The Xavánte in Transition

The book series, Human-Environment Interactions, is designed to publish the best research being produced on the reciprocal interactions between human populations and the biophysical environment. This scholarship is inherently multidisciplinary, combining contributions from the physical, biological, and social sciences, coalescing around investigations into the human dimensions of global environmental change. Because this is an integrated science agenda, the books in the series will reflect this multidisciplinary complexity.

The series publishes work that addresses issues both timely and urgent in nature. They speak to our future, our present, and our past. In order to better address the ways in which feedbacks operate between the biophysical environment and the social and political context of human choice-making, series books provide a nuanced understanding, solidly backed by quantitative science, of the interactions of population and environment, their changing spatial and temporal distribution, their consumption patterns and their impact, and their changing production and reproduction strategies. Other issues that concern the series are the health impacts of global warming and the distribution of disease vectors; the impact of El Niño events upon coastal areas and upon places as varied as the desert Southwest and the Amazon; and how people forest and reforest their landscape. How might we anticipate these changes? How might we respond and prepare for these changes? Who is most vulnerable? These and other related topics will drive the series as we seek to stimulate readers to think, and to act.

The Xavánte in Transition: Health, Ecology, and Bioanthropology in Central Brazil, Carlos E. A. Coimbra Jr., Nancy M. Flowers, Francisco M. Salzano, and Ricardo V. Santos

The Xavánte in Transition

Health, Ecology, and Bioanthropology in Central Brazil

Carlos E. A. Coimbra Jr.
Nancy M. Flowers
Francisco M. Salzano
and
Ricardo V. Santos

Ann Arbor

THE UNIVERSITY OF MICHIGAN PRESS

First paperback edition 2004

Published in the United States of America by
The University of Michigan Press
Manufactured in the United States of America
♾ Printed on acid-free paper

2007 2006 2005 2004 5 4 3 2

A CIP catalog record for this book is available from the British Library.

Library of Congress Cataloging-in-Publication Data

The Xavánte in transition : health, ecology, and bioanthropology in central Brazil / Carlos E.A. Coimbra, Jr. ... [et al.].
p. cm. – (Linking levels of analysis)
Includes bibliographical references and index.
ISBN 0-472-11252-X (cloth : alk. paper)
1. Xavante Indians—Population. 2. Xavante Indians—Health and hygiene. 3. Xavante Indians—Anthropometry. I. Coimbra, Carlos E. A., 1952– II. Series.

F2520.1.A4 X39 2002
306′.089′984—dc21 2002074320

This paperback edition introduced into the Human-Environment Interactions series
ISBN 0-472-03003-5 (pbk. : alk. paper)

To the Xavánte

Series Introduction

The Human-Environment Interactions series is designed to publish the best scholarship on the reciprocal interactions between human populations and the biophysical environment. Inherently multidisciplinary, combining the physical, biological, and social sciences, this scholarship focuses on "the human dimensions of global environmental change." A measure of its importance is the existence over the past decade of a standing committee at the National Research Council/National Academy of Sciences that has identified and provided direction to research on the human dimensions of global change. The U.S. Global Change Research Program has been funded at significant levels by the National Science Foundation (NSF), the National Institutes of Health (NIH), the National Aeronautics and Space Administration (NASA), the National Oceanic and Atmospheric Administration (NOAA), the Department of Energy (DOE), the Environmental Protection Agency (EPA), and many other U.S. agencies. This effort is also international, with notable funding from the European Union and many other nations, such as Japan, Russia, and Brazil.

As the importance of human interactions with the earth system began to be recognized, a large community of scholars undertook a vigorous research agenda, combining methods from the biophysical and the social sciences to understand these interactions. Human-environment interactions research has addressed such concerns as tropical deforestation; the societal impacts of climate change; the reciprocal interactions of population-environment-consumption; and mega-urbanization dynamics; it has also provided large-scale monitoring of changes in vegetation and historical reconstructions of human interactions at local and regional scales. Topics such as these will give scope and structure to the series. The series will also reflect the multidisciplinary fashion in which the field developed, due to the complex nature of these issues. Over the coming decade, research in this field is sure to grow.

The authors of the present book, the first in the series, bring together sophisticated understanding of indigenous South America, medical anthropology, biological anthropology, and social anthropology within an explicitly political-economic context that makes the local human conditions richer and more complex than if they had been only treated ethnographically, medically, or economically. The research underlying the book pulls together work over decades by several investigators and collaborations by the authors for more than a decade. It is a perfect example of longitudinal research and personal commitment to the subjects of study.

The Xavánte of Central Brazil have been of interest to outsiders for many years. They resisted contact for a very long time, gaining fame for their fierceness, and for the past fifty years have reluctantly engaged in interethnic relations, with high, even devastating, costs. As with so many other indigenous populations, the mortality from introduced diseases left them a shadow of themselves both biologically and culturally. Thus, the Xavánte give us a well-documented window into the experience of many other ethnicities throughout lowland South America that are much less well chronicled. Most indigenous groups are poorly known, particularly their health status and their demographic history. This study of a Xavánte population begins to correct this situation. It documents the changing demography, ecology, and economy of one of the surviving populations. They have managed better than many others in coping with rapidly changing conditions, and they provide insight into the more effective ways to survive.

The book is also an important contribution of the "new" biological anthropology. In *Building a New Biocultural Synthesis* (1999), a volume in the Linking Levels of Analysis series that preceded this one, Alan Goodman and Thomas Leatherman collected studies that integrated considerations of political economy with those of human biology to show that much of what had passed for adaptability (or not) was a product of the poverty and exploitation experienced by the subjects of study, rather than inherently a biological process of adaptation. The authors of the present book were represented in that volume; *The Xavánte in Transition* is more complete in its analysis of ecology, health, disease, and demography in their full political and economic contexts. This is human biology with a social conscience.

It is my hope that this book and the volumes to follow in the series will inspire readers to consider how complex human environmental relations are, how heavy the costs to a population can be, and what we might do to make this transition less fraught with the danger of biological and cultural extinction.

Emilio Moran, Series Editor

Contents

Figures

Tables

Foreword

On 22 and 23 July 2002, Carlos E. A. Coimbra Jr. and Ricardo V. Santos went to Etéñitépa to present The Xavánte in Transition: Health, Ecology, and Bioanthropology in Central Brazil *and to discuss it with the community. Tsuptó, the village chief, made the following comments.*

In former times, in those days, our health and the foods we ate were more traditional than they are today. There were not so many sicknesses. We were very strong and resistant because of the way we ate. At that time there wasn't the interference that there is today, the interference of the whites. At that time there was no tuberculosis, no diabetes; at that time there were none of the sicknesses that we see today. Now we have them all. These sicknesses are not from our village; they are not from other Indian populations. I think they are because of interference from outside. I think that the diseases that the Indian populations are suffering from are brought by the whites. Many things have changed. So, compared with former times, there are many more kinds of diseases. Our organisms, our bodies, the bodies of our children, can't stand up to all this; they can't resist. Because these are not our sicknesses. They come from outside.

About the work of research, I think the research team has to relate to the community; the team has to like what it is doing. We don't see that very often. The health teams that come to the village, they come to treat a particular sickness, just that one time; then they go away. I think that what is affecting our health, what is damaging it, these things that come into the village from outside, have to be investigated. It is very important to understand what is going on. The things that are brought in make garbage that builds up and takes time to clean away. We don't know how to deal with this. About those influences that come from away, those sicknesses from away that are prejudicing the village, the community—it is about those things that research should be done.

The research that has already been done has helped a lot, even if it is over the long term. But we see this as an example of how we can find out what is damaging the health of the village, what may damage it in the future. So that instead of letting it do damage, we can avoid it. We also see research as a way for people outside to learn about our lives. We have health problems that people outside don't know about, so they don't look for solutions. Because at present, the diseases that Indian populations suffer from are not being dealt with. Well, I think these things should be studied in more depth so that if solutions for these things that are happening can be found, they will be.

Tsuptó Buprẽwen Wairi, chief of the village
Etéñitépa, Terra Indígena Pimentel Barbosa

Preface and Acknowledgments

Sometimes fortuitous encounters in our lives may start us along paths that take us a great distance, involve relationships that last for years, and, if we are fortunate, give us great personal and intellectual pleasure. This book is the result of such a meeting. The idea for a multidisciplinary research project among the Etéñitépa Xavánte bringing together perspectives from biological anthropology, human ecology, and public health was born in 1988 from a casual conversation between Carlos Coimbra Jr. and Nancy Flowers during a break between sessions of a seminar on biological anthropology in Belém, Pará. Soon after, the idea was discussed with Ricardo Santos and Francisco Salzano, who immediately joined the project. Authors' names in this book appear in alphabetical order.

For some twenty-five years Flowers has been doing research, especially in human ecology and anthropological demography, among the Xavánte. In 1976 and 1977, Flowers lived at Etéñitépa for fourteen months. Flowers's research was part of a project, Human Ecology in Central Brazil, planned and coordinated by Daniel Gross. According to the design of the project, Flowers, with two other graduate students from the City University of New York, spent a year in different Indian villages of Central Brazil, Dennis Werner with the Kayapó-Menkrangnotí and Madeline Ritter with the Canela (Ramkókamekra), collecting data for the comparative project as well as their dissertations.

Throughout the 1980s, Coimbra and Santos were involved in research in medical-biological anthropology and epidemiology among indigenous peoples in the Brazilian Amazon. Both based their doctoral dissertations, defended at Indiana University, Bloomington, on research carried out among the Gavião, Surui, and Zoró, Tupí-Mondé-speaking groups who live on reservations near the boundary between the states of Mato Grosso and Rondônia. Both Coimbra and Santos are interested in understanding how the health and biology of Amazonian indigenous peoples are affected by contact and increasing interaction with the Brazilian national society.

Francisco Salzano has been working since the 1950s with indigenous peoples in Brazil and other South American countries. He initially worked with the Kaingáng in southern Brazil and shortly after, in the early 1960s, did research among the Xavánte. In the following years this research was extended to a number of different groups, chiefly from the Amazon region, in collaboration with a large number of colleagues from both Brazil and abroad. Particularly noteworthy was a joint project that involved the Department of Human Genetics, School of Medicine, University of Michigan, and the Departamento de Genética, Universidade Federal de Rio Grande do Sul. These studies were conducted over a period of four decades, contributing in a significant way to knowledge of the genetics of Amerindian tribal populations.

In our early discussions of our Xavánte project it became clear that we all had a particular interest in studying the ways through which socioeconomic and environmental changes affect the health, biology, and ecology of indigenous peoples. Aspects of the Xavánte experience, including their history and patterns of interaction with non-Indians, offered a unique opportunity for this kind of research. An unusual and interesting aspect of the Xavánte is the considerable time depth of historical references to them. Based on these documents we can draw a reasonably detailed picture of interaction between the Xavánte and the Brazilian national society over the past two and a half centuries.

Another advantage was that some members of the team had already done fieldwork among the Xavánte. Both Flowers and Salzano did research with the group that now lives on the Pimentel Barbosa reservation. Work could be done in the same community where previous research was carried out, making it possible to compare data collected at different times in the recent history of the group. Moreover, two anthropologists, David Maybury-Lewis and Laura Graham, had made ethnographic studies of the group at different periods.

The first fieldwork of Coimbra and Santos among the Xavánte was in May and June of 1990, when they made a field trip to Etéñitépa with Flowers. At that time they collected many of the data that we present here. But they and Flowers returned to Etéñitépa many other times, and over the period of their research they developed very warm relations with the Xavánte. During the years of the project it attracted a growing number of people. At the beginning of the 1990s Coimbra and Santos both began to work actively as teachers and advisers in the master's and doctoral programs of the Escola Nacional de Saúde Pública (ENSP, National

School of Public Health) of the Fundação Oswaldo Cruz (FIOCRUZ), in Rio de Janeiro. Rui Arantes, Silvia Gugelmin, Rubens Ianelli, Maurício Leite, Silvana Pose, and Luciene Souza, who were students in that program, chose to do their graduate research among the Xavánte, focusing on a number of different topics, such as the epidemiology of infectious and parasitic disease, nutrition, demography, and human ecology. Working with these collaborators, as well as other students doing research on indigenous health (Ana Lúcia Escobar and Eliana Diehl) at the Escola Nacional de Saúde Pública, has been an extremely enriching experience, both in personal terms and from the opportunity given us to build a collective body of knowledge. Looking back, we can see that research training has become an important aspect of the project, which we did not anticipate when it began.

The building of modern laboratory and computing facilities at Salzano's Genetics Department at the Federal University of Rio Grade do Sul in Porto Alegre has made possible the detailed genetic study of the samples collected in the 1960s and 1990s at both the protein and DNA levels. Molecular investigations of these samples are still being performed while this book is being written.

We certainly would not have been able to carry out our research among the Xavánte without financial assistance from many different sources. For the 1990 field trip we had the support of the Wenner-Gren Foundation for Anthropological Research (grant to Nancy Flowers and Carlos Coimbra Jr.). Activities between 1992 and 1995 were largely made possible through a fellowship granted to Ricardo Santos from the John D. and Catherine T. MacArthur Foundation. Throughout the project we had the constant support of the Fundação Oswaldo Cruz, which both directly and indirectly provided resources. The FIOCRUZ program PAPES—Programa de Apoio à Pesquisa Estratégica em Saúde (Program to Support Strategic Research on Health)—provided important financial support, often at times when we had no other funding.

Writing a book with eight hands naturally requires many, many meetings. Flowers came to Brazil for various lengths of time in 1992, 1994, 1996, and 2000 for further fieldwork and data analysis as well as for writing articles and the book itself. Her trips were funded by the Conselho Nacional de Desenvolvimento Científico e Tecnológico (CNPq, National Council for Scientific and Technological Development), the Fundação Oswaldo Cruz, the Fulbright Commission, and the Fundação de Amparo à Pesquisa do Rio de Janeiro (FAPERJ, Rio de Janeiro Foundation for the

Support of Research). Throughout the Xavánte project, Coimbra, Salzano, and Santos held research fellowships from the CNPq. Postdoctoral fellowships awarded to Coimbra and Santos by the Coordenação de Aperfeiçoamento de Pessoal de Nível Superior (CAPES, Council for Postgraduate Study) made it possible for Coimbra to spend twelve months in 1998 and 1999 at the Anthropology Department of the University of Massachusetts at Amherst, and for Santos to spend the same period of time at the University of Massachusetts and the Program in Science, Technology and Society of the Massachusetts Institute of Technology. Finally, CNPq and CAPES gave fellowships to the graduate students of the Escola Nacional de Saúde Pública who carried out research with the Xavánte. The studies in population genetics of the research group coordinated by Salzano at the Universidade Federal do Rio Grande do Sul, in Porto Alegre, were supported by the Programa de Apoio a Núcleos de Excelência (PRONEX, Program in Support of Excellence), CNPq, the Fundação de Amparo à Pesquisa do Estado do Rio Grande do Sul (FAPERGS, Rio Grande do Sul Foundation for the Support of Research), and the Financiadora de Estudos e Projetos (FINEP, Agency of Studies and Projects).

Beyond direct support for our research, we received indirect support in a number of ways. The laboratories of the Instituto Oswaldo Cruz, of the Escola Nacional de Saúde Pública and the Hospital Evandro Chagas, all linked to the Fundação Oswaldo Cruz, analyzed biological samples collected at Eténñitépa and provided consumable supplies. We thank Cláudio Daniel-Ribeiro, Ricardo Lourenço de Oliveira, Wilson Souza, Bodo Wanke, and Clara Yoshida. Analyses were also made at the Instituto Evandro Chagas, the Instituto de Medicina Tropical de São Paulo, and the Hospital da Universidade de Brasília. We thank the following colleagues from these institutions: Monamaris M. Borges, Alexandre C. Linhares, Ronan Tanus, and Amélia P. A. Travassos da Rosa.

As must be the case with a project that involves so many researchers, the list of people whose assistance and collaboration we should like to acknowledge is a long one. Collectively, we should like to thank Adauto Araújo, Michael Brown, Regina Lana Costa, Carlos Fausto, Maj-Lis Fóller, Laura Graham, Daniel Gross, Paulo Sabroza, and Sheila Mendonça de Souza. As well as their friendship, some of these colleagues gave us comments that have greatly enriched our analyses. The support of Emilio Moran, editor of the Linking Levels of Analysis series with the University of Michigan Press, was indispensable throughout the process of bringing this book to light. Ingrid Erickson and Ellen McCarthy, at the University of Michigan Press, gave us invaluable help in the preparation of the book. There are also many people who, as individuals, we should like to thank. Flowers thanks Dan and Judith Bates, Elizabeth Butson, Warren DeBoer, Brian Ferguson, Bill Fisher, Denny Gilmore, Ken Kensinger, Susan Lees, Sally McLendon, Debra Picchi, Sydel Silverman, Sara Stinson, and Dennis Werner. Santos and Coimbra jointly thank a number of friends from Amherst and Boston (Michael Fischer, Alan Goodman, Lynnette Leidy, Debra Martin, Lynn Morgan, Adriana Petryna and João Biehl, Alan Swedlund, Jim Trostle, and R. Brooke Thomas) as well as from Rio de Janeiro (Ângela Cançado, Bruna Franchetto, Dora Chor, Reinaldo Santos, Marília Facó Soares, and Antônio Carlos F. do Valle).

José Levinho and Ana Paixão provided support to our archival research at the Museu do Índio in Rio de Janeiro. Sula Danowski and Adriana Cataldo Silva helped us in preparing maps and illustrations. Jussara Long at the library of the Escola Nacional de Saúde Pública helped us to locate important bibliographic materials.

At the Fundação Nacional do Índio (FUNAI, National Indian Foundation) we thank Fábio Oliveira and Tizuko Tsumori from the Cuiabá regional office and Jô Cardoso de Oliveira from the Brasília headquarters.

The research on which this book is based would not have been possible without the collaboration and enthusiasm of the Xavánte themselves. At each of our visits to Etéñitépa we came before the *warã,* or men's council, to explain what we intended to do and also to present the results of research to date. This dialogue was always very helpful to us. We hope that our research and the parallel activities that are developing from it have benefited and will benefit the Xavánte. We have not designed our research to be only of academic value. Among the Xavánte that we should like to thank in particular are Suptó, Roberto, Sereburã, Barbosa, Agostinho, Paulo, and Jamiro. We have decided to assign the royalties from this book, and from any future edition in Portuguese, to the Etéñitépa community.

Abbreviations

AIH	Autorização de Internação Hospitalar (Authorization for Hospitalization)
ARI	acute respiratory infection
ASFR	age-specific fertility rate
BCG	vaccine against tuberculosis, prepared from a weakened strain of tuberculosis bacteria named BCG—bacille Calmette-Guérin, for the French scientists who developed the product
BMI	body mass index
CAPES	Coordenação de Aperfeiçoamento de Pessoal de Nível Superior (Council for Postgraduate Study)
CBR	crude birth rate
CDR	crude death rate
CEDI	Centro Ecumênico de Documentação e Informação (Ecumenic Center for Documentation and Information)
CIMI	Conselho Indigenista Missionário (Missionary Council for Indigenous Affairs)
CNPq	Conselho Nacional de Desenvolvimento Científico e Tecnológico (National Council for Scientific and Technological Development)
CPI	Centro de Pesquisa Indígena (Center for Indigenous Research)
DMFT	decayed, missing, filled teeth
DSEI	Distrito Sanitário Especial Indígena (Special Indigenous Health District)

DTP vaccine	vaccine to protect against diphtheria, tetanus, and pertussis
ENSP	Escola Nacional de Saúde Pública (National School of Public Health)
EPF	endemic pemphigus foliaceus, or fogo selvagem
EVS	equipe volante de saúde (mobile health team)
FAPERGS	Fundação de Amparo à Pesquisa do Rio de Janeiro (Rio de Janeiro Foundation for the Support of Research)
FAPERJ	Fundação de Amparo à Pesquisa do Estado do Rio Grande do Sul (Rio Grande do Sul Foundation for the Support of Research)
FBC	Fundação Brasil Central (Central Brazil Foundation)
FINEP	Financiadora de Estudos e Projetos (Agency of Studies and Projects)
FIOCRUZ	Fundação Oswaldo Cruz (Oswaldo Cruz Foundation)
FUNAI	Fundação Nacional do Índio (National Indian Foundation)
FUNASA	Fundação Nacional de Saúde (National Health Foundation)
HBV	hepatitis B virus
HLA	human leucocyte antigen
IBAMA	Instituto Brasileiro do Meio Ambiente e Recursos Naturais Renováveis (Brazilian Institute of the Environment and Renewable Natural Resources)
IBGE	Instituto Brasileiro de Geografia e Estatística (Brazilian Institute of Geography and Statistics)
IBP	International Biological Program
IMR	infant mortality rate
ISA	Instituto Socioambiental (Socioenvironmental Institute)

MF	Médicins sans Frontières (Doctors without Borders)
NCHS	National Center for Health Statistics
NCI	Núcleo de Cultura Indígena (Center for Indigenous Culture)
NGO	nongovernmental organization
NIDDM	non-insulin-dependent diabetes mellitus
ORT	oral rehydration therapy
PAHO	Pan American Health Organization
PAPES	Programa de Apoio à Pesquisa Estratégica em Saúde (Program to Support Strategic Research on Health)
PEM	protein-energy malnutrition
PNSN	Pesquisa Nacional Sobre Saúde e Nutrição (National Survey of Health and Nutrition)
PRODEAGRO	Projeto de Desenvolvimento Agro-Ambiental de Mato Grosso (Project for the Agro-Environmental Development of Mato Grosso)
RIHGB	*Revista do Instituto Histórico e Geográfico Brasileiro (Journal of the Brazilian Historical and Geographical Institute)*
SPI	Serviço de Proteção aos Índios (Indian Protection Service)
STD	sexually transmitted disease
SUS	Sistema Único de Saúde (Unified Health System)
SUSA	Serviço de Unidades Sanitárias Aéreas (Service of Airborne Health Units)
TFR	total fertility rate
UNI	União das Nações Indígenas (Union of Indigenous Nations)
WHO	World Health Organization
WWF	World Wildlife Fund

Guide to Pronunciation of Xavánte Words

We have not attempted to reproduce the phonetics of the Xavánte language. The orthography is currently under discussion. In the pronunciation of Xavánte words as they are written in the text, the consonants *p, b, t, d, m, w,* and *h* are pronounced as they are in English. To pronounce the affricate sounds *ts* and *tz,* raise the tongue to the roof of the mouth just before making the English sounds *s* or *z*. The letter *r* is a tap; the tongue momentarily touches the roof of the mouth.

a is pronounced like the *a* in *father*
e is pronounced like the *e* in *eight*
é is pronounced like the *e* in *men*
i is pronounced like the *ee* in *meet*
o is pronounced like the *o* in *open*
ö is pronounced like the *oeu* in the French word *boeuf*
ó is pronounced like the *ou* in *bought*
u is pronounced like the *oo* in *shoot*
~ indicates the nasalization of a vowel
Two vowels together indicate a long vowel, as in *ii*
ñ is pronounced like the *ny* in *canyon*
An apostrophe (') indicates a glottal stop

The spelling of the names of indigenous peoples and languages in Brazil follow the Associação Brasileira de Antropologia (Brazilian Association of Anthropology) 1953 convention, according to Rodrigues (1986).

Chapter 1

Introduction

This book presents the results of research in an indigenous community of Central Brazil. Our principal aim, in collecting and analyzing demographic, biological, epidemiological, and ecological data, has been to produce a diachronic view of the long and complex interaction between the Xavánte people, especially those of the Etéñitépa community, and the surrounding Brazilian national society. On a broader scale, our research may be seen as an attempt to understand how local systems interact with larger social, economic, and political institutions and processes.

Although the most recent contact, which must now be considered permanent, between the Xavánte and the national society began only fifty years ago, in the 1940s, historical sources show that they have been resisting Western expansion for centuries. Accounts of the Xavánte in the eighteenth and nineteenth centuries refer to epidemics, armed conflicts, and attempts of the Portuguese colonial government to "pacify" and settle them. The Xavánte responded to successive fronts of European penetration with increased mobility and eventually escape through westward migration. In the mid–twentieth century they were again overtaken, as the Brazilian government implemented the "March to the West," opening Central Brazil to colonization and economic development. We argue, therefore, that there is no way to regard the Xavánte as a society until recently isolated from the political and economic processes that have been taking place for centuries in Brazil and beyond.

The consequences of European expansionism for native societies of the Americas over the last five centuries are well documented. As the New World was conquered, they suffered from epidemic disease, forced labor, violent death, and finally extinction or marginalization. The fate of indigenous peoples in Brazil has been no different. While the exact magnitude of the demographic reduction of native populations in Brazil over the last five centuries may still be debated, there is little doubt that it was catastrophic.

In our case study of the Etéñitépa Xavánte, we document some of the biological, epidemiological, and ecological consequences of contact with the Brazilian national society in the second half of the twentieth century. However, our intention is to reach beyond merely documenting effects to show the importance of particular circumstances. We argue that extrinsic events and situations, as well as ecological, social, and biological characteristics intrinsic to Xavánte society, have played important roles in shaping their contact and postcontact experience. Social transformations imprint "signatures," or biological marks, upon individuals in a population, and one of the intents of our study is to highlight some of these.[1]

Moreover, there can hardly be any region of the world today where a small-scale or indigenous society may continue to enjoy true political and economic autonomy. In most cases such societies subsist as subordinated and impoverished segments of the national society. We believe that our documentation of the Xavánte experience has broad significance, exemplifying the process through which small-scale societies become disadvantaged populations without access to services and benefits available to other social segments.

At present, some two hundred indigenous societies, with a total population of nearly 280,000, survive in Brazil. This corresponds to less than 0.5 percent of the total Brazilian population. Most are settled on reservations[2] and interact at different levels of intensity with the Brazilian national society. They range from a few still relatively isolated groups in the Amazon to many that are deeply involved in regional markets and a growing number that live on the outskirts of Brazilian cities. A common denominator that unites them is their marginalization within Brazilian society, reflected in poor health and economic conditions and the difficulty they experience to obtain access to health care, education, and other social services. While some groups still remain at risk for epidemic diseases capable of decimating their populations, others are experiencing high rates of population growth even while they face the simultaneous presence of both infectious and chronic noncontagious diseases. The prevalence of accidental injuries, violence, and social suffering (alcoholism, suicide, and so on) also varies widely from group to group. What can be generally stated, however, is that for the great majority of indigenous peoples in Brazil morbidity and mortality rates are unequivocally higher than overall national rates, while life expectancy at birth is disturbingly low.

There is relatively little reliable information about the health and demography of indigenous peoples in Brazil. Even less is known about how these dimensions are changing as Indian societies increasingly interact with wider systems. It is important to recognize that these experiences may lead to health/disease profiles radically different from that of the Brazilian national society, which over recent decades has undergone major demographic and epidemiological transitions. Knowledge about the differences is highly relevant to the planning, implementation, and evaluation of public services for indigenous peoples. In this case study of the Etéñitépa Xavánte, we document the changing state of demography, ecology, and health/disease of a contemporary Indian society in Brazil. We also point to their efforts to cope with these rapidly changing conditions. Our hope is that the knowledge gained may help to provide a firmer base for decision making and action affecting them as well as other indigenous peoples. We believe that, while some aspects of our research apply more directly to the Xavánte, much of it, at a higher level of generality, is relevant to the situation of other small-scale societies, whether characterized as "indigenous" or not.

Human Biological and Epidemiological Research in Amazonia

Knowledge about many aspects of the biology and epidemiology of indigenous Amazonian peoples in Brazil has been accumulating since the 1960s. Research has been most intense in the areas of population genetics and epidemiology/tropical medicine. This research was spearheaded by a relatively small group of investigators. Among the most influential have been James V. Neel and Francisco M. Salzano in the field of population genetics and Francis L. Black and Roberto G. Baruzzi in research related to the epidemiology of infectious and parasitic diseases.

Research in population genetics has focused primarily on the mechanisms involved in the production and maintenance of genetic variability. The general orientation of this research agenda was made explicit by Neel.

> The general thesis behind the program was that, on the assumption that these people represented the best approximation available to the conditions under which human variability arose, a systems type of analysis oriented toward a number of specific questions might

> provide valuable insights into problems of human evolution and variability. We recognize, of course, that the groups under study depart in many ways from the strict hunter-gatherer way of life that obtained during much of human evolution. . . . We assume that the groups under study are certainly much closer in breeding structure to hunter-gatherers than to modern man; thus they permit cautious inferences about human breeding structure prior to large-scale and complex agriculture. (1970, 815)[3]

This research program in population genetics was not oriented by a search for genetic adaptation per se, in the sense of reproductive advantage in particular environmental circumstances. Instead, the focus was on population structure and the application of methodologically sophisticated biometric analyses to genetic processes at the microevolutionary level. The data collected included medical examinations, anthropometric and dental measurements, vital statistics (mortality, fertility and migration), and blood genetic typing. These studies have proved to be important in revealing how random processes (e.g., genetic drift and founder effects) result in high levels of interpopulation genetic variability. The major human biology projects were carried out among the Xavánte, Makiritáre, Yanomámi, and Kayapó (Neel 1994; Salzano and Callegari-Jacques 1988).

In a work entitled "A prospectus for genetic studies on the American Indians," published in 1967, Neel and Salzano set down three specific themes that should receive particular attention in this research program: (1) questions related to the early peopling of the Americas, (2) questions related to the microevolutionary dynamics of these populations from which models might be formulated that would help in understanding human evolutionary history, and (3) questions about the emergence of noninfectious diseases and "genetic adjustments" deriving from contact of these populations with Western society (Neel 1968; Neel and Salzano 1967a, 246–57; Neel et al. 1977; Salzano 1991). Of these topics, the first two received the most attention in the investigations of the 1960s and 1970s.

A good example of theory building resulting from these Amerindian studies is the "fission-fusion model," which was first proposed by James Neel and Francisco Salzano based on observations made among the Xavánte and further developed through the Yanomámi studies (Neel and Salzano 1967a; Smouse et al. 1981; Thompson 1979). This model, which was originally formulated to explain the large differences that were sometimes observed in gene frequencies among South American Indian villages, even when located close to one other, was later expanded to elucidate patterns of genetic microdifferentiation and population structure in small-scale human populations in general. Neel et al. (1977, 121) summarize in general terms the model and its implications.

> New villages (populations) do not arise by some random sampling of a large "parental" gene pool, but by a fissioning of a pre-existing village, usually to some extent along lineal lines. . . . This is because village alliances (and so migration matrices) are subject to sudden changes and because the dictates of war and disease also result in village fusions, which probably occur with sufficient frequency to equal in importance small-scale intervillage migrations. . . . Social structure thus sets the stage for isolated demes with unusual combinations of gene frequencies and this may play a role in the rapid evolution which seems to have characterized our species.

Given these motives for the research programs in human biology, it is not surprising that emphasis fell on the study of populations considered to be relatively isolated or little affected by external historical and political processes. The predominant view when the program began was that the expansion of economic and demographic frontiers was occurring at such a rapid pace that conditions suitable for studying native peoples in a "relatively undisturbed condition" were vanishing.[4] It is important to emphasize, however, that the researchers were aware of the extent of the impacts of Western expansion on New World populations. For example, in 1967 Neel and Salzano wrote that "there is no Indian group completely untouched by the discovery of America and subsequent contacts, direct or indirect, with the Western World"; moreover, "we have no way of knowing with certainty to what extent the surviving groups of . . . Indians have had their way of life seriously disrupted by the events of the past 450 years" (1967b, 246, 248). Referring to his attempts to use data from indigenous groups in order to gain insights into broader processes related to human biological diversity, Neel drew on the metaphor of a "mirror cracked and dusty, but as accurate as any we have" (1994, 139).

These investigations in human biology carried out in Amazonia, while incorporating "history," envisioned history on a macroevolutionary scale. Their aim was to explain the biological trajectory of humans over long periods of time. Within this line of research, it was thought appropriate to use biological data collected from populations relatively unaffected by "modern" living conditions to generate analytical models for human biological evolution. For indigenous groups to be used as models, producing results that could be generalized beyond the regional scale, it was considered important to assure that they were influenced as little as possible by local historical, political, and economic processes. In order to guarantee the validity of conclusions on a macrohistorical scale, it was necessary to control (in the sense of excluding) microhistorical dynamics related to the interaction of the native groups with the surrounding national societies.

Another area of research in human biology in Amazonia that received considerable emphasis was the epidemiology of infectious disease. We can identify two main lines of investigation.

The first line was particularly associated with work in human genetics, where research on infectious diseases was primarily aimed at characterizing the "disease pressures" to which indigenous populations were submitted. In these populations, infectious diseases were the most important causes of death. Researchers were also interested in finding out which infectious diseases affected native Amazonian populations before Europeans reached the Americas (Black et al. 1974; Neel 1968, 1977, 1982a; Neel and Salzano 1967b).

The second line of investigation in Amazonian epidemiology was related to the broad interests of worldwide biomedical research in the 1960s. Through research on populations held to be "isolated," epidemiologists were attempting to understand the relationships between population size and the persistence of certain contagious diseases. More specifically, discussions turned on the concept of the "critical population size" necessary to maintain at endemic levels viral infections that confer lasting immunity, such as measles and smallpox (Black 1966, 1975, 1990).

Perhaps it was in the study of the dynamics of infectious disease that research in human biology in Amazonia came closest to considering the processes of interaction between indigenous populations and the surrounding national society. In any case, these studies placed their emphasis on the first moments of contact.

The high mortality observed during outbreaks of viral diseases, especially measles epidemics, led to heated discussions about the relative weight of biological/immunological determinants versus sociocultural factors (Coimbra 1987; Neel 1974). One hypothesis was that the population collapse following epidemics resulted from genetically based susceptibilities to novel pathogens (Black 1990, 1992; Black et al. 1982). According to this explanation it is not that native peoples have "inappropriate" genes to cope with diseases, but rather that their greater genetic homogeneity may lead to relatively restricted diversity of immune reaction and therefore to impaired survival potential. Other explanations emphasized the impacts of socioeconomic and ecological breakdowns experienced by indigenous people hit by epidemics and attributed high mortality largely to these effects (Neel 1977, 1982a; Neel et al. 1970).

These lines of epidemiological research involved questions associated with central discussions in the field of human biology, like the early peopling of the Americas and the origins of diseases in the New World, the demographic dynamics and population size of indigenous peoples, and parasite loads and disease pressures as they relate to genetic processes.

It is important to mention another line of biomedical research that was consolidated during the 1970s and 1980s. This research focused on the clinical description and epidemiological analysis of so-called tropical diseases with little-known etiologies and unclear transmission cycles. Roberto Baruzzi, his students, and his associates made a number of these studies among Upper Xingu indigenous populations. An example is their study of the epidemiology of toxoplasmosis (Baruzzi 1970). At the end of the 1960s little was known about the modes of transmission and geographical distribution of this cosmopolitan parasitic disease, which is caused by the protozoan *Toxoplasma gondii.* Another concern of tropical medicine that led to studies in indigenous populations at this time was the "syndrome of tropical splenomegaly" and its association with endemic malaria (Baruzzi et al. 1976). There was also interest in studying certain infectious diseases considered to be rare or exotic, such as lobomycosis and mansonelosis (Baruzzi et al. 1973, 1979). At the same time, and over many years, Baruzzi and his associates provided health care to the Indian groups living in the Xingu Park (Baruzzi et al. 1978).

Some of the genetic and epidemiological studies on native Amazonian populations have reached a remarkable degree of methodological and theoretical sophistication. On the basis of specific case studies, researchers have been able to generate important theoretical formulations in human population genetics and health/disease processes related to infectious diseases. While there have been discussions on the epidemiological aspects of the interaction between indigenous peoples and national societies, the main focus has been upon recently contacted and/or semi–isolated populations, and, although the importance of a research agenda that includes the study of change has been stressed (see Neel 1991; Salzano 1985, 1991), relatively little attention has been given to this topic. The theme in itself—the increasing connectedness of indigenous societies in Brazil to regional and macroregional economic and political networks—has yet to become an autonomous and productive domain of knowledge in the areas of biological anthropology and the epidemiology of indigenous peoples in Brazil.[5] We feel that it is crucial to increase this knowledge as indigenous peoples struggle to overcome their marginalization within the national society and to widen their opportunities and choices. The primary focus of our study, therefore, is not on the interaction of the Xavánte population with its "natural environment," important as this is, nor on the biological impact of contact as an event, but on the Xavánte as a society in transition, with emphasis on the demographic, ecological, and epidemiological effects of ongoing change.

Survival, Permanence, and Changing Research Agendas

The experience of indigenous peoples in Brazil that came into contact with the national society during the first half of the twentieth century led to pessimistic forecasts about their ultimate chances for survival, with special fear for groups that were still "isolated." Toward the middle of the century Brazilian anthropologist Darcy Ribeiro (1956) analyzed a large body of ethnographic information and demographic data that demonstrated the magnitude of the depopulation affecting indigenous peoples, largely due to epidemics of infectious diseases. According to Ribeiro (3), "the history of relations between Indians and whites in Brazil teaches that the weapons of conquest were . . . bacteria and viruses, especially viruses." Ribeiro (1977, 239) pointed out that of the 230 Indian groups known in Brazil in 1900, 87 were extinct by 1957. When only groups defined as "isolated" in 1900 were considered, the percentage of population loss in such groups between that date and 1957 was estimated to be around 73 percent, showing the dimensions of extermination to be even more appalling (434–35). However, Ribeiro did not foresee the physical extinction of all groups, and he thought that "those that survived [would remain] Indian—no longer in their habits and customs but in their self-identification as people different from the Brazilians and as victims of their domination" (8).

Around the middle of the 1960s the military government of Brazil initiated a series of development projects in the Amazon. This period was characterized by huge economic enterprises and infrastructure development programs that involved the construction of highways and hydroelectric dams, mining projects, and agricultural settlements, all of which stimulated migration into the Amazon (Barbira-Scazzocchio 1980; Cleary 1990; Hemming 1985; Moran 1981; Schmink and Wood 1984; Smith 1981; Velho 1981). The repercussions for a great number of native peoples, many of them still living in relative isolation, were disastrous epidemics, social breakdown, and in some cases physical extinction (Davis 1977; Ramos 1984; Santos and Coimbra 1998).

In various areas of Amazonian anthropology, the combined threat of the disappearance of indigenous peoples and the accelerated rate of environmental destruction seems to have stimulated—as we have seen in the case of human biology—lines of research that focused on indigenous societies as yet little affected by the social and economic impacts of Western expansion. About the field of ecological anthropology Hames and Vickers (1983, 26) wrote, "As scientists, we also recognize that tropical environments everywhere are now experiencing rapid change due to deforestation and the impact of development. . . . This knowledge gives our research a sense of urgency and a concern that scientific findings be incorporated into the ongoing process of policy formulation and planning for these regions." In ethnology as well there was a sense of urgency: "Reviewing the recent literature [on lowland South American ethnography] brings into focus the enormous amount remaining to be done, and shows how rapidly opportunities for future work are being eliminated by the far-reaching effects of 'progress' and 'development' on all fronts, although seemingly most seriously in Brazil" (Jackson 1975, 330–31). This pessimism seems to have been pervasive among anthropologists studying indigenous societies at the time. Wright (1988, 371) writes that "between 1952 and 1968 there rose in the international academic community a plea for 'urgent' anthropological studies to be done among tribal societies because they were considered to be in the process of 'disappearing.' Hence, anthropologists felt the need to "save" aboriginal cultures in the face of what was called the 'tragedy of anthropology.'"

By the end of the 1960s the accumulating evidence that the survival of indigenous peoples in Brazil was threatened became known far beyond the country's borders. The international press used terms like *genocide* and *ethnocide* to describe what was happening (Davis 1977, 11–12, 74). Several international human rights organizations sent "fact-finding missions" to Brazil to ascertain whether these accusations were justified. The decade of the 1970s saw the growth of indigenous advocacy organizations throughout the world. At this time a number of groups also formed in Brazil for the defense of Indian rights, often criticizing prevailing models of "development," which seemed to result, for affected populations, only in increased poverty and inequality as well as environmental destruction (Ramos 1998; Wright 1988). By the 1980s, as news of widespread deforestation in the Amazon spread in North American and European countries, many human rights organizations added environmental protection to their agendas, thus strengthening their public appeal (Conklin and Graham 1995).

The development programs launched in Amazonia brought major social and demographic consequences for indigenous peoples. However, the more pessimistic forecasts of population extinction were not confirmed. By the 1980s the demographic future of Brazilian Indians already looked brighter. According to anthropologist Mércio Gomes,

> What comes out as most surprising and extraordinary in relations between Indians and Brazil is the possible historic turnaround in indigenous demography. It would certainly be bold to affirm that the Indians, at last, have survived, and that this is a concrete and permanent reality. . . . But the fact is that there are strong indications that the surviving Indian populations have been growing in the last three decades, confounding the alarming prognostications . . . of a short while ago. (1988, 16–17)

By the 1990s perceptions about the future of Indians had changed. In Ricardo's words, "the hypothesis of the physical disappearance of Indians in Brazil is gone, and therefore we are not facing a 'lost cause,' as was sometimes said a few years ago" (1996, xii). This reversal of the pessimistic picture of the future of indigenous peoples was based on the realization that many populations had been growing continuously for quite a long period of time.

Also, indigenous peoples in Brazil were increasingly making their voices heard. In part through their experience with support groups, the

Indians learned of the power of the "indigenous image." The Xavánte wore paint and carried bows and arrows when they appeared at government agencies in Brasília, winning media attention for their struggle to reclaim their land base. After democracy returned to Brazil, a new constitution, promulgated in 1988, "revolutionized relations between the state and indigenous peoples and terminated five centuries of integrationist policy. . . . It increased enormously the rights of the Indians, recognizing their social organization, their practices, religions, languages, and beliefs. Above all, it called the Indians 'Indians' and gave them the right to continue as such" (Souza 1994, 218–19). This constitution also gave the Indians broader rights to organize themselves. A number of Indian organizations and support groups were founded at this time (Ramos 1998, 259–60; Souza 1994). As Urban (1985) points out, the number of distinct indigenous societies in Brazil, formerly politically autonomous, brought into contact with national society at different times and under different circumstances makes it difficult to define common goals. Nevertheless, Brazilian Indian leaders consistently define self-determination to include control over their lands' natural resources and the right to use them as they see fit. This may not always be in accordance with the ideal of resource sustainability upheld by their environmentalist supporters (Conklin and Graham 1995, 705).

Eduardo Viveiros de Castro argues that the newfound demographic and sociocultural permanence of indigenous peoples obliges Amazonian anthropologists to give some deep thought to their analytical tools as well as their research topics. According to him, the result of the new situation is a (welcome) blurring of the division of labor between those specializing in so-called "pure" societies and those studying "acculturated" societies. Ethnographic examples show that the trajectory of indigenous peoples in contact is not necessarily from "adaptive integrity" to a future characterized by social breakdown and anomie. Nor are their paths to that future inevitably those laid down by the heavy hand of Western expansionism. Castro calls attention to the need to recognize "the historical agency of native peoples" (1996, 192).

In the 1990s, an emphasis on history, justified by the requirement to better understand the diachronic dimensions of interaction between indigenous peoples and national sociopolitical systems, has become central to anthropological discourse about Amazonian peoples.[6] Recent edited volumes on Amazonia and indigenous peoples in Brazil, covering the various subfields of anthropology, agree in stressing the historical perspective.[7] Anna Roosevelt (1994, 11) calls for greater emphasis on postcontact history.

> Traditional ethnographies customarily explain the present configurations and adaptations of Amazonian Indians without reference to the many marked changes that have taken place in indigenous lifeways and their political, economic, and social context in the last 500 years. . . . In order to gain a clearer understanding of the reasons for the patterns of native lifeways in specific regions today, it is important to evaluate the specific impact of the European conquest.

Manuela Carneiro da Cunha (1992a, 22) argues that rethinking the history of indigenous people has the potential to help us reflect on the position of these societies in the national context and more widely on the future of interethnic relations in Brazil.

> During nearly five centuries the Indians were thought of as ephemeral beings in transition, transition to Christianity, to civilization, to assimilation, to disappearance. Today, we know that indigenous societies are part of our future and not only of our past. Our common history . . . was a rosary of iniquities committed against them. We can only hope that the relations established from now on will be more just, and perhaps the sixth centenary of the discovery of America will be something to celebrate.

About the Book

We see this book as a case study that may help us to better understand not only the situation of Indians in Brazil but also that of indigenous and other marginalized peoples encapsulated in national societies in many other parts of the world. Although Western expansionism and colonialism have taken on different tonalities under different historical, political, and economic circumstances, the end result has usually been that indigenous societies have fallen to the lowest rung of the social ladder. Toward the end of the book, when we explore the biological and health transitions that the Xavánte are undergoing, the parallels between our case study and the situation of indigenous peoples in other parts of the world will become evident.

There are two rather unusual things about this book: one is the long time period over which the field data on the Etéñitépa Xavánte that we have analyzed and compared in the book were collected, and the other is the book's interdisciplinary breadth. David Maybury-Lewis, in 1957 to 1958, carried out the first anthropological study of the group, resulting in a book, *Akwẽ-Shavante Society* (1967), on their social structure. One of the authors of the present book (Salzano) took part, with James Neel, in the extensive study, carried out in 1962, of the human biology and genetics of the Etéñitépa group. Another author (Flowers) did research in 1976–77 covering the human ecology and demography of the same group. In the 1980s an anthropological linguist, Laura Graham, studied aspects of Xavánte oratory and leadership. The most recent field studies, by Coimbra, Santos, Flowers, and their students, were made over five years, from 1990 to 1995. Although the village site had changed, the community that we saw in the 1990s was largely made up of the descendents of the 1960s population and even included a number of individuals seen in the 1960s. Chapters were not written by the authors separately; all authors worked (with many heated but friendly arguments) on the writing of each chapter. The expertise of each author contributed to the whole book.

In chapter 2 we introduce the geographical area where the Xavánte live, describing the climate, vegetation, soils, and fauna of the *cerrado* region of Central Brazil. Xavánte reservations are located along with population numbers and transportation links to regional urban centers. The chapter goes on to place the Xavánte as a Jê linguistic group and then describes some aspects of Xavánte social structure, including the age set system and the exogamous moieties that regulate marriage. A description of the site and physical layout of Etéñitépa village follows.

Chapter 3 describes the history of the interactions of the Xavánte with the Brazilian national society over a period of nearly three centuries. In the early eighteenth century the Xavánte clashed with gold seekers in colonial Goiás. We describe how the Indians raided mining towns and how the colonial governors eventually made peace with some Xavánte and settled them in secular mission villages. Nineteenth-century accounts tell how the Xavánte abandoned the settlements and migrated west across the Araguaia River, where they maintained their isolation and hostility until the middle of the twentieth century, when they again came into contact with the national society. From this time on we can follow the particular history of the Etéñitépa group over a number of village splits and moves. In recent times the Etéñitépa

Xavánte have struggled to include more of their former territory in their reservation and have been involved in the rise and fall of a government project to mechanize rice growing on their reservation. While maintaining many traditional values, the Etéñitépa Xavánte are forming an increasing number of political, economic, and cultural links to outsiders. The history of the Xavánte shows that, although the intensity of the interaction has varied, their society has been connected to the world system for centuries, deeply affecting its internal functioning and structural relations.

In chapter 4 we review genetic information concerning the Xavánte. We compare data from the Xavánte at the morphological and biochemical levels with similar data from other South American groups. We refer to data derived from research conducted at Etéñitépa in the 1990s as well as that obtained in previous research on the Xavánte. Our analysis indicates that no significant changes in the gene pool of the Etéñitépa Xavánte have occurred in the last thirty years.

In chapter 5 we describe the demographic dynamics of the Etéñitépa Xavánte from the time of contact in 1946 to the present, based on data collected in 1957–58, 1962, 1976–77, and 1990. We point out some of the methodological difficulties of doing demographic analyses of indigenous groups in Brazil because of the absence of a system for collecting vital statistics on a regular basis. Data from censuses collected in the Etéñitépa Xavánte population show that, after a period following contact when the population declined due to epidemic disease and social disruption, it has been increasing up to the present. Based on the reproductive histories of women, we show that during the 1960s fertility dropped and childhood mortality rose sharply. While infant mortality began to decline in the 1970s, it is still very high relative to national rates. In the last part of the chapter we describe how adult mortality after contact was due not to disease alone but also to violence provoked by accusations of sorcery that fell harder on the weaker faction, and we show the resulting effects on marriage patterns a generation later. An important point in this chapter is the interplay between aspects of Xavánte social organization and the impacts of the depopulation that followed contact.

Chapter 6 begins by showing how the Xavánte have traditionally made their living in the *cerrado* regions of Central Brazil. They have learned to gain their subsistence from its resources by hunting, gathering plants for food and manufacture, fishing in the rivers and streams, and planting crops in the gallery forest. We describe recent changes in their subsistence activities, in particular those deriving from the influence of government-sponsored development projects. During the late 1970s and early 1980s the Indian agency FUNAI implemented a project on the Xavánte reservations intended to make the Indians large-scale producers of rice for the regional market. We show how the program failed, and, comparing time allocation data from 1976–77 to data collected in 1994, we find that the Xavánte at present are pursuing a more "traditional" subsistence strategy. However, because of the need for resources from the outside even to practice "traditional" subsistence, the line between tradition and "modern" has become blurred. In recent decades much of the *cerrado* has been irrevocably changed by intensive agriculture; Pimentel Barbosa reservation is one of the largest areas of relatively nondegraded *cerrado* vegetation remaining in Brazil. The Etéñitépa Xavánte, with the assistance of environmentalist nongovernmental organizations (NGOs), are fighting to block a proposed inland waterway that would have a negative impact on the fish and other wild resources of the reservation. The Etéñitépa Xavánte have formed many links in recent years with outside organizations, most of which are NGOs concerned with environmental issues.

In chapter 7 we briefly examine the history of Indian health services in Brazil, pointing out that they have often been badly organized and inefficient. Our analysis shows that health care available for the Xavánte far from meets their needs. We describe recent structural changes in the national health system and discuss how these are affecting health services for indigenous peoples.

In chapter 8 we present epidemiological data for the Etéñitépa Xavánte in order to discuss their current health status. Because of the lack of reliable statistics on Indian health in Brazil, we rely for our analyses almost entirely on primary data collected by our research team. We describe the Etéñitépa village and its surroundings from a sanitary point of view. Data from local hospitals and clinics show high morbidity and death rates among young children, primarily from gastrointestinal and respiratory diseases. The epidemiological profile of the Xavánte is still dominated by infectious and parasitic diseases. We also present information on other diseases affecting Xavánte health, including intestinal parasites, malaria, and tuberculosis. We used anthropometry to evaluate the nutritional status of children, pointing out that growth deficits may often be due to a synergy between infection and malnutrition. High rates of anemia were found in young children and women of reproductive age.

Protein-energy malnutrition is present at Etéñitépa at rates similar to those found in the poorest regions of Brazil.

In chapter 9 we discuss the epidemiological transition that the Xavánte are experiencing. We begin by reviewing some of the major formulations of the theory of epidemiological transition. We argue that, although infectious and parasitic diseases are still the leading causes of sickness and death, it is clear that chronic noncontagious diseases, including hypertension and diabetes, are becoming increasingly important. Among the Xavánte, changes that have taken place in settlement, levels of physical activity, and dietary patterns, are affecting health conditions. We compare anthropometric and other health-related data collected in the 1990s to data collected in the 1960s. The trend has been toward changes in body composition (weight gain) and the emergence of hypertension. Although the Etéñitépa Xavánte clearly seem to be going through a health transition, the pace is slow compared to what is happening on other Xavánte reservations, where rates of obesity and diabetes are already alarming. These comparisons suggest that the different rates of transformation in health conditions that one observes in the various Xavánte reservations are related to political and economic aspects of their postcontact experience.

Finally, in chapter 10 we bring together the topics looked at in previous chapters to discuss whether the Xavánte case fits any of the theoretical frameworks of health transition that have been proposed to describe the experience of other societies. Our argument is based on the demographic, ecological, and health/disease data presented throughout the book. We conclude that the Xavánte transition has characteristics that differentiate it from previously formulated models. We stress the importance of recognizing diversity and paying attention to local contexts when talking about the current situation and patterns of transition that indigenous societies in Brazil are experiencing today.

Chapter 2

Geographical and Social Setting

The Geographical Setting

The Xavánte live today, as they did in the past, on the Brazilian Plateau, on uplands that are drained by northward-flowing rivers: the Tocantins, the Araguaia, and the Xingu (fig. 2.1). Most of this southeastern region of the Amazon river basin, often loosely referred to in the literature as Central Brazil, is covered by the distinct type of vegetation known as *cerrado.*[1]

The first human inhabitants of Central Brazil were hunter-gatherers who appeared around 6000–9000 B.C. and are known from their flaked tools and numerous painted rock shelters (Roosevelt 1992, 58). After about 800 A.D. sites attributed to agricultural peoples are found, with pottery and houses set in a circular arrangement reminiscent of the typical village layout of contemporary *cerrado* peoples (Wüst 1994, 332). Since many of the indigenous groups of Central Brazil speak languages related to that of the Xavánte and share subsistence practices and social structures that seem relevant to living in the region it has been argued that these similarities indicate a long period of familiarity with the *cerrado* environment.

The total Xavánte population is approximately eight thousand, according to a recent FUNAI census. This population is distributed over seventy villages of various sizes located within the boundaries of six federal reservations (*terras indígenas*)—Pimentel Barbosa, Areões, São Marcos, Sangradouro, Parabubure, and Marechal Rondon—all in eastern Mato Grosso state (see table 2.1 and figure 2.1). Nearly half of the total Xavánte population lives on the Parabubure reservation, where there are some forty-one villages. A seventh reservation, Marãi-wasede, in a region occupied by Xavánte groups prior to contact, has been proposed but not surveyed, and the Xavánte have not yet been able to return there. Our research was carried out at the Pimental Barbosa reservation.

The Pimentel Barbosa reservation is located in northeastern Mato Grosso (approximately 13° 20′ S, 51° 40′ W) (fig. 2.2). The total area of

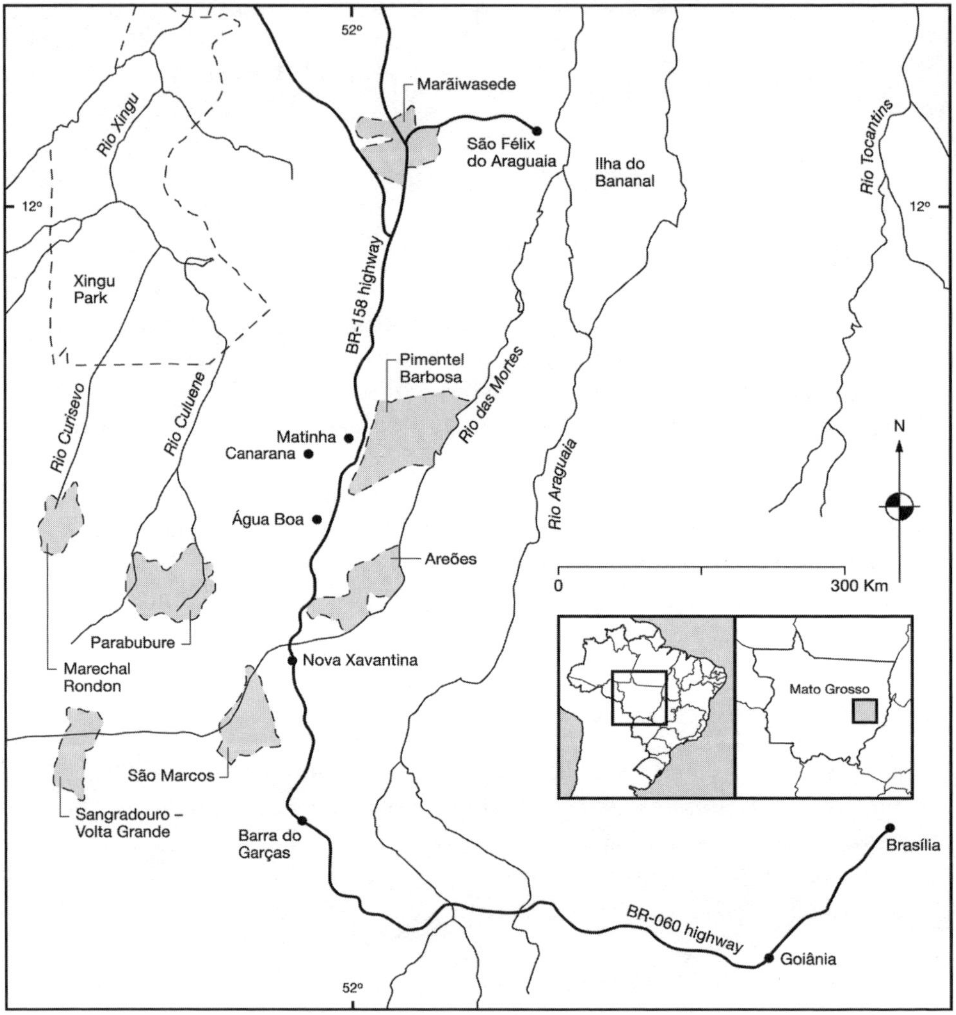

Fig. 2.1. Location of Xavánte reservations, state of Mato Grosso, Brazil

Pimentel Barbosa is 328,966 ha. At present there are five villages on the reservation: Etéñitépa, Tanguro, Caçula, Água Branca, and a new village, Pe'azarupré, recently split from Etéñitépa. All of our research on Pimentel Barbosa reservation was carried out at Etéñitépa village, which has been continuously occupied since 1972 and has an attached FUNAI (National Indian Foundation) post.

The reservation is bounded on the east by the Rio das Mortes, a tributary of the Araguaia, which flows north to join the Tocantins, a great river that empties its waters into the Atlantic about 200 km east of the mouth of the Amazon. On the west, Pimentel Barbosa is bordered by the Serra do Roncador (the Snoring Mountain), consisting of irregularly shaped hills and flat mesas that may rise to altitudes of 400 to 600 m at the highest points. The northeastern slope of Serra do Roncador falls gently to the Tanguru and Suiá-Miçu Rivers, tributaries of the Xingu. The region is crisscrossed by various smaller rivers and streams, which though never drying up completely, alternate seasonally between low and high water. There are also small lakes and marshes within reservation boundaries, especially near the Rio das Mortes. Overall, Pimentel Barbosa's topography is flat or rolling, with outcroppings concentrated in the region of the Serra do Roncador.

The average annual temperature in the region is 20 to 22°C. Average maximum temperatures in the hottest month, September, may reach 30 to 36°C, and extremes of 40°C have been reported (Caldeiron 1993). Average daily temperatures in the coldest months (June and July) range between 15 to 24°C maximum and 8 to 18°C minimum. Relatively cooler daytime temperatures resulting from polar cold fronts (*friagens*) may last from three to four days and often occur at various times during the

TABLE 2.1. Xavánte Reservations

Reservations (Terras Indígenas)	Total Area (hectares)	Population	Year of Information	Population Density (pop/area)	Legal Situation
Pimentel Barbosa	328,966	1,060	1995	0.0032	*Homologada*[a]
Areões	218,515	800	1995	0.0037	Demarcated[b]
São Marcos	188,478	1,813	1996	0.0096	*Homologada*
Sangradouro/ Volta Grande	100,280	920	1997	0.0092	*Homologada*
Parabubure	224,447	3,100	1995	0.0138	*Homologada*
Marechal Rondon	98,500	376	1996	0.0038	Demarcated
Marãiwasede	168,000	—	—	—	Delimited[c]
Total		8,069			

[a] *Homologada* means that the registration of a reservation has been officially ratified. It is the final stage of territorial recognition.

[b] *Demarcated* indicates that a reservation has been surveyed and the boundaries marked.

[c] *Delimited* indicates that a reservation has been proposed and preliminary boundaries identified. Marãiwasede is a precontact village site from which the inhabitants were removed after contact to make way for a large ranch. They have been living at a village called Água Branca on the Pimentel Barbosa reservation, waiting until they can return to their ancestral lands.

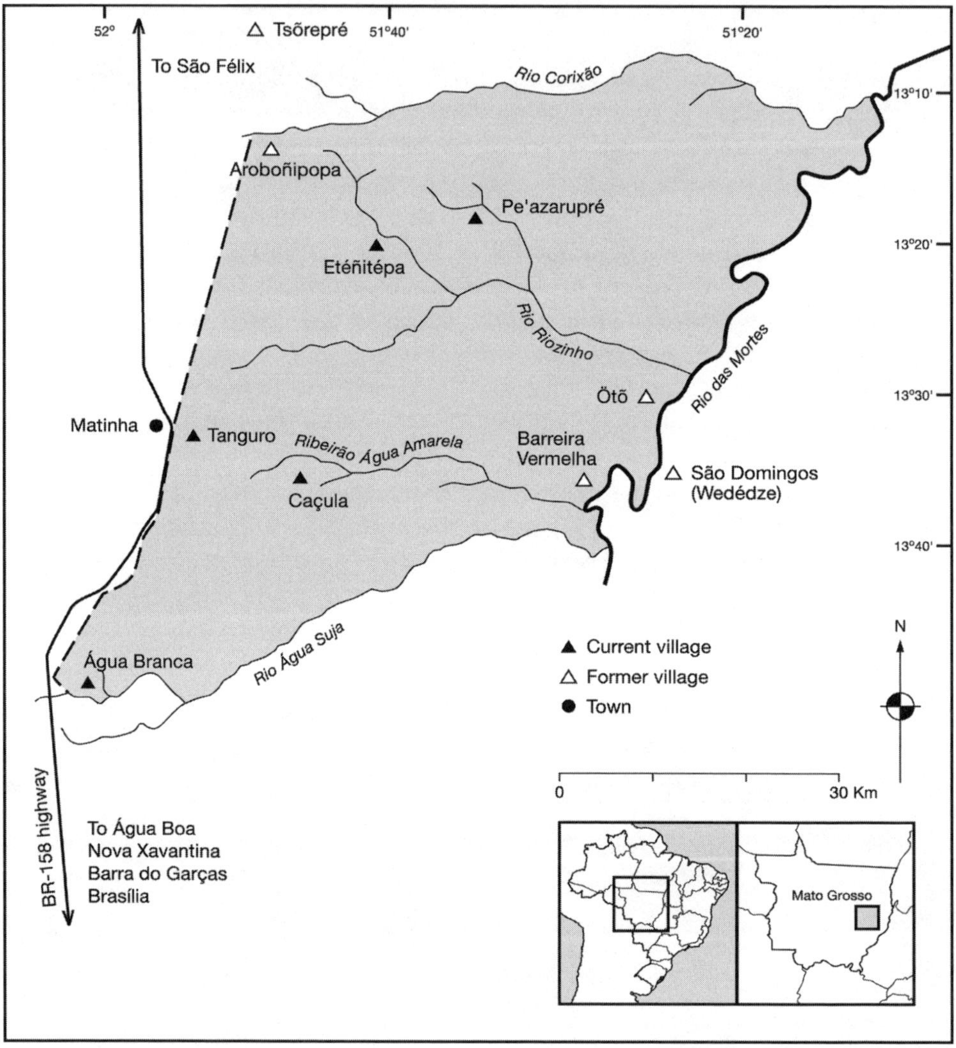

Fig. 2.2. Location of Pimentel Barbosa reservation, state of Mato Grosso, Brazil

dry winter months (Caldeiron 1993; Nimer 1989). Ratter et al. (1973) report that the daily temperature range may be over 19°C in the dry season and 12°C in the wet season. Thus, daily variation in temperature is much greater than seasonal variation.

Mean annual rainfall in the region ranges from 1,750 mm to 2,000 mm. Rainfall, however, is not evenly distributed throughout the year. In the driest months (July through August) average rainfall may be as low as 10 to 30 mm or even less (Eiten 1972). The rains last from late September (though in some years they may not start until late October) until the end of April or early May; the rainiest months are January and February. During the dry season relative humidity may drop to 25 to 30 percent or less. Nevertheless, considering the relatively short dry season and intermediate overall rainfall averages, northeastern Mato Grosso may be considered a "humid savanna" (Harris 1980a, 1980b). In spite of the short duration of the dry season, the arid months leave their mark on the landscape: "All the herbaceous and low bushy vegetation generally dries up and disappears. The leaves and withered stems of the grasses turn a characteristic golden yellow. The sparse bushes lose their leaves, retaining only their dry branches" (Joly 1970, 38).

The *Cerrado* Habitat

The vegetation that covers most of the region inhabited by the Xavánte is referred to as *cerrado*. This is the second-largest vegetation type in Brazil, covering approximately 25 percent of the country's land area (Eiten 1972; Ratter et al. 1973). It is a highly diversified vegetational form that may range from closed canopy forest to grassland, with or without scattered trees and shrubs. This vegetational mosaic offers a wide range of resources, which are seasonally exploited by the Xavánte. Relatively dense, low scrub predominates, with the ground covered by hard, narrow-leaved grasses and sedges. In general, the trunks and branches of the trees have thick and often deeply fissured corky barks. Their curved and twisted shapes are a characteristic feature of the *cerrado* landscape. Compared to other Brazilian ecosystems, the *cerrado* has relatively lower aboveground biomass, only 9 percent of that of the Amazon rain forest. In the *cerrado,* more than 70 percent of total plant biomass may be below ground (Castro and Kauffman 1998) (fig. 2.3).

Four basic structural types of *cerrado* may be distinguished at Pimentel Barbosa: (1) *campo limpo,* mostly grassland with low shrubs and/or palms emerging here and there; (2) *campo cerrado,* open scrub dominated by grasses and herbs, with scattered medium-tall or low trees and shrubs; (3) *cerrado* (*sensu stricto*), usually composed of 30 to 40 percent (or less) semi-open scrub or low arboreal forms, with scattered trees ranging from 3 to 4 m in height; and (4) *cerradão,* a taller arboreal

Fig. 2.3. *Cerrado* vegetation covers the region where the Xavánte live. *Cerrado* is highly diversified, ranging from open grasslands to closed canopy forest. The twisted shape of the trees is a characteristic feature of the *cerrado* landscape. (Photo by C. E. A. Coimbra Jr., 1995.)

form, with trees rising 10 to 15 m and on occasion forming a closed canopy (see Cole 1986; Coutinho 1978; Eiten 1972, 1975, 1994; Ratter et al. 1973; and Rizzini 1979). It is important to note that, despite the botanical and ecological peculiarities of each of these domains, the range of structural forms in *cerrado* is "completely continuous in the sense that stands can be found in any region which may be ranged in a series from arboreal, through all grades of scrub and structural savanna, to (usually) pure grassland" (Eiten 1972, 231).

The structural continuity of *cerrado* vegetational types is only broken at valley bottoms or by the margins of the many winding clear-water rivers and streams that crisscross Xavánte territory. In these humid areas *cerrado* gives way to broadleaf mesophytic evergreen gallery forest (fig. 2.4). These are narrow strips (100 to 200 m wide) of closed canopy arboreal formations with tree heights reaching 15 to 40 m. Alternatively, in areas where the water table reaches the surface during most of the

Fig. 2.4. Gallery forest grows along streams where more moisture supports taller tress. The Xavánte make their gardens in gallery forest. (Photo by C. E. A. Coimbra Jr., 1995.)

year and hydromorphic soils tend to be permanently saturated, sometimes even forming small pools, gallery forest is replaced with palm stands. These are characteristically dominated by *buriti* palm (*Mauritia vinifera*), sometimes combined with *açaí* (*Euterpe* sp.) and are usually separated from the *cerrado* by strips of grass and sedge marshlands.

Fire resistance is a distinctive feature of the *cerrado.* This evolved in response to a history of frequent fires (anthropogenic or not) and resulted in the development of an incredible variety of belowground biomass, which includes various kinds of large roots, tubers, and other organs of water and nutrient storage (see Castro and Kauffman 1998; Eiten 1972; and Rizzini and Heringer 1961). These adaptations allow the vegetation to grow back vigorously after burning, even before the first rains. According to Eiten (1972, 303), "annual fires over many years will not destroy a cerrado . . . [fires] may lower and open out the woody layer but hardly affect the herbaceous cover except for species composition."

Although natural fires may occur every once in a while, most often triggered by lightning, man-made fires were the most likely cause of recurrent burning even before the arrival of Brazilian farmers and ranchers in the region. Eiten estimated in the late 1960s that, on the average, a given *cerrado* area would be burned over by indigenous peoples of the Serra do Roncador region every three to five years (1975, 132). In the dry season the Xavánte set fire to extensive areas of *cerrado,* either to surround game (Leeuwenberg and Robinson 2000, 391) or, as noted by Flowers (1983a, 45), to make it easier to travel. According to Harris (1980b, 32), "fire set during the dry season after tuberous plants have accumulated starch in their underground organs, also makes foraging for roots, tubers, and other plants easier."

In the Serra do Roncador region, *cerrado* vegetation grows on very deep red and/or red-yellow latosols, well drained, acid, and nutrient poor, often showing patches of podzols (mostly alfisols) (Askew et al. 1970a, 1970b; Eiten 1975; Ratter et al. 1973). The amounts of clay and sand vary substantially, and quartz stones or laterite pebbles may be present in various amounts and at various depths. Scattered throughout Xavánte territory one often sees blocks of spongelike laterite outcroppings; on occasion the soil surface may be made up entirely of laterite pebbles and small stones (Askew et al. 1970a, 1970b; Eiten 1975).

These soils have low cation exchange capacity, low base exchange, and relatively small amounts of organic matter (Freitas and Silveira 1977; Goodland and Ferri 1979; Haridasan 1994). High aluminum saturation is a major limiting factor, which *cerrado* flora has evolved to endure (Eiten 1972). Under prevailing soil conditions of high acidity and aluminum toxicity, coupled with the poor content of other key plant nutrients such as nitrogen, potassium, and phosphorous, commercial farming in the *cerrado* requires the utilization of large amounts of expensive fertilizers. Traditional Xavánte horticulture, however, takes place on the humus-rich podzolic soils that support the gallery forests. The higher humus content of these forest soils prevents nutrient ions from being leached so quickly, thus making them more available for absorption by plants.

The variety of *cerrado* landscapes and vegetation types is matched by the diversity in animal life (Hershkovitz 1972). Notwithstanding its diversity, the biomass of each particular species tends to be low in relation to the overall faunal biomass (Cerqueira 1982). Available faunal listings for the Central Brazilian *cerrado,* which are far from complete, include approximately 950 species of birds and 300 species of mammals (Costa et al. 1981; Pine et al. 1970; Sick 1965) (see also chapter 6).

Transportation and Connections

Different parts of the Pimentel Barbosa reservation are connected by tracks, some only footpaths and some usable by trucks and other vehicles, although most of the latter may become impassable in the rainy season. Etéñitépa is about 40 km by dirt road from BR–158, a partially paved federal highway that leads to Barra do Garças, a commercial center 340 km to the south (fig. 2.1). The Xavánte of Etéñitépa maintain economic and political connections with towns along the highway to the south, starting with Canarana and Água Boa, commercial towns that are centers for local agribusiness where the Xavánte shop, get medical attention, and collect their social security benefits at local banks. The FUNAI regional office that has jurisdiction over their reservation is also located in Água Boa. Nova Xavantina is another town where the Xavánte go for shopping and medical care. Until 1994, Nova Xavantina was the location of the office of the Associação dos Xavantes de Pimentel Barbosa, an organization set up by the Xavánte for political representation. Another 140 km south of Nova Xavantina is Barra do Garças, where BR–158 intersects BR–070, leading eastward to Goiânia and ultimately to Brasília.

The Etéñitépa Xavánte go to Goiânia, where they have numerous social connections, for schooling and medical attention. Brasília, the capital of Brazil, is 1,000 km distant from the reservation, but it is a frequent destination for Xavánte leaders, as they travel there by bus to gain benefits for their communities by applying political pressure at FUNAI headquarters and other federal government offices. But Xavánte connections to the country and the world do not end at Brasília. Several young leaders from Etéñitépa were educated at Ribeirão Preto in São Paulo state, where they sometimes visit, and the community has formed close ties with nongovernmental organizations and people in academia and the arts in both Rio de Janeiro and São Paulo. Most recently, with the sponsorship of various international NGOs and academic institutions, a number of Xavánte have traveled nationally and internationally (see chapters 3 and 6).

The Xavánte are distributed over six reservations, which are socially and politically independent. Sometimes Xavánte leaders cooperate for political purposes, as when they are involved in territorial struggles or other major issues that may affect more than one community. Also, because of conditions associated with contact, a complex network of kin relationships connects the various Xavánte groups (see chapter 3). Xavánte families from Pimentel Barbosa sometimes visit relatives on other reservations.

Unlike many other Indians living on reservations, the Xavánte from Etéñitépa are not socially or politically limited by their reservation boundaries, even though they derive their subsistence primarily from reservation resources. They identify very strongly with their home territory, as the region where Pimentel Barbosa is located is where the Xavánte settled after they migrated westward and crossed the Rio das Mortes in the nineteenth century (see chapter 3). It was from here that, after the population increased, some groups spread out to found new villages.

The Social and Cultural Setting

The Xavánte as a Jê Group

In the early nineteenth century the naturalist Karl von Martius (1867, 1:256–60) recognized the linguistic and cultural similarities among a number of Central Brazilian groups and introduced the general term *Gê,* now written *Jê* (Rodrigues 1986, 11), to characterize them. The modern classification of Jê groups generally follows Nimuendaju (1942, 1–2), who divided them into three main branches.

1. The Northern Jê groups are divided into the Eastern Timbíra (including the Krahô, the Ramkókamekra or Canela, the Parakatejé or Gavião, and other groups), the Western Timbíra (Apinayé), the Northern and Southern Kayapó (Kreen-akarôre or Panará), and the Suyá.
2. The Central Jê have two subbranches. One, the Akwẽ, includes the Xavánte, Xerénte, and Xakriabá. The other was comprised of the Akroá (often divided into the Akroá-Assu and the Akroá-Mirim), with the closely related Gueguê, all now extinct.
3. The Southern Jê consist of the Kaingáng, the Xokléng (Aweikoma), and the Botocudo.

Greenberg (1956, 794) and Mason (1950, 287) placed the Kaingáng and Botocudo languages within the wider classification of Macro-Jê, rather than considering them as a branch of Jê proper, and Mason (290) added the Jaikó as a separate branch of the Jê. All the languages of the

Macro-Jê linguistic family are spoken by peoples living in the eastern and central part of the Brazilian Plateau. According to Urban's (1992) reconstruction based on historical linguistics, separation between the Macro-Jê and Jê languages may have begun five or six thousand years ago. Central Jê is considered to be the parent branch of the Jê languages and probably originated in the headwaters regions of the São Francisco and Tocantins Rivers, near the former territory of the Xakriabá. Separation of the Southern Jê languages could have taken place around three thousand years ago as some groups migrated south. As other groups expanded north and west into the Amazon Basin a further division, one to two thousand years ago, separated the Central from the Northern Jê. Finally, linguistic differentiation within groups occurred over the last five hundred years.

Besides their linguistic relationship, the groups that occupy the uplands of Central Brazil share many similarities in social structure and subsistence strategy. They build large, nucleated villages, with houses forming a circle (or a semicircle in the case of the Central Jê) around a central plaza where communal events take place. They have elaborate social structures, with organizational units that may include lineages, clans, nondescent moieties, age sets, age grades, and men's societies. Individuals may have special ascribed roles and "formal friendship" links with other individuals. Many Jê groups practice ceremonial log racing, in which two teams carry heavy logs over a distance of several miles, rolling the log successively from the shoulder of one team member to that of another.

Internal factionalism is often intense, and the multiple forms of organization offer possibilities for cleavage that frequently result in village splitting. A disaffected segment, sometimes a group of extended families, sometimes a clan or men's society, may leave to found a new village or join another already established (Maybury-Lewis 1967; Neel and Salzano 1967a).

Although all Jê groups of Central Brazil have long practiced agriculture and rely on it for subsistence during a large part of the year, most of them until recently had a period of dispersal, usually in the dry season, when families went "on trek," that is, took long hunting and gathering trips during which they lived mainly on wild foods. In the past these treks often covered long distances, leading to familiarity with the natural resources, as well as the human occupants, of an extensive territory (see chapter 6). All groups are now confined to reservations of various sizes and have either modified trekking by making much briefer trips or abandoned it altogether.

The social organization of Central Brazilian Jê groups tends to maintain relations among villages, even though these relations are often hostile. The varied linkages provide a basis for political alliances between autonomous communities and weave a social network around each individual that connects him to other communities to which, if conditions become unfavorable in the home village, he may eventually move.

It is possible, as Gross (1979) has suggested, that Central Brazilian Jê villages of the past became larger and their social organization more elaborate as a result of colonial warfare. Some villages had populations of one thousand or more, and at times the Indians were apparently able to organize a large group of warriors to attack settlements in force. At other times they could evade the expeditions sent to punish them by scattering into smaller groups that, when harassed, might wander for several years in the *cerrado* without planting crops. These wandering periods were undoubtedly times of stress and hardship, but the tactic must have allowed some to survive catastrophic events, which became more frequent after the invasion of their territory by colonists (see chapter 3).

Relationships among the Central Jê

Since all the Central Jê groups except the Xavánte and Xerénte have disappeared,[2] we must go to earlier accounts to investigate relationships among the Central Jê up to the time of the Xavánte migration across the Araguaia River around 1850. In the early nineteenth century Martius (1867, 2:134–46) compiled word lists obtained from informants identified as Xavánte, Xerénte, Akroá-Mirim, and Xakriabá. Comparison of these word lists suggests that differences among Central Jê languages were little more than dialectical (Flowers 1983a).

While the different Central Jê groups may be distinguished, it appears that during the eighteenth and into the nineteenth century a flexible network of their base villages covered the region from the western tributaries of the São Francisco to the Araguaia. Alliance, fission, and recombination among these villages took place in response to varying pressures that resulted from attacks by the colonists, the extinction of "pacified" groups, opportunities for predation on the colonists' cattle, and hostilities that arose from rivalries in leadership.

The Xavánte and the Xerénte may have been one people until the nineteenth century, when the Xavánte crossed the Araguaia and the Xerénte remained near the Tocantins. In 1994, 1,552 Xerénte were living in Tocantins state on two reservations with a total area of 183,245 ha (ISA 1996, 635–36).

Xavánte Social Organization

The social organization of Jê societies is remarkably complex. This complexity has attracted the attention of anthropologists for a long time, and it has been a major topic of ethnological research in Central Brazil (for reviews see Cunha 1993 and Maybury-Lewis 1979).

The interrelation of the Xavánte age grade and age set systems exemplifies this complexity. While the age grades refer to life stages for men and women (small children, adolescents, young adults, and mature adults), the age set system is made up of eight named age sets, which rotate in a forty to fifty year cycle (table 2.2).

Young boys, between the ages of eight and thirteen, become *wapté.* They then leave their family households and go to live as a group in the *hö,* or "bachelors' hut," a house that is built for them at one end of the village semicircle. A boy begins to belong to an age set when he moves into the *hö,* where he will remain for approximately five years, until the time when his age is initiated into manhood and allowed to marry. According to Maybury-Lewis:

> The bachelors' hut is . . . the cornerstone of the age set system. It is the place where a Shavante boy first feels what it is to belong to an age set and participates in the comradeship, cross-cutting distinctions of clan and lineage, which such membership implies. . . . Before that time he is, in a sense, not a full member of society. He has no place in a system where social and ceremonial activities are largely carried out by the age sets. . . . The Shavante do not keep count of the actual ages of their children before they are old enough to belong to an age set; after that their relative age, i.e. the age set to which they belong, are all that matters. (1967, 105–6)

Table 2.2 shows the age set cycle and the initiation dates of *wapté* at São Domingos[3] and Etéñitépa during the second half of the twentieth century. Since there are eight age sets and they are five to six years

TABLE 2.2. Xavánte Age Grades and Age Sets at Etéñitépa

	Age Set Positions[b]								
Age Grades[a]	1953	1958	1963	1970(?)	1978	1987	1990	1996	2002(?)
	Abare'u	Tsada'ro	Anorowa	Hötörã	Ai'rere	Étẽpa	Tirowa	Nodzö'u	Abare'u
	Tsada'ro	Anorowa	Hötörã	Ai'rere	Étẽpa	Tirowa	Nodzö'u	Abare'u	Tsada'ro
iprédu (mature man) *pi'õ* (mature woman)	Anorowa	Hötörã	Ai'rere	Étẽpa	Tirowa	Nodzö'u	Abare'u	Tsada'ro	Anorowa
	Hötörã	Ai'rere	Étẽpa	Tirowa	Nodzö'u	Abare'u	Tsada'ro	Anorowa	Hötörã
	Ai'rere	Étẽpa	Tirowa	Nodzö'u	Abare'u	Tsada'ro	Anorowa	Hötörã	Ai'rere
ritai'wa (young man) *abada* (young woman)	Étẽpa	Tirowa	Nodzö'u	Abare'u	Tsada'ro	Anorowa	Hötörã	Ai'rere	Étẽpa
wapté (bachelor) *soimba* (girl)	Tirowa	Nodzö'u	Abare'u	Tsada'ro	Anorowa	Hötörã	Ai'rere	Étẽpa	Tirowa
waptébremi (little boy) *ba'õno* (little girl)	Nodzö'u	Abare'u	Tsada'ro	Anorowa	Hötörã	Ai'rere	Étẽpa	Tirowa	Nodzö'u

[a] For a man, transition from one age grade to another is closely associated with his place in the age set cycle. For a woman, change in age grade depends not only on her place in the age set cycle but on betrothal, childbirth, the naming ceremony, and so forth. Therefore, it may happen that a woman may change her age grade at a different time than a man in the same age set (for more information see Maybury-Lewis 1967, 150–52; and Graham 1995, 95–96).

[b] The years refer to the time when *wapté* were initiated and graduated to the *ritai'wa* age grade and another group of boys entered the bachelors' hut (*hö*). Maybury-Lewis (1967, 164) states that the Étẽpa were initiated in 1953 (they became *ritai'wa*). When he carried out research in 1958, the Tirowa were initated and the Nodzö'u entered the *hö* (338). The Nodzö'u were probably initiated in 1963 (154). In 1976–77, when Flowers lived at Etéñitépa, she observed that the Tsada'ro were in the *hö* and they would be initiated in 1978 (1983a, 163). Graham (1995, 91–93) writes that the Tsada'ro became *iprédu* in 1987, which indicates that the Anorowa graduated to *ritai'wa* and the Hötörã entered the *hö*. In 1990 the Hötörã were initiated and the Ai'rere entered the *hö*. Finally, in 1996 the initiation of the Ai'rere took place and the Étẽpa entered the *hö*. The remaining initiation date (1970) is estimated because it was not formally recorded. We estimate that the next initiation will take place around 2002.

apart, the full cycle lasts four to five decades. For example, in 1953 the Tirowa age set entered the *hö* and in 1958 it was initiated, that is, its members moved to the *ritai'wa* age grade (young men). The age set that entered the *hö* next was the Nodzö'u. In 1963 the members of the Tirowa age set who had entered the *hö* a decade before became *iprédu* or mature men. Because the Xavánte recycle age set names over time, a new age set with the name Tirowa will enter the *hö* around the beginning of the twenty-first century, possibly in 2002.

At any given time the Xavánte population is divided into eight groups containing individuals of approximately the same age.[4] In 1990, the Tirowa, who were between eight and thirteen years of age in 1953, when they entered the bachelors' hut, were forty-five to fifty years old in 1990.

Boys living in the *hö* are not separated from village life. They may visit their family houses and may work with their fathers in the gardens. They may attend school if one is functioning at the FUNAI post. But they usually sleep and eat in the *hö,* sharing the food that their mothers bring them. In her recent study of the Etéñitépa Xavánte, Graham writes:

> During this time of collective residence, under the guidance of their senior sponsors, they engage in various solidarity building activities. One of the most important of these is *da-ño're* singing. The pre-initiates [*wapté*] also practice various skills that require group coordination, such as hunting, fishing, and gardening, which will be important in their adult male lives. These activities, particularly singing, foster sentimental attachments between age set members. (1995, 93)

Graham's account, based on observations made in the 1980s, is not very different from Maybury-Lewis's description of *wapté* life at São Domingos in the late 1950s (see Maybury-Lewis 1967, 105–15).[5]

The elaborate initiation rituals for a boys' age set take up most of the last year the *wapté* spend in the *hö* and culminate in the ear-piercing ceremony. After this they wear short white ear sticks symbolizing maturity, enter the *ritai'wa* age grade, and are eligible for marriage.

The age set system operates for women as well. As pointed out by Maybury-Lewis, "at the time when small boys are first inducted into the bachelors' hut . . . the small girls who are about the same age are also said to belong to the same age set. Thus the respective ages and seniority of women as well as of men may be given by simply naming the age set" (1967, 149). There are important differences, however. The girls of the same age set as the *wapté* are not separated from their families and may already be married. Through the years from late childhood into adolescence, when their male peers engage in special activities while living in the *hö,* they are helping their mothers and female kin with household chores.

The clan and lineage systems are two other fundamental dimensions of Xavánte social organization (table 2.3).[6] The Xavánte have a patrilineal clan system (i.e., the child belongs to his or her father's clan). Three clans are recognized by the Xavánte (Poridza'õno, Öwawẽ and Topdató), and each of them may be represented in a given village by one or more named lineages (Maybury-Lewis 1967, 340 and genealogy 3). As Maybury-Lewis points out, although the clan is a fundamental unit of Xavánte social organization, and one that tends to persist over time, the political system operates mostly through the lineages.

> It is the lineage which is expected to act as a corporate group. A member of it cannot dissociate himself from the others on an important issue and still remain part of it. In these circumstances, he must found a separate lineage or get himself adopted into another one. Sometimes a lineage becomes so small that it disintegrates and its surviving members are all incorporated into other lineages. But the expectation that members of a lineage will not actively oppose each other is much stronger than in the case of the clan. (1967, 169)

The relationship between Xavánte lineages tends to be conflictive, as there is an ongoing struggle for political power, women, and resources. As Maybury-Lewis points out (1967, 190), "The factions are in perpetual competition for power and prestige."

According to Maybury-Lewis (1967, 75), the Xavánte at the time of his research considered it incestuous for members of the Öwawẽ to

TABLE 2.3. Xavánte Clans and Lineages

Clans:	Poridza'õno	Öwawẽ	Topdató
Lineages:	Wamãri	Uhö	Aiuté'mañãri
	Tebe	Dzutsi	Wahi

marry members of the Topdató clan. The accepted marriage was between members of the Poridza'õno and the two other clans; thus one could say that the Etéñitépa Xavánte had "a system of exogamous moieties with two clans in one moiety and one clan in the other." In one moiety was the Poridza'õno clan and in the other the Öwawẽ and the Topdató. Flowers observed in 1976–77 that these marriage practices were still operating in Etéñitépa, despite some changes. In the mid-1970s the Xavánte had a symmetrical moiety system because there were no longer Topdató men in the village, and the remaining Topdató women were all married to Poridza'õno men.[7]

The Xavánte practice polygynous marriage. Among them polygyny is largely sororal, that is, if a young man marries into a household where there are several daughters he may marry first the eldest, then the younger sister or sisters as they come of marriageable age. Postmarital residence is matrilocal, that is, the husband moves into the wife's parents' household. Brothers often marry into the same household, so a frequent arrangement is for two or more brothers to be married to the daughters of the household. Some men, after years of monogamous marriage, or after the death of the first wife, may acquire a much younger wife. In this case the young wife is usually brought to live in her husband's house, thus not conforming to the practice of matrilocality that applies to the first marriage of a young man.

The Etéñitépa Xavánte do not seem to be abandoning polygyny. Comparing the censuses of 1977 and 1990 we observed that the frequency of polygynously married men remained almost the same. In 1977, 24 percent of married men had more than one wife. In 1990, the frequency was 23 percent. In 1990, 41 percent of women were in polygynous unions, hardly varying from the 1977 figure of 40 percent. In other Xavánte reservations polygyny is not as common as at Etéñitépa. At Sangradouro, Souza and Santos (2001) found that only 4.1 percent of the married men and 8.3 percent of the married women were in polygynous marriages. This decline in polygyny may be due to the influence of Salesian missionaries at Sangradouro.

In 1990 we collected census data that illustrate changes in marital status over the life cycle of men and women. In figure 2.5 one can see that most men only marry after the age of twenty. Although a few may marry a second or third wife while in their late twenties, polygyny is more common among men in their thirties and over. The pattern for women is very different from that of men (fig. 2.6). A good number

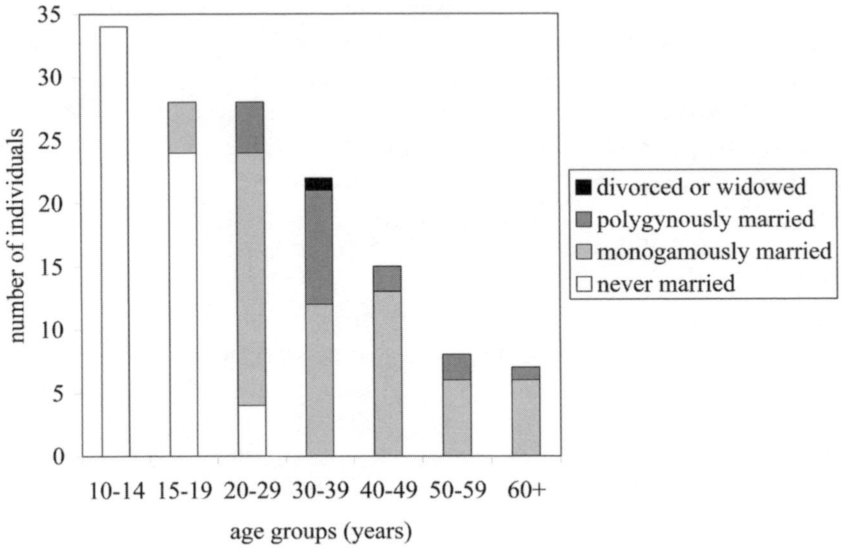

Fig. 2.5. Marital status of adult men at Etéñitépa, according to age group, 1990

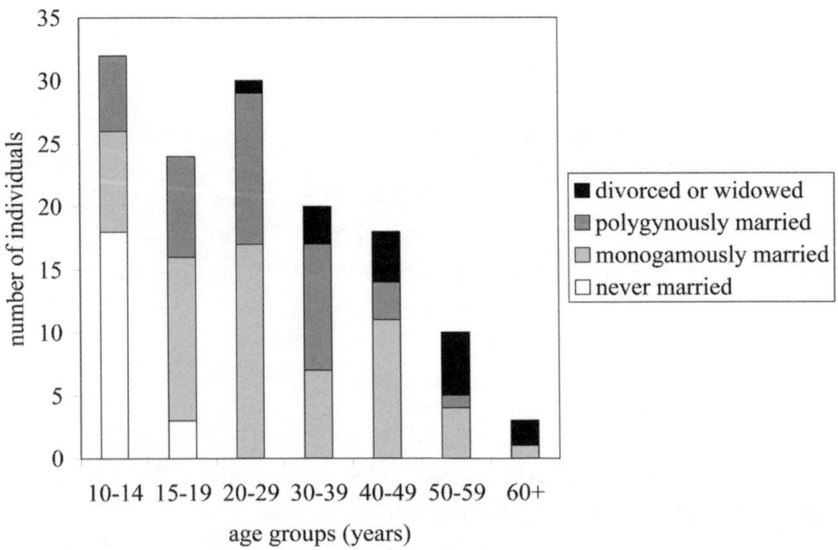

Fig. 2.6. Marital status of adult women at Etéñitépa, according to age group, 1990

marry when they are between ten and fourteen years of age; by their twenties they are all married. Often young girls marry an older man as his second wife. Another difference between men and women is that older women who become widows often do not remarry, whereas older men do.

Polygyny leads to a situation of constant scarcity of women to marry the young men. Xavánte men are not permitted to marry until their age set is initiated, when most of them are fifteen to eighteen years old. At the end of the initiation period each male youth is formally betrothed to a girl of the proper clan affiliation. However, it may be a long time before the marriage is consummated because often a young man is betrothed to a small girl and it will be several years before she reaches puberty. In 1958, Maybury-Lewis witnessed a betrothal ceremony at São Domingos, about which he wrote: "all the brides were immature, and some of them were carried out to be married on the mothers' hips" (1967, 78).

As Graham (1995, 69) points out, marriage among the Xavánte is a gradual process that begins with betrothal and proceeds, when the girl is considered old enough, with the young man making nocturnal courtship visits to his bride. When the young couple has cohabited for several months to a year, and it seems likely that the marriage will work out, the bridegroom goes hunting with a number of his age mates and presents the meat to the household of his future father-in-law. Only after the couple's first child is born does he take up residence in her father's house (see Graham 1995, 66–70; and Maybury-Lewis 1967, 75–90).

Xavánte Ceremonial and Social Life

At Etéñitépa social life is intense, and it is ordered and enlivened by ceremony. Some rituals recur daily, such as the men's evening council meeting, or *warã* (fig. 2.7), and the young men's singing and dancing around the village. Others are seasonal, such as log races (*uiwede*), which are held in the rainy season, and ceremonial preparations for hunting with fire in the dry season (*du*). Still others celebrate events in the life cycle such as the boys' initiation and weddings.

Most log races are between two opposing teams made up of the most recently initiated age set and the next oldest, distinctively painted (see fig. 9.7). In 1976 these were the Abare'u and the Nodzö'u. Both men and women log race, and races give women the rare opportunity to paint and perform as members of their age set.

Fig. 2.7. The *warã* (men's council) meets every day, morning and evening, in the center of the village, to discuss the day's affairs and make decisions. (Photo by N. M. Flowers, 1977.)

The *wai'a* ceremony, usually lasting three days, in which all initiated men participate, is held several times in the dry season (fig. 2.8). This is a complex ritual during which the men seek to get in touch with, and gain the protection of, spiritual powers. Although missionary influence has suppressed the *wai'a* in a number of other Xavánte villages, it has been held continuously at Etéñitépa.

It is not our intent in this book to describe Etéñitépa ceremonial life in detail, as this is a rich and complex topic that has already been well covered by Maybury-Lewis (1967) in the late 1950s and by Graham (1995) in the 1980s. What is striking is that, despite the many changes in the life of the community over the thirty-year interval between these accounts, ceremonial remains an essential part of Xavánte life. There is no doubt that the maintenance of ceremonial at Etéñitépa, much of which has been suppressed or has died out in other Xavánte communities, has been important in preserving sociocultural and political identities. Graham demonstrates this clearly in her analysis of the dream narrative told by Warodi, the chief of Etéñitépa, in which he frames his appeal to the villagers to "remain Xavánte forever" as a communication

Fig. 2.8. The *wai'a* ceremony expresses the progression of Xavánte men toward spiritual knowledge and power. Here a senior man is performing an "attack dance" toward a young initiate, who must maintain a ceremonial stance with eyes downcast. (Photo by N. M. Flowers, 1977.)

from the immortals. Moreover, Warodi produced an innovative performance of his dream in which the entire village took part and which Graham photographed and tape-recorded. As she points out (9), "this performance illustrates the Xavánte ability to preserve continuity through creative response to change."

Xavánte Leadership and Factionalism

Every evening and often in the early morning as well, the mature men of the village hold a formal meeting of the men's council, called the *warã,* at which the affairs of the village are discussed. Subjects that may be addressed are plans for hunting or fishing expeditions or for holding ceremonies, family disputes, strategies for dealing with territorial incursions, dealings with FUNAI, and any other matters that affect the village as a whole. When leaders return from a trip to Brasília they report to the council. If outsiders, such as anthropologists, wish to work in the community, they must be introduced to the council and explain their plans, which the council either accepts or rejects.

As Graham (1993) explains, the *warã* offers a forum for the open discussion of conflicting viewpoints. However, the style of debate differs from that found in Western meetings. When men speak, they do not frame their speech as representing their individual viewpoint but rather that of their faction. When making formal speeches, elders speak in a special way, which differs in style and vocabulary from conversational Xavánte. Others join their speech to that of the orator so that discourse becomes collective rather than individual. According to Graham (718), "The discursive practices of *warã* meetings physically and acoustically blur the boundaries among individuals to promote social cohesiveness and counteract the factionalism that constantly threatens to tear the community apart."

The political ambitions of mature men, involving competition for power, prestige, and material resources, are expressed through factionalism. A man usually relies most heavily on his sons, his brothers, and his brothers' sons, who are men of his clan and lineage, but his leadership may attract other allies. As Maybury-Lewis (1967, 168–69) writes, "clanship by itself is meaningless. It becomes significant only in context, and the context is supplied by the factional structure of the community." The village chief is head of a faction powerful enough to dominate the village. Political relations always threaten to become tense, and when rivalries and disagreements reach a certain point the weaker faction may leave to join another village or found a new one. In former times the stronger faction might resort to bloodshed to purge the community of its rivals.

At present the power and prestige of a leader depend to a great extent on his ability to deal effectively with outsiders and to obtain concessions and benefits for his community from outside sources, including FUNAI, owners of land bordering the reservation, neighboring communities, and national and international NGOs that may render material assistance or offer opportunities for community development. If a leader and his faction fail to produce the anticipated results they may lose control of the community. The factionalism that pervades internal Xavánte politics may have helped to make leaders aware of the value of public opinion when dealing with outside institutions, for it is only by swaying public opinion that they can keep their own positions. Xavánte leaders quickly became aware of the importance of the media, taking advantage of the value of "image" by appearing in full regalia when confronting the authorities. So effective have Xavánte leaders become in using these tactics that they have aroused the resentment of other indigenous groups, which accuse them of aggressive lobbying and skimming off more than their share of government resources.

Village Site and Household Structure

Etéñitépa is located on a height of land and commands a broad view over gallery forest and beyond a stream to the *cerrado* and distant rocky outcroppings. The Xavánte call the site by this name in reference to a flat-topped hill east of the village that glows a dull red when it is lit by the setting sun on dry season evenings. They occupied the same site on at least two previous occasions, as they liked its high position and the good garden land nearby. In the 1930s, before contact, there was a Xavánte village about three kilometers northwest of the present village location. At that spot Flowers picked up shards of the pottery that the Xavánte made before they obtained manufactured articles. From around 1950 to 1956 the same group had a village located about one kilometer east of the present village. One could still trace the outlines of houses and find on the ground scattered remains of trade goods such as bits of rusty wire screen from a sieve.

Some 350 m from the village are located the buildings of the government Indian post. During most of the 1990s these buildings, which included a schoolhouse, an infirmary, and a house for the Indian agent, were abandoned. The schoolhouse had actually fallen down. In 1996 a new schoolhouse was built according to the model of a municipal rural school. Behind the village houses is an airstrip, which is now rarely used (fig. 2.9). From the post, a dirt road leads to BR–158, the federal highway, which is more or less 40 km distant.

While the village is built on open *cerrado,* the stream where the villagers bathe and from which they get water flows through gallery forest. As the village, which is higher than the stream, has no form of sanitation, there are increasing problems of water pollution from runoff, especially in the rainy season (see chapter 8).

For a few years after 1972, when the village was founded, the Xavánte cut all their gardens in the nearby strip of gallery forest. As time when on and the land was used up, they opened gardens at a greater and greater distance from the village. In the 1990s, most families had gardens that were as far as five kilometers, or an hour's walk, away.

Flowers lived in Etéñitépa in 1976–77 and returned on several occasions in the 1990s. Not only did she find the village larger than in the 1970s, it had acquired an air of permanence. In contrast to past mobility the sedentism of the present had allowed mango trees to grow up, forming a continuous, deep green canopy that dominated part of the village. The soccer field, which in 1977 was marked only by two goalposts in the village center, now occupied its own location between the village and the stream.

Etéñitépa follows the traditional layout of a Xavánte village, which graphically expresses in many ways both the social organization and the ideology of the community. Wherever the Xavánte build a village, or even a temporary camp, the disposition of the houses follows the same pattern: a semicircle with the open side facing a stream used for household water and bathing.

The area in the center of the village is "public space" and includes the place where the *warã* meets twice daily. Every activity that takes place in this central area concerns in some way the social life of the village: a party of hunters returns with game, the afternoon soccer game is played, log-racing teams arrive and drop their logs with a flourish, young men form dancing circles and "sing around the village," or a man shouts for everyone to hear a decision reached in the *warã.* On moonlit nights, after the council meeting, women may join the men in the dancing circle. The center of the village is in a way the stage on which the

Fig. 2.9. Aerial view of Etéñitépa village. (Photo by N. M. Flowers, 1977.)

social life of the village is performed. Surrounding the village center is a semicircle of houses. Each house has only one entrance, facing the village center, so that to leave one's house is to enter the public arena. Domestic life takes place mostly within, and in the immediate vicinity of, each house.

A Xavánte house is associated with the clan and lineage to which the senior male member is affiliated, but since Xavánte society is both patrilineal and matrilocal in each generation the affiliation of a household changes. As we explained, if a man is Poridza'õno his daughters must marry men of the Öwawẽ moiety. The young men who marry the daughters and move in are often brothers, for there is a strong preference for brothers to stay together by marrying a group of sisters. When the father-in-law dies, the sons-in-law become joint heads of the household, which then changes in clan affiliation. At this point, the brothers often decide to build houses of their own, but these are usually adjacent in the village circle. As Maybury-Lewis (1967, 172–78) noted, this often results in segments of the village being associated with different lineage-based factions. For example, when Flowers lived in the village in 1976–77, a house was built for her near that of the chief, Apöwẽ, at the left side of the village circle. Next to his house were the houses of two of his sons. In fact, all but two of the houses on this side of the village belonged to men of Apöwẽ's lineage. Of the two exceptions, one belonged to a man married to Apöwẽ's daughters. The right side of the village was less homogeneous, but the two houses on the far end were occupied by Apöwẽ's two oldest sons; one son (Warodi) succeeded him, while the other left to found a new village. This arrangement expresses the traditional importance of the "end houses," known as *amrã,* which are usually occupied by politically prominent families (172). The only other named house is the *hö.* This house is placed outside the village circle, and for each succeeding age set a new one is built, the position to left or right alternating. In recent years, as leadership has passed to a younger generation, the political expression of the village layout is less clear. For example, the present chief (Suptó) is a young man still living in his father-in-law's household, and the families living in the *amrã* are not particularly prominent in the political life of the village.

In 1990 there were thirty-two houses at Etéñitépa. Most of them were still of the traditional beehive shape, known as *ri uptabi,* or "true houses." A traditional Xavánte house is built with a sturdy tree trunk as the center post. On the circumference of a circle approximately five to six meters in diameter flexible saplings are planted to form the perimeter of the house; their tops are bent to the center post and tied together with vines. The house is thatched with palm leaves from the top to the ground, with no distinction between walls and roof. In a traditional house there is but one doorway, and "windows" are made when needed by pulling the thatch aside. Houses are separated by a distance of five to twelve meters. The shelters built by groups on trek are essentially miniatures of village houses, only big enough for people to crawl into. According to Maybury-Lewis (1967, 58) the relative positions of the shelters in the camp were kept the same as in the village, so that people had the same neighbors.

Just as the layout of the village expresses the public life of the community, the internal layout of a house expresses the domestic organization of a Xavánte extended family. Inside the house the section at one side of the door is usually occupied by the sleeping mats of the senior man and his wife or wives, while on the other side of the door is the place of the oldest daughter and her husband. The sleeping mats of other family members are distributed around the house walls. Each nuclear family has a designated space; if a man is married to more than one wife, one sleeps on his left and the other on his right, each with her children (Sá 1983; Silva 1983). The interior of a Xavánte house is an open space without partitions except when a daughter of the house marries; then a matting screen is erected around the sleeping area of the young couple to give them some privacy (see Graham 1995, 66–69).

Prior to contact the Xavánte did not use hammocks, as they sleep on mats cushioned with palm leaves, which are laid either on sleeping platforms or directly on the house floor. There are now usually some hammocks of Brazilian manufacture used casually for resting or naps. The cooking fire was formerly built in the center of the floor, with smoke drifting out through the palm fronds of the roof. One innovation that has become general is that the cooking fire is no longer inside but in a kitchen lean-to at the side of the house, thus eliminating the smoke that formerly blackened the interior (fig. 2.10). Belongings are stored in

Fig. 2.10. Xavánte house with a kitchen lean-to attached. This has become a more common house type in recent years. A woman is hulling rice. (Photo by C. E. A. Coimbra Jr., 1991.)

Fig. 2.11. The interior of a Xavánte house with the fire in the center. The fire gives warmth at night as well as being used for cooking. Belongings are kept in baskets hung on the walls of the house. (Photo by N. M. Flowers, 1977.)

baskets hung on the walls. A Xavánte house is dark and relatively cool during the day, a welcome relief from the sun's heat, although the thatch may be pushed aside in places to provide light or to observe what may be taking place outside (fig. 2.11).

Although the Etéñitépa Xavánte have experimented with various exterior house forms, often building palm-thatched houses with separate walls and roofs that resemble the houses of rural Brazilians, the open interior arrangement is usually preserved. Recently there has been a movement to return to the traditional house, which by the late 1990s had become almost universal at Etéñitépa.

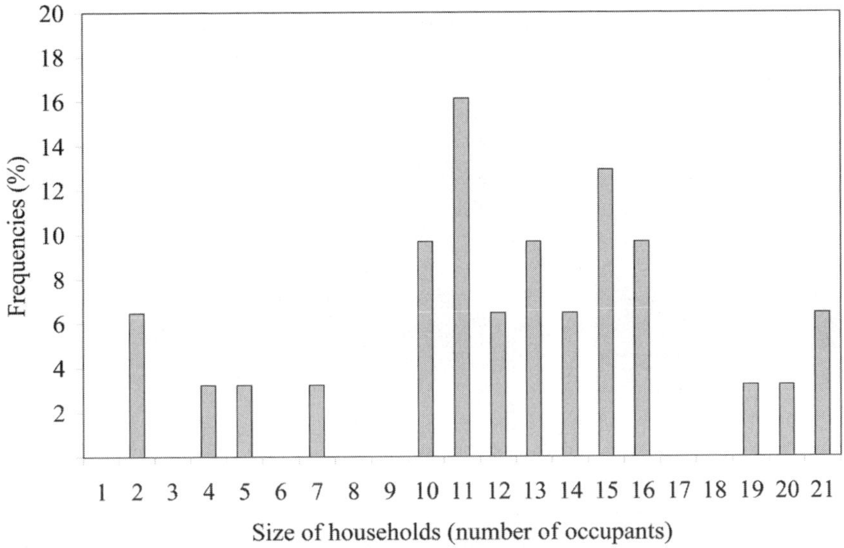

Fig. 2.12. Frequency distribution of households at Etéñitépa in 1990, according to the number of occupants

In 1990 most of the houses at Etéñitépa were inhabited by extended families. The average number of persons per house was 18.3, ranging from 2 to 21, and most households had between 10 and 16 (fig. 2.12). Few households were either very small (fewer than 5 occupants) or very large (more than 20). An old couple constituted the smallest household, while several households included four generations. The large households were generally those of older polygynous men who had several daughters and a number of grandchildren.

The diagram of a large Xavánte household will help us to understand the kin relationships that structure both the household and its transformation over time. In figure 2.13 we show the household headed by Agostinho as it was constituted in 1996. He is married to three of Warodi's daughters. Warodi was the oldest son of Apöwẽ, who headed the group at the time of contact. Warodi was chief from 1978 (when Apöwẽ died) until 1985; he died in 1988. Agostinho first married Araci and later her two sisters, Marilda and Zerra. In 1977, when Flowers was living at Etéñitépa, Agostinho, who had moved in according to the rule of matrilocal postmarital residence, was a junior member of Warodi's household. Two of Agostinho's daughters (Cecita and Pore) are married

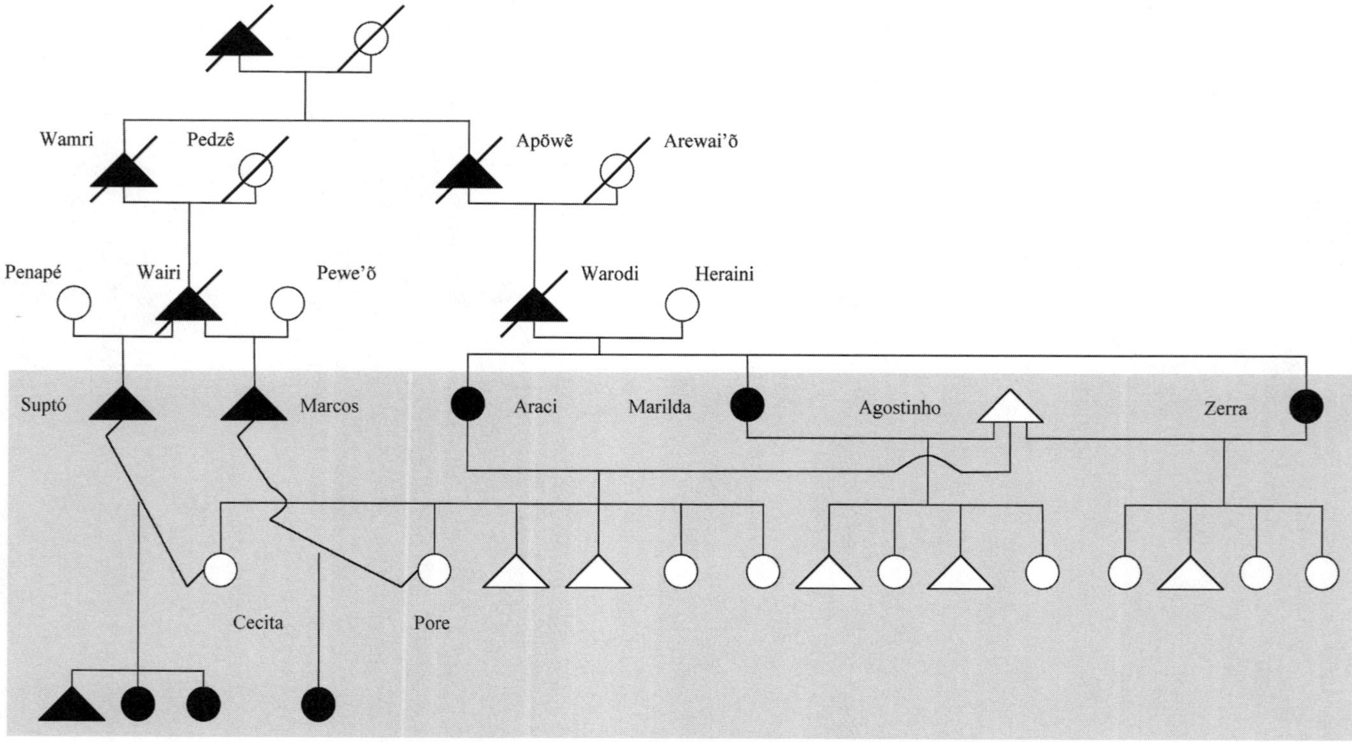

Fig. 2.13. Kin relationships in a large household at Etéñitépa in 1996. The gray area indicates the people actually living in the household. Triangles refer to males and circles to females. Black symbols indicate people of the Poridza'õno moiety and white symbols the Öwawẽ. A line across the symbol indicates that the person has died.

to two half-brothers, Suptó and Marcos.[8] Both live in Agostinho's house and are a few years older than their wives. The genealogical diagram also shows that Suptó and Marcos are distant kin to their wives and that marriages follow clan exogamy.

Groups of brothers often marry groups of sisters. This can be seen in Agostinho's household, as Suptó and Marcos have married two sisters. In the future, it is possible that Suptó and Marcos may marry other daughters of Agostinho's or one or more brothers may move in and marry daughters of the household. Over time, the tendency will be for the influence of Suptó and his brothers to grow. Eventually, Suptó or one of his brothers will become the head of the household.

A Day at Etéñitépa

If you look across the village in the early morning of a day in the dry season, you often see mist hanging low over the trees, making the beehive shapes of the houses on the other side dim and ghostly. The air is chilly, for dry season nights are cold on the Central Brazilian Plateau. People emerge slowly from their houses and may build a small fire to warm themselves and their children, as the sun takes some time to break through the mist. In the center of the village, at the *warã,* half a dozen men stand, blankets over their shoulders, discussing the affairs of the day. The morning meeting brings together only a few of the leading men; the main council meeting is in the evening. Soon women are gathering up their water tins, and their children and are on their way to the stream below the village to bathe and bring water for cooking and drinking.

As the sun rises higher people go about the activities of the day. In a Xavánte village, where houses are only a few meters apart and much activity takes place outdoors, one is constantly surrounded by what Graham (1995, 65) calls "socially organized sound," voices in conversation or choral song, children laughing or crying, dogs barking, the thump of rice being hulled in wooden mortars. People leave for the gardens in family groups, the women carrying on their backs large baskets suspended by the handle across their foreheads, perhaps with a baby inside or a toddler perched on top. The men carry tools and invariably a rifle in case game is sighted. In the dry season, a formal collective hunt, called *du,* may be planned. The men taking part, painted and with their hair bound up, assemble in the center of the village with their weapons; these are mostly rifles, but a few hunters carry bows and arrows. Since the dry scrub will be fired to drive game, one hunter climbs to the top of a house and shoots an arrow straight up to establish wind direction. The men climb into the back of the small community truck and are off to the distant hunting ground.

In the heat of noon the village becomes quiet, as those who are not away at the gardens or hunting seek the dark coolness of their houses to nap. By midafternoon activity resumes. Families return, grimy and tired, from the gardens, bringing threshed rice, to be hulled as needed. The pounding of mortars resumes, and cooking fires are lit. Young boys start a game of soccer. Toward dusk the hunters return, with game if they have been fortunate. There is much going and coming from house to house as the meat is shared out, and before long pieces are roasting over the fires. The dry season sunset is red and smoky, and looking from the village across the *cerrado* after dark you can see brush fires flickering in the distance. The evening council circle forms, and when it breaks up one of the men shouts out the decisions to inform the women, who are gathered with the children at the doorways of their houses. A woman may respond in disagreement or mockery, starting an exchange that amuses the listening village. If there is a full moon, the women may join the men in the center of the village in a round dance and singing.

By early evening people are in their houses, not necessarily to sleep but to talk quietly as they lie on their sleeping mats. The dim light from a lamp made by dipping a cotton wick in a can of kerosene throws shadows on the palm leaf walls. As the night draws on, those who are sleeping may be awakened briefly by the deep voices of the men as they sing around the village and be reassured that it is being kept safe from all human or supernatural threats. Then they can fall asleep again.

Chapter 3

History: Confrontations and Connections

The Xavánte have a legend telling how their ancestors once crossed a wide river on the backs of river porpoises. In a version of this legend told by an old man from Etéñitépa, the Xavánte crossed this river on their way west, coming from where the sun rises and fleeing places where there were many white settlers.[1] They founded a village called Tsõrepré. Times were good for the Xavánte when they lived in this village. The land was red and fertile, the population grew, the village was very large, and the people lived to be so old that they were bent like armadillos. When the *warã* (men's council) met in the center of the village, there were so many men that, instead of one ring, as we see today in Xavánte villages, men had to arrange themselves in several rings. The white-haired elders sat in the middle. Behind them, each according to his age group, sat the younger men: a ring of middle-aged men, another ring, and another, and finally the youngest in the outside ring.

Versions of this story are told in every Xavánte village and have been recorded by missionaries and anthropologists.[2] It has been interpreted as an account of the experience of the Xavánte with colonizers. The region from which the Xavánte were fleeing was the province of Goiás, east of their present territory, where the Xavánte were located in the eighteenth century and where for a long time they fought with the settlers (fig. 3.1). This story, with its mythical elements, refers to historical events, and it is emblematic of the experience of the Xavánte with Western society.

In this chapter we will show that the history of indigenous peoples of Central Brazil over the past three centuries is connected to economic and political processes whose causes were complex and whose consequences were felt far beyond the region. Gold washed from river gravel of the central plateau in the eighteenth century brought wealth to some miners, but most Brazilian gold went overseas to make the king of Portugal one of the richest monarchs of Europe, and large amounts ended up in the

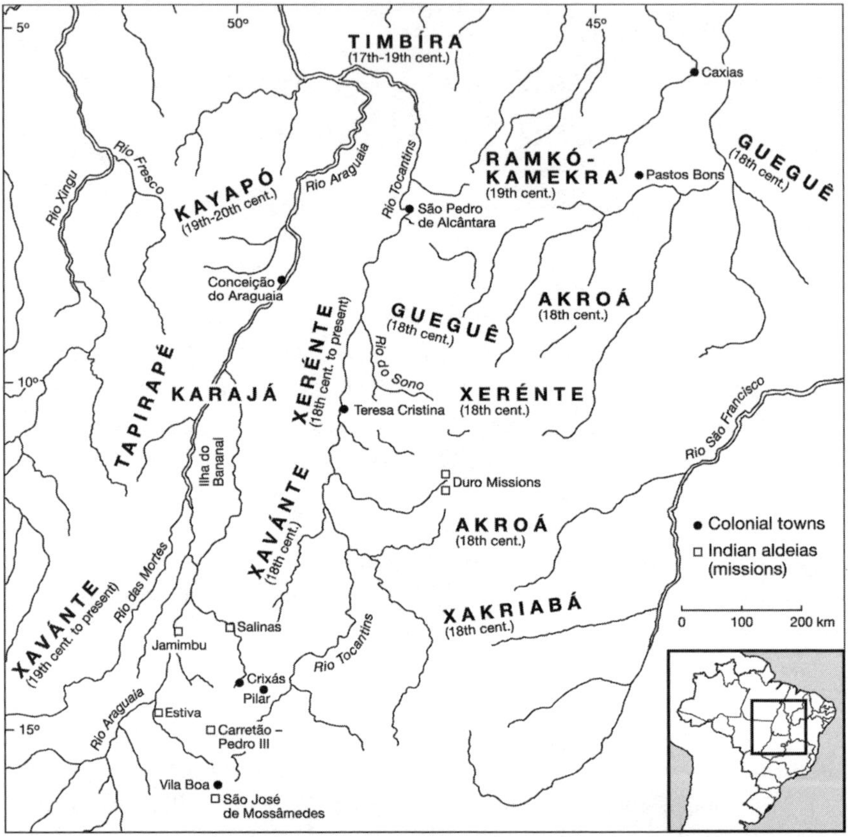

Fig. 3.1. Historical map of the Tocantins-Araguaia frontier, showing Indian groups, colonial towns, and secular missions in the eighteenth century. (Based on Hemming 1987.)

pockets of English merchants. At about the same time cattle raisers began to penetrate the hinterlands in search of pasture for vast herds. These herds supplied the sugar plantations and towns of the coast with meat and draft oxen to haul cane to the sugar mills. The export trade in sugar throve as Europeans increasingly demanded sweetening for their coffee and tea. The need for plantation labor in turn fueled the Atlantic slave trade. The historical processes in which indigenous peoples from Central Brazil were involved were far reaching.

Inevitably the advance of European settlement into the hinterlands affected Indian subsistence, and they reacted against the loss of access to fishing rivers, agricultural land, and hunting grounds. But the indigenous groups of the central plateau of Brazil had opportunities for resistance that were not available to those on the Atlantic coast that bore the brunt of European colonization and exploitation in the sixteenth and seventeenth centuries. Since settlement of the interior was not based on a plantation economy but on extensive cattle raising and gold mining, occupation of the land was not continuous but was concentrated according to the needs of these activities. The colonists were geographically tied to their resources, mines and cattle, while the Indians had the advantage of mobility.

One response of the Xavánte and other indigenous populations to interaction with Western society and resulting loss of access to resources was a change in settlement pattern. In several periods we find Central Brazilian groups moving into regions left open by the decline of other populations; in most cases these were native groups, but occasionally they were colonials. For the Indians this strategy often implied increased mobility and warfare, both for defense against the colonists' aggression and for raiding the settlers' cattle and crops, which came to replace in part the loss of game and garden land. Some indigenous groups accepted peaceful relations and settlement in mission villages, but they either abandoned them after a short period, perhaps with some advantage from having acquired knowledge of the whites' customs and weapons, or they declined in numbers and eventually succumbed to stress and disease.[3]

Increased mobility also provided some protection against the spread of disease. It is generally agreed that the most important factor in causing the catastrophic depopulation of native peoples in the Americas was the impact of epidemic diseases carried from Europe and Africa, diseases to which the indigenous populations had not been previously exposed. Epidemic diseases could be transmitted from village to village without direct contact with Europeans, but their results tended to be most lethal when susceptible Indians were concentrated in mission and other settlements where they were in sustained contact with Europeans and the effects of disease were often exacerbated by malnutrition and hard labor. Population decline was then unrelenting, usually ending with the extinction of the group.

In the early eighteenth century the discovery of gold led to a tumultuous influx of Europeans and their African slaves into the province of Goiás. The Indians were caught within the two arms of a vast pincer movement of settlement: to the east the cattlemen and to the west the miners. Historical documents of the period report Xavánte raiding cattle and fighting gold miners. Later, in the second half of the century, the colonial government settled a number of indigenous groups in "secular missions." During the time when some Xavánte groups lived in these settlements they suffered the effects of epidemic diseases. Within a few years most of these settlements were abandoned. In the nineteenth century, as Goiás became more densely populated, the Xavánte migrated west into the region where they live today. Xavánte history is closely tied to the expansion of these demographic and economic frontiers in Brazil over the last three centuries, expansion that has been continuous, though based on the exploitation of different resources.

The Eighteenth Century

The first gold strike in Goiás was made by a stubborn and flamboyant old backwoodsman nicknamed Anhangüera ("Old Devil" in Tupi). When he was a boy he went with his father on a slave-raiding expedition to the country of the Goyaz Indians, where it was said that the women wore gold ornaments. More than fifty years later Anhangüera was determined to rediscover the source of this gold. After several unsuccessful attempts, he again reached Goiás. Anhangüera had been given up for dead when in 1725 he returned to São Paulo bringing with him, according to one account, the extraordinary amount of eight thousand drams of alluvial gold (Boxer 1962, 36, 267–68).[4]

A gold rush into Goiás followed the news of Anhangüera's discoveries in the 1720s (see table 3.1): "From Minas Gerais, Rio de Janeiro, and the whole Captaincy of São Paulo, flocked many people, leaving their houses, goods, wives and children, pouring into these backlands as if they were the Land of Promise or the hidden Paradise in which God placed our first parents" (Barbosa de Sá, quoted in Boxer 1962, 255). Mining camps sprang up along the tributaries of the Araguaia and Tocantins Rivers. By the middle of the eighteenth century Goiás had some thirty thousand nonindigenous inhabitants, of whom ten thousand worked in the mines (Ravagnani 1978, 6). But this population was unstable. Placer mining was the only technique used, and when miners heard rumors of a new strike they packed up their few tools and possessions and moved on in the unending hope of the fortune that would make them rich. As a result, "mining camps were likely to appear suddenly, flourish briefly, and to vanish overnight" (269).

TABLE 3.1. Time Line of Historical Events Affecting the Xavánte

1720s	Discovery of gold in Goiás
1748	Goiás is made an independent captaincy.
1749–55	Dom Marcos de Noronha, Conde dos Arcos, first governor of Goiás
1751–53	Settlement of the Akroá and Xakriabá at the Duro missions
1751	"Land of the Xavánte Heathen" appears on a map dated 1751.
1756–58	Conde de São Miguel, governor of Goiás
1759–70	João Manuel de Mello, governor of Goiás
1757	Implementation of the Directorate legislation
1760	Governor Mello reports that the Duro missions are abandoned and the Akroá and Xakriabá are raiding settlements.
1761	King of Portugal authorizes "offensive war" on the Indians of Goiás.
1762	Xakriabá destroyed. Xavánte are reported to be attacking mining camps.
1764	Punitive expedition against the Xavánte is raised by settlers.
1772–78	Dom José de Almeida Vasconcelos, Baron of Mossâmedes, governor of Goiás
1775	Founding of the Mossâmedes secular mission
1780–17??	Luiz da Cunha Menezes, governor of Goiás
1784–17??	Tristão da Cunha Menezes, governor of Goiás
1784–88	Expedition to pacify the Xavánte, led by José Rodrigues Freire and Miguel de Arruda e Sá
1788	A group of Xavánte is settled at Carretão.
1819	The Austrian naturalist Johann Emanuel Pohl visits Carretão and writes that it is almost abandoned.
Circa 1840–60	Split between Xavánte and Xerénte. Xavánte cross the Rio das Mortes.
Circa 1870–1920	Xavánte live at Tsõrepré, their "mother" village, after they arrive in the Roncador region.
1938	President Getúlio Vargas launches the March to the West.
1943	Roncador-Xingu Expedition begins exploration of Central Brazil.
1946	First peaceful contact of the Xavánte with SPI agents at São Domingos
1950–60	Xavánte groups settle at missions and SPI posts.
1956	Apöwẽ's group moves to São Domingos.
1957–58	Maybury-Lewis's fieldwork at São Domingos
1962	Field study by James Neel, Francisco Salzano, and colleagues at São Domingos
1966	Xavánte at Marãiwasede lose their land and are taken to São Marcos.
1967	SPI is investigated and replaced by FUNAI.
1972	Apöwẽ's group settles at Etéñitépa.
	Pimentel Barbosa reservation is demarcated, with an area of 224,000 ha.
1976–77	Nancy Flowers's fieldwork at Etéñitépa
1978	Death of Apöwẽ. Warodi becomes chief.
1980	Pimentel Barbosa reservation is redemarcated with an area of 328,966 ha.
1980–85	Xavánte Rice Project

(*continues*)

TABLE 3.1.—*Continued*

1981–91	Laura Graham does fieldwork at Etéñitépa at various times during this period.
1988	Death of Warodi. Associação dos Xavantes de Pimentel Barbosa is formed.
1988–94	Jabiru Project at Pimentel Barbosa
1990–2000	World Wildlife Federation game management project at Etéñitépa
1991	Suptó becomes chief at Etéñitépa.
1995	Compact disc of Xavánte singing at Etéñitépa is issued.
1998	*Wamrêmé Za'ra: Nossa Palavra,* describing Xavánte history and myths, is published.
1996–2000	Xavánte campaign against the proposed Tocantins-Araguaia Hidrovia (inland waterway) project

In their eagerness to work the placers, the miners often neglected to plant adequate crops, and as a result in the early years the settlers in Goiás suffered great privation, to the point of starvation (Alencastre 1864, 65). Because everything had to be imported, prices were astronomical, and probably, as in the other mining regions, gold very quickly slipped out of the hands of the miners into those of merchants, who were able to take advantage of the shortages (Ayres de Casal [1817] 1945–47). In 1824, Brigadier Raymundo Cunha Mattos (1874, 303), in his *Chorographia Histórica da Provincia de Goyaz,* wrote that throughout the eighteenth century epidemics of smallpox and measles frequently swept through the mining camps, resulting in many thousands of deaths. Endemic diseases were also rife, including goiter, syphilis, dysenteries, and "dropsies" (possibly beriberi). Around the camps conditions were filthy, with stagnant water, decaying vegetation, and the carcasses of dead animals left putrefying in the sun. As for the thousands of black slaves who worked the mines, Mattos (1874, 302) wrote: "Hard work, bad treatment, poor food, and lack of medical care took these poor devils so early to their graves that they left no descendants."

The discovery of gold in Goiás attracted considerable attention from the government in Lisbon, for the collection of the royal fifth (a tax due to the Crown from all gold-mining operations) was of prime importance. In 1748, Goiás was separated from São Paulo and made an independent captaincy, with a governor appointed by Lisbon (Boxer 1962, 270). The seat of government of Goiás was established at Vila Boa, which was central to the mining district (fig. 3.1).

Letters to Lisbon from the early governors of Goiás make constant reference to Indian raids on the mining camps and efforts to deal with the "Indian problem." The settlers favored "offensive war," which justified destroying Indian villages and making slaves of the captives.[5] But the Crown promoted Indian settlement in the hope of increasing the number of industrious subjects populating Portugal's overseas dominions. The governors were often frustrated as they attempted to reconcile opposing interests.

The first effort to settle Indians in Goiás was in the 1750s. By that time, the Akroá and the Xakriabá, both Central Jê groups closely related to the Xavánte (see chapter 2), were reported to be "devastating and depopulating" the gold mining camps. In 1752, Noronha, the first governor of Goiás, wrote to the king:

> They [the Akroá] are killing blacks and whites, destroying the crops, eating the cattle; and lately, so that they can continue their insults with greater ease, the Indians have taken to stealing horses from their pastures, and have entered the cattle ranches on horseback, where they have made great destruction, not only by killing cattle, but by driving cattle off to their villages where they keep them for their consumption. (letter quoted in Chaim 1974, 181)

The settlers were protesting bitterly and wanted the governor to petition the king to declare offensive war, but the governor was caught between the demands of the colonists and directives received from Lisbon. The governor sent emissaries to the Akroá, telling them that if they made peace they would be "civilly treated, and would receive lands, and neither they nor their kin would be subject to any law of captivity" (182–83).

Two mission villages were built between 1751 and 1753 at Duro (São José do Duro and São Francisco Xavier do Duro), on a tributary of the Tocantins River (fig. 3.1). Several hundred Akroá and Xakriabá were settled there, in the region where they had previously been raiding. A pictorial map of the period, possibly drawn by a Jesuit missionary, shows one of the Duro missions (fig. 3.2). The map shows that not all the Indians in the region were in the mission settlement, as it depicts two very large villages some distance to the north. The way from the mission to the villages is drawn, indicating that the journey requires crossing three rivers and takes thirty-one days to reach the first village and three more days to reach the second village. These villages have beehive-shaped houses arranged in a Central Jê semicircular formation. Near the

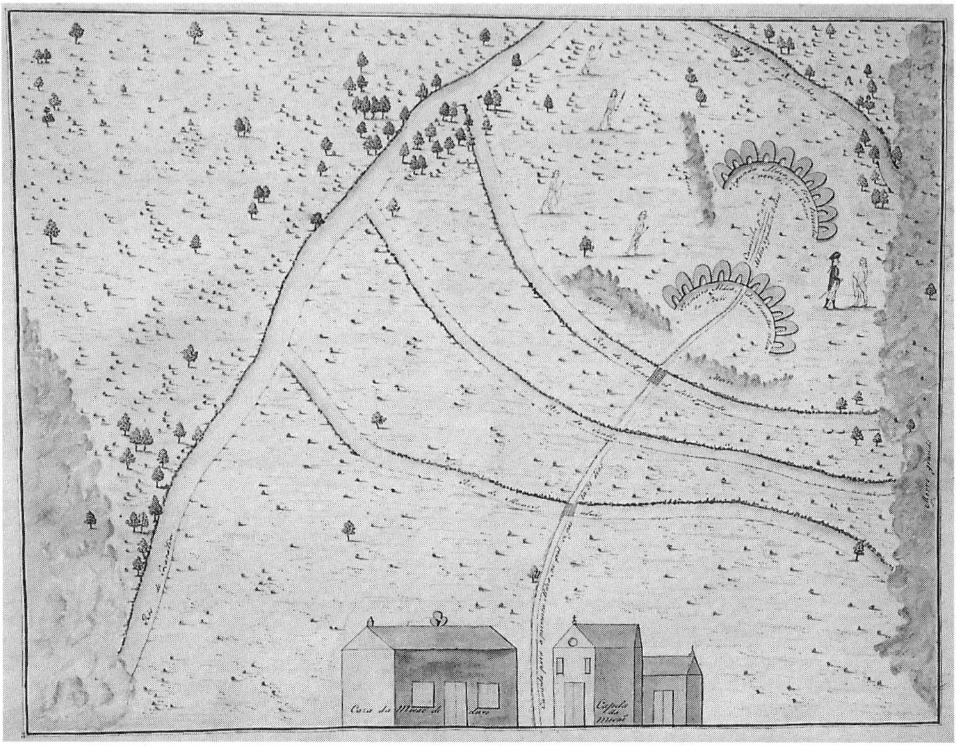

Fig. 3.2. Eighteenth-century map of one of two Duro missions on a branch of the Tocantins. The mission and the church are shown at the bottom. The road to two large semicircular Jê villages is shown, and there are small figures of Indians, one of whom is talking with a Portuguese officer. (Courtesy of the Map Library, Itamaraty, Ministério das Relações Exteriores, Rio de Janeiro.)

villages several Indians carrying bows and arrows are depicted, as well as a Portuguese officer apparently conversing with an Indian. The caption for the first village reads that it has 286 houses; the second village has 396 houses, indicating a population for each of perhaps three to four thousand (see also Chaim 1974, 111–12).

Over the next few years a series of epidemics and uprisings at the Duro missions discouraged succeeding governors of the Captaincy from further attempts to make the Akroá and the Xakriabá peaceful subjects of Portugal. By 1760, one settlement was abandoned, and the other had only sixty Indians, "of both sexes, but all either very old or very young"

(RIHGB 1918, 67). The colonists claimed that the situation was worse than before the attempt at settlement, for the Indians had killed more than two hundred people since the first uprising and were now carrying firearms, which they had learned to use while at the missions (67). Governor Mello wrote to the king in 1760: "Experience has shown that the settlement of the Akroá and Xakriabá is useful neither for Religion nor for the State; from the former they have become apostates, and from the latter rebels" (61). He added, "It seems to me that the best thing is for Your Majesty to authorize offensive war, obliging the Indians to retreat into their former territory" (61). Soon the king agreed to sanction offensive war.

In 1762 the governor again wrote to Lisbon, this time describing the fate of the Xakriabá. This document is particularly interesting because it shows how disputes among indigenous groups worked to the advantage of the colonizers. Settlers from the region of the now deserted missions organized a punitive expedition to attack the Xakriabá and then pursued them into Akroá territory. The Akroá received the Xakriabá in their villages but behaved treacherously, massacring them while they slept. "Only forty escaped, and these came to our *bandeira* asking to make peace.[6] This was granted them. The women and boys were shared out among the most wealthy households of the surrounding settlements to serve in bondage, while those capable of bearing arms offered to join our *bandeira,* saying that they wished to avenge the deaths of their kinsmen" (RIHGB 1918, 83). The governor estimated that some six hundred Xakriabá had been killed and "few remain of this inconstant nation."

The Campaign against the Xavánte

The first known reference to the Xavánte in colonial documents appeared in a map dated 1751, which showed the location of the "Land of the Xavánte Heathen" (Sertão do Gentio Xavánte) east of Ilha do Bananal (Bananal Island; see Chaim 1974, 43).

In 1762, shortly after the defeat of the Xakriabá, a letter written by Governor Mello described in detail a Xavánte attack on the mining camp at Crixás (RIHGB 1918, 83–84).[7] According to this letter, the Indians' advance on Crixás was only halted when the settlers armed three hundred slaves who were employed in the mining operations. The Xavánte withdrew but after a few days again began raiding the plantations with a larger number of warriors, many of them armed with muskets. The miners were on the alert and went out to drive the Indians back, but they feared to attack as the Xavánte were now so numerous and well armed. The miners were reluctant to risk the lives of the slaves on whose labor they depended. The opposing forces faced each other for two days, without an attack on either side, until the Xavánte finally withdrew.

Because of the Xavánte attacks the settlers appealed to the governor for help, arguing that they were too few to resist such large numbers of Indians. Mello (RIHGB 1918, 84) proposed that the miners should raise a *bandeira.* He also noted that sending a *bandeira* into unexplored territory was doubly advantageous, since they could prospect for gold as well as seek out and punish the Xavánte. At this period, about forty years after the discovery of gold in Goiás in the 1720s, the known gold sites were rapidly becoming depleted. As the mines became less and less productive the miners were increasingly convinced that there must be rich and unexplored sites in regions where they dared not venture because of the presence of hostile Indians. Most *bandeiras,* therefore, had a double purpose: to harass Indians and search for gold.

The *bandeira* finally set out in 1764, after being delayed for two years because of disputes about how much of its cost should be borne by "spontaneous contributions of the people." In the end it was made up of only two hundred men instead of the five hundred originally planned. After many days of searching, the Indian trackers who accompanied the *bandeira* found a trail that led to a populous Xavánte village on which the *bandeirantes* made a dawn attack. The Indians were caught by surprise; a number were killed, and the rest fled, leaving some children in the village. But they did not go far, and when the Xavánte saw that their attackers were few they prepared to counterattack. João de Godoy then sent an Indian who spoke Xavánte to carry a message to the headman, offering to make peace if the Xavánte would declare their allegiance to the king of Portugal and embrace the Catholic faith. The Xavánte chief pretended that he could not hear the message and called the messenger nearer. When the man did so, the chief and his warriors shot and killed him (RIHGB 1918, 89).

The Xavánte then retreated slowly, drawing the *bandeira* in pursuit through difficult terrain, until they reached a wide and swift river, which the Indians crossed, halting on the opposite bank. Any enthusiasm remaining among the troops to continue the campaign was now washed away by the rains, which were breaking as it was the end of September.

As Mello (RIHGB 1918, 89–90) pointed out, in the rainy season muskets got wet and would not fire, giving the advantage to adversaries armed with bows and arrows. So the *bandeira* retreated, and João de Godoy informed the governor that in the next dry season they would attempt to destroy two other Xavánte villages in the region. According to Mello this was the only effective way of dealing with the Indians: "For experience has shown that when we burn their villages and capture their children all these heathen seek out more distant wildernesses in which to live and cease to invade our lands" (90).

Rather than ceasing, hostilities with the Xavánte would only increase in the following years. Ten years later, in 1774, Mello's successor, Jose d'Almeida e Vasconcellos (RIHGB 1918, 96–97), reported that in the previous dry season the Xavánte had raided four mining towns along the Tocantins. This indicates that the Xavánte at that time, far from seeking out "more distant wildernesses," had expanded their activities eastward. It appears that at this time, perhaps because they could muster large numbers of warriors, the Xavánte attacked openly and in force.

The "Secular Missions" of Goiás

In 1757 a new system imposing secular control on the Indians was extended to all of Brazil. This legislation, known as the Directorate,[8] suppressed Indian slavery, encouraged mixed marriages, and envisioned the eventual conversion of the *aldeias* into townships. The aim was the gradual incorporation of Indians into the general population. With this legislation the Portuguese Crown placed its own interest, which was to populate its vast dominions in Brazil with Portuguese subjects, whether Indian or not, above that of the colonists, who regarded pacified Indians only as a supply of manual labor (Almeida 1997; Hemming 1987, 4–5; Prado 1963, 88–89).

In Goiás, however, there was no attempt to put the new system into effect until the 1770s, because, after the defection of the Akroá and the Xakriabá from the Duro *aldeias* in the 1760s, there were no pacified Indians in Goiás. Mello (RIHGB 1918, 62), writing in 1760, maintained that the new legislation was intended to apply to peaceful village cultivators and not the "nomadic pirates" that "infested" Goiás. Mello supported the colonists in their contention that any attempt to pacify and settle the Indians of Goiás would be futile and that peace could be attained only by killing them off or driving them far from settled regions.

But Vasconcellos, who succeeded Mello as governor in 1772, received from Lisbon a firm directive to enforce the Directorate. By the third quarter of the eighteenth century gold production in Goiás was well into the decline from which it never recovered. The new governor was urged not only to seek out new mines but to stimulate agriculture and the exploitation of new resources such as salt deposits. In his instructions, Vasconcellos was told that "the civilizing of Indians is an objective much more important in its consequences than all other wealth . . . since the mining of gold, diamonds, and other precious metals involves considerable expense, while the civilizing of Indians may be easily accomplished as soon as the correct method of proceeding is arrived at" (quoted in Chaim 1974, 94).

Vasconcellos wrote to Lisbon of his resolve to carry out the royal instructions, although he anticipated little cooperation from the colonists (RIHGB 1918, 95). He took up the task with enthusiasm. Near the capital he built a large model village (São José de Mossâmedes), which took its name from his own royal title: baron of Mossâmedes. For a frontier captaincy the design of the settlement was grandiose. The plans show a large church and a palacelike building, which was to serve as the governor's summer residence, as well as workshops with looms for weaving cloth, a kitchen garden, a large banana plantation, and long barracks to house the Indians (fig. 3.3).[9] There are quarters for a schoolmaster and a schoolmistress. One of the directives of the Crown concerning the new *aldeias* was that Indian children should be taught in Portuguese and their native languages prohibited (Almeida 1997, app. 4). The first inhabitants of Mossâmedes were Akroá (Alencastre 1857; RIHGB 1918, 99–100). In the following year Vasconcellos sent a mission to the Karajá on Ilha do Bananal. The Karajá agreed to sign a treaty of peace with the Portuguese in return for a promise of protection against the Xavánte who were harassing them by fording the river in the dry season and raiding their plantations and villages (RIHGB 1918, 116–17). Soon a number of Karajá were persuaded to settle at Mossâmedes. Eventually a total of eight thousand Indians, including Xavánte, were settled there (Chaim 1983, 120–21).

Under succeeding governors the effort to settle Indians continued. By the early 1780s the Xavánte were one of the few groups in Goiás that remained hostile. The Xavánte were most active in the north of the captaincy, where several cattle ranches had been abandoned because of

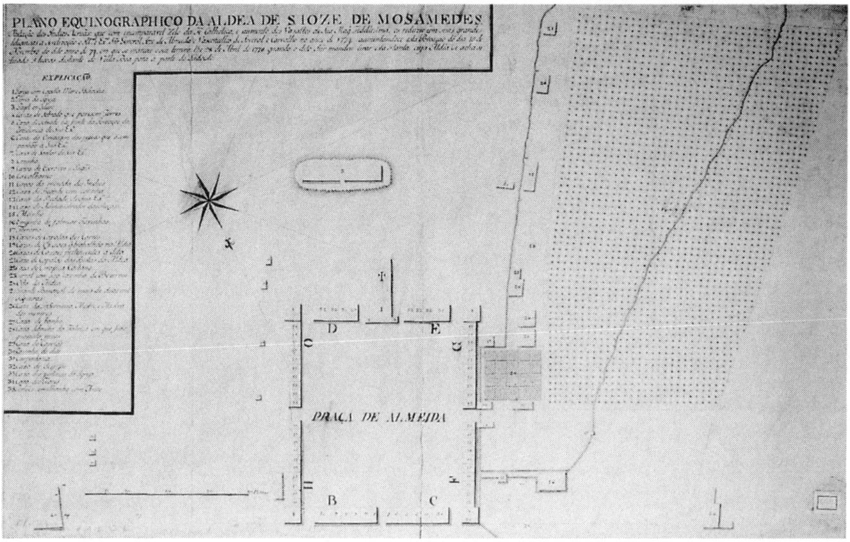

Fig. 3.3. Plan of São José de Mossâmedes, an eighteenth-century secular mission built by the baron of Mossâmedes, governor of Goiás, for "pacified" Indians. (Courtesy of Arquivo Histórico do Exército, Rio de Janeiro, series CO, subseries GO, 10-01-2085.)

their depredations, and because of Xavánte raids on crops the mining camps were running short of food supplies (RIHGB 1918, 156).

In 1784 Governor Tristão de Menezes raised a *bandeira* to pacify the Xavánte. In 1790 the document *Relação da Conquista do Gentio Xavánte* (*Report of the Conquest of the Xavánte Heathen*), attributed to Lt. of Dragoons José Rodrigues Freire, was published in Lisbon (Freire [1790] 1951). It gives an eyewitness account of the events that followed. It is worthwhile to look at this report in detail because it gives us an excellent view of the motivations as well as the strategies utilized by the colonial government in dealing with indigenous groups in Goiás (see also Karasch 1992).

As reported by Freire, the *bandeira* succeeded in reaching a village of the "barbarous assassins," who firmly rejected all peace proposals. Freire speculated that the Xavánte, since they had recently killed a number of Portuguese, feared that these peace proposals were simply a trap intended to deceive them and deliver them into the hands of the Portuguese. When the commander realized that his peace proposals were going nowhere, he gave the Kayapó who accompanied the *bandeira* orders to take prisoners, which they were only too pleased to do, as the Xavánte were their traditional enemies. They captured one warrior, four women, and several children and returned to the capital. The governor treated the captured Xavánte man with such "kindness and affability" that he promised to attempt to persuade his fellow villagers of the whites' good intentions. This Xavánte, who was now called Tristão after the governor, was allowed to return to his village a few months later to carry the governor's peace message: "to share with the whites the sweet fruits of civil society in order to forget forever all their hostilities against us, and all the crimes they had perpetrated" (Freire [1790] 1951, 14–15).

After long debates, during which Tristão barely escaped being killed by his own people as a traitor, the village leaders were apparently convinced by his arguments, for they allowed him, with his brother and several women, to return to the whites with the message that the Xavánte were ready to make peace and that a settlement should be prepared and crops planted for their maintenance. At the beginning of the next dry season an escort should meet them, and they would go to live there.

The governor began preparations to provide an *aldeia* for the Xavánte at a place not far from Vila Boa called Carretão (Fig. 3.1).[10] He had crops planted for the subsistence of the newcomers. But the settlement process did not proceed as smoothly as the governor would have wished. At the designated meeting place the escort found only a Xavánte hunting party, and the soldiers soon realized that they were surrounded by a large number of warriors who had come armed and ready for battle, showing that they had little confidence in the good faith of the Portuguese. According to Freire, an Akroá tribesman with the escort party had put the Xavánte on guard against the whites. It was the headman of the Kayapó who finally persuaded the Xavánte to lay down their bows and arrows, for he declared that if they did not do so the Kayapó, along with the Portuguese, would seek out and destroy all the Xavánte villages, no matter how distant.

At the beginning of the following dry season, the governor was startled to receive word from the north that over two thousand Xavánte were marching toward the capital. Tristão de Menezes had never anticipated such a large number, and the idea that the Indians might arrive at Vila Boa in force filled him with apprehension, for it was a drought year and the inhabitants had little food for themselves, much less enough to feed two thousand extra mouths. After conferring with the town officials, the governor decided to divide the Xavánte into two groups, one to settle at Carretão and the other at Salinas near the Araguaia River. Not only would it be easier to provide for the Xavánte at two *aldeias,* but the potential threat of so many warriors, who had only recently adopted the ways of peace, would be lessened (fig. 3.1).

The governor sent an escort with orders to guide the Xavánte on back roads and avoid the towns, but the soldiers found it easier simply to take them from town to town. The inhabitants were terrified at first, hiding their families and closing up their shops, but when they found the Indians friendly they showered them with gifts and even contributed food supplies for the future *aldeia.* However, when the Xavánte came to the place where the road forked, the officer who was charged with splitting the group met with such resistance that he was forced to abandon the plan. When they left their wilderness, the Xavánte declared, they intended to live among the whites, and not in such a distant place as Salinas, which was known to have an unhealthy climate and to be infested with mosquitoes. In January of 1788, after six months on the road, for they traveled slowly, burdened as they were with children, the old, the lame, and the sick, the Xavánte finally entered Carretão to the sound, as Freire tells us, "of their disagreeable instruments, clamor, and dances, according to their customary festivities" ([1790] 1951, 18).

The group settled in at Carretão, and only the chief and a few other leaders came to the capital, where the governor gave the chief so many gifts that when he returned to the *aldeia* he made a fervent speech to persuade his followers of the government's good intentions and to encourage them to work at farming with the tools that the governor had given him to distribute. Tristão de Menezes had reason to be pleased with the apparent success of the Xavánte settlement, for it was now possible to prospect for gold in regions where no one had dared to venture when the Xavánte were hostile. Several miners who had been taken prisoner by the Xavánte were released at the time of settlement, and they assured the governor that they had seen some likely spots during their captivity but had been unable to prospect because they lacked the necessary tools.

Freire concluded his *Report of the Conquest of the Xavánte Heathen* in a highly optimistic vein. He wrote that a local priest had baptized 412 Xavánte children on one visit, in spite of the fact that many families were absent from the *aldeia,* having fled into the forest because of an epidemic that attacked them almost immediately after their arrival, killing over 100 people. Since the epidemics had ceased, conditions, according to Freire, were improving, as the Xavánte were becoming involved in agriculture. The governor could have the satisfaction of having freed "the people of this captaincy [Goiás] . . . from the wild beasts that devoured their entrails, at the same time earning the incomparable glory of bringing to the Church an equal number of faithful sons and an equal number of vassals to the Portuguese empire" ([1790] 1951, 19).

Although the *aldeia* appeared to be flourishing at the time that Freire wrote his report, within a few more years Carretão was practically deserted (Ravagnani 1978, 73). In the 1780s there must have been well over ten thousand Indians settled at the various *aldeias* of Goiás. By the end of the century only a handful remained, and any attempt to "civilize" them in accordance with Directorate guidelines had collapsed. The situation was much as it had been fifty years before: there were no pacified Indians in Goiás.

What we must find puzzling is not so much why the Xavánte who were settled at Carretão abandoned the *aldeia* as why some groups accepted settlement in the first place. It must have been a difficult decision, given all the reasons they had to be suspicious of the whites' motives. The governor used methods, liberal distribution of gifts with affirmations of paternalistic benevolence and goodwill, similar to those first employed successfully by Jesuit missionaries. To this day FUNAI agents use the same strategies to attract and "pacify" isolated Indian groups. The allure of trade goods, including tools that facilitated many tasks, and muskets, which apparently the Xavánte were already acquiring in raids, must have been great, as was perhaps the prospect of living at peace with their former enemies. The prestige of the leaders could also have been strengthened by the many gifts that they were receiving. Whitehead (1993, 214) notes that the introduction of steel tools, which make possible the cultivation of larger areas, has the political effect of permitting village chiefs to produce food surpluses for feasts, thereby increasing their prestige. This was "a point well understood by the Europeans who actively reformed native leadership using the lure of such items to reinforce the position of those native leaders most receptive to their needs" (216).

Giraldin (1997, 94) suggests more specific factors for the temporary success of the governors in settling a number of different tribes around this time. The 1780s were difficult years for the Goiás colonists. Not only was the productivity of the mines declining, but several drought years in succession had caused crop failures. A smallpox epidemic in 1772 that affected the entire captaincy was followed in 1782 by an epidemic of "putrid fever" (malaria?) that frustrated attempts to open a promising new mining site on the Tocantins, as over eighty miners died each day (Pohl [1837] 1951, 316). Indians, as well as the colonists, must have been affected by the droughts, and the colonists' constant harassment may have forced them to abandon their fields before the crops were ripe. The Indians themselves may have suffered from the epidemics, weakening their ability and will to resist and giving the advantage to the "peace faction" in their councils. If that was the case, they were soon disillusioned, as within a few years the *aldeias* were largely deserted.

For over 150 years after the failure of their brief experience with coexistence the Xavánte would hold to their separate path, rejecting all attempts of the whites to make peaceful contact.

The Nineteenth Century

In Goiás, as mining declined, the colonists failed to exploit new resources that might have led to the development of internal trade. Most of the population survived by means of subsistence agriculture, and what little gold they could wash from the gravel of the rivers (RIHGB 1918, 278). River transport was both difficult and dangerous, as it was not easy to assemble boat crews, and there were few river settlements from which to obtain supplies. The constant risk of Indian attacks made a troop escort necessary (Barata 1848, 343). Also, life on isolated farms was hazardous because of Indian raids, which often caused them to be abandoned (Mattos 1875, 14). Therefore, as the mines ceased to be productive, much of the population simply left the captaincy (Prado 1963, 53). According to Mattos (1875, 14), "The parish of Crixás once had four thousand inhabitants; today there are scarcely one thousand. . . . In the district of Pilar there were once more than nine thousand slaves; today the total population, in continuous decline, is less than three thousand souls . . . [and] the land is full of the ruins of abandoned houses."

A number of European men of science traveled through Central Brazil in the first half of the nineteenth century and wrote accounts of their experiences and observations. Among these were Johann Emanuel Pohl ([1837] 1951), Johann Baptist von Spix and Carl Friedrich von

Martius (1824), Francis de Castelnau (1850–61), George Gardner ([1846] 1970), and Auguste de Saint-Hilaire (1847–48). From them we have eyewitness reports of the economic decline of the region and the hostile relations between colonists and Indians.

In 1819, when the Austrian naturalist Pohl visited Carretão, he wrote ([1837] 1951, 2:35): "at the beginning the settlement was inhabited by 3,500 Indians brought from the Araguaia and Tocantins Rivers, but most of them died soon after they arrived. At present the *aldeia* is still inhabited by 227 Indians, including children." They had a recognized leader, the *capitão,* but he and the other Indians were subjected to the authority of the director and his soldiers. The *capitão* complained bitterly to Pohl of mistreatment and the heavy burden of work imposed on the Indians. By 1824, when Raymundo Cunha Mattos (1874, 245–46) visited Carretão, there were only 199 Indians left.

The most powerful and active Indian group in Goiás at this time was the Xavánte, who were apparently expanding eastward. Around this time the Xerénte first begin to be mentioned in the literature. According to Pohl ([1837] 1951, 2:132), one of whose informants was a colonist who was captured by the Xavánte and lived among them for several years, the Xavánte, the Xerénte, and the Akroá-Mirim, formerly distinct groups, were now united. They ranged over a wide territory on both sides of the Tocantins, from Porto Real downstream to São Pedro de Alcântara and eastward into the region of Pastos Bons in Maranhão. Mattos (1875, 10), writing in 1824, calculated that the Xavánte had four thousand fighting men; this would indicate a total population of at least twelve thousand, including women and children.

After over a century of warfare, resettlement, and epidemics, the Indians of Goiás were still numerous and were vigorously defending their territories. All groups retaliated against slaving raids, which were again semilegalized, by raiding and destroying farms and taking cattle (Prado 1963, 93–95). The economic deterioration of the region and the resulting population decline were also relevant factors. Relations between Indians and whites had deteriorated sharply since the eighteenth century, when the governors of Goiás had celebrated the success of the *aldeias* in keeping the peace. The few Indians who remained in the *aldeias* vegetated miserably, subject to the authority of ignorant and often brutal directors.

Francisco de Paula Ribeiro, a Portuguese army officer who served along the frontier from 1800 to 1823, gives us an eyewitness account of Indian-colonist relations at this time (Ribeiro 1841, 1849, 1870). The government gave little support to either military defense against Indians or attempts to pacify and settle them. Rather, the colonists were encouraged to take matters into their own hands. According to Ribeiro, Indian campaigns organized by the settlers were not really intended for the pacification of hostile natives or the opening up of new lands for cultivation but rather for the capture and enslavement of Indians, mainly children. Ribeiro (1841, 449) saw Indian captives sold at public auction both at São Pedro de Alcântara on the Tocantins and at Caxias in Maranhão. As Ribeiro wrote, this policy and treatment could only have the effect of perpetuating hostilities.

Ribeiro's (1841, 304–13; 1849, 41–45; see also Hemming 1987, 185–86) account of the fate of the Capiekran (now known as the Ramkókamekara or Canela), a Timbíra tribe that in 1815 accepted peace, shows how disastrous to the Indians any relations with the whites could be. The Capiekran, as had been agreed, came to settle at a place near the colonists' farms. But, since no provision had been made for the Indians' support, they soon began killing the settlers' cattle and raiding their crops for food. The settlers, in order to rid themselves of their unwelcome neighbors, plotted to attract them to the town of Caxias, where the Indians were not only left hungry but many were whipped or thrown into prison. When the Indians found themselves betrayed and defenseless they fled, pursued by troopers, who cut them down. Worse, in the town they were infected with smallpox, so those who escaped the troops began to fall sick along the way of their retreat. "Any who became ill on the march would lie down with his head on a rock, and then a friend or relative would crush his head with another large rock, thus giving him rest and respite from his pain" (Ribeiro 1841, 313). According to Ribeiro, only scattered remnants of the tribe remained.

> The sickness that they carried was not fatal to them alone, because it was transmitted to all the native populations of the captaincy, and even to some of our own settlements that were so distant they had never been affected since they were first founded. Principally among the Indians it caused such horrifying slaughter and was carried so far that in October of 1817 it was raging among the tribes that live in the wilderness beyond the Tocantins. . . . We can hardly guess how many thousand souls may have perished, since we know the extravagant means through which these ignorant people attempt to cure themselves: by lying in the rivers to cool themselves

> as soon as they feel the heat of fever or by immediately taking the lives of any who show the first symptoms of disease. (1849, 45)

Both the Xavánte and the Xerénte could well have been affected by this epidemic, as it spread over a wide region both east and west of the Tocantins.[11]

Westward Migration

Around the middle of the nineteenth century, the French naturalist Castelnau traveled across Minas Gerais to Goiás. It seems clear that at this time the Xavánte were already moving westward, for Castelnau (1850–61, 2:114–16) heard of a large Xavánte village west of the Araguaia. As the Xavánte withdrew across the Araguaia, they became separated from the Xerénte, and Brazilian settlement began to surround the Xerénte on all sides. By the 1860s it appears that the separation between Xavánte and Xerénte was complete. Those who were now known as Xerénte were pacified and settled along the Tocantins, particularly in the Rio do Sono region. The Xavánte were distinguished as those who had migrated beyond the Araguaia and were mostly free. In the 1860s Brazilian settlers who lived along the Araguaia told Gen. Couto de Magalhães (1934, 99) that a column of smoke seen to the north indicated the camps of the Xavánte along the Rio das Mortes (fig. 3.1).

There seems little doubt that the Xavánte separation from the Xerénte and the withdrawal across the Araguaia were due to pressure from the growing Brazilian population. By 1824 the population of Goiás had increased by 25 percent since the 1804 census (Castelnau 1850–61, 2:121; Pohl [1837] 1951, 1:293). In the eighteenth century Indians settled in the *aldeias* remained in contact with their free relatives whose villages were located in the same region (RIHGB 1918, 73). It appears that free Indians visited the *aldeias* when on trek and then left again. The authorities deplored this custom, but as long as the region was not settled by colonists, and the Indians were free to move about, they could do little to control it. These contacts became increasingly difficult as the colonist population increased. But in the west of Goiás there were still relatively few settlers, and beyond the Araguaia there were none. It was therefore in that direction that the Xavánte migrated.

The Xerénte who remained east of the Araguaia suffered gradual depopulation. By 1927 they were living in twelve scattered villages along the tributaries of the Tocantins and numbered only about 1,400 (Vianna 1927, 135). In 1930 Nimuendaju (1942, 11) found only nine small Xerénte villages and in 1937 only seven. Many Xerénte by this time were apparently no longer living in their villages. Nimuendaju (8) found the Xerénte demoralized and living conditions in their villages deplorable: "Economically and socially ruined, hemmed in by Neobrazilian settlers. . . . His native village has turned into a place of scarcity." The Xerénte did not forget, nor completely break off, their contact with the Xavánte. Coudreau (1897, 211), in his travels on the Araguaia in the 1890s, heard that parties of Xerénte from the Rio do Sono occasionally visited their "brothers" on the Rio das Mortes.[12]

By midcentury the imperial government of Brazil had largely abandoned its attempts to control Indians by means of punitive expeditions. In 1857 a decree returned the administration of *aldeias* to the Church. On the Araguaia the provincial authorities were seeking to promote navigation by encouraging missionary settlement of pacified Indians. Italian Capuchins founded missions on the Araguaia and succeeded in settling a number of Karajá and Xavánte Indians. It was hoped that these missions, protected by small military garrisons, would become the kernels of flourishing settlements that would supply river navigation with fuel and provisions (Hemming 1987, 385–404).

Beyond the Rio das Mortes

Some Xavánte groups, however, had now crossed the Araguaia and were settled in the region of the Rio das Mortes, a major tributary of the Araguaia. In 1856 Friar Sigismundo de Taggia, with a party of mission Indians, soldiers, and settlers, ascended the Rio das Mortes, attempting to make contact with these Xavánte (Hemming 1987, 390–91; Karasch 1992, 398). They reached a village but found only women, children, and old men, for the young men were off hunting. In spite of the offers of peace and gifts that the mission Indians brought, an old man told them that he had no use for the whites, who he knew to be evil, for when the Xavánte were at Carretão they had suffered mistreatment and torture. The party prudently retreated before the return of the village hunters, but before they reached the canoes on the river they stopped for the night and were ambushed. In the fight that ensued the Xavánte were driven away by gunfire, and some were killed. Back at his mission on the Araguaia, Friar Sigismundo petitioned the authorities for weapons so that he could try to contact the "wild" Xavánte again, but there is no record that he did so.

Not all the Xavánte crossed the Rio das Mortes.[13] The Xavánte legend of the river porpoises tells how whites began to cut the forest and plant their crops on the eastern side of the river. An elder, Sereburã, from the village of Etéñitépa tells the story (Sereburã et al. 1998, 89).

> Only a few crossed the Rio das Mortes to the side where we live now; the rest of the people stayed behind. We thought that everybody would make it. But suddenly the porpoises appeared and crossed in front. Our people were afraid of the spray that came out of the water; we were really afraid. . . . The group that crossed asked their relatives to take care of their nieces and nephews. "Don't deny anything to my nephew! Give him whatever he asks for!" Many people stayed on the other side of the river—many more than those who crossed.

The history of Xavánte expansion into eastern Mato Grosso around the middle of the nineteenth century is difficult to unravel. Much of what is known of this period comes from reconstructions on the basis of oral histories collected by missionaries and anthropologists in the twentieth century, thus several decades after the actual events took place (see Giaccaria and Heide 1984, 27–49; Graham 1995, 29; Leeuwenberg and Salimon 1999; and Silva 1992, 365–67). Epidemics, political disputes, population movement, fission, and recombination were frequent events. According to some Xavánte accounts (Leeuwenberg and Salimon 1999), after Xavánte migrants crossed the Araguaia they remained for a number of years between the two rivers until some crossed the Rio das Mortes. The first village the Xavánte founded after the crossing was abandoned when an epidemic killed all the old people (Giaccaria and Heide 1984, 27–49). Later they founded a village called Tsõrepré, located about one hundred kilometers north of the present village of Etéñitépá, which they still consider to be their ancestral village. According to Silva (1992, 366), Tsõrepré was occupied from the end of the nineteenth century to the 1920s. When Tsõrepré broke up, some groups moved west and north, while the group that now occupies the Pimentel Barbosa reservation remained in the vicinity of the Rio das Mortes (see fig. 2.2).

It appears that when the Xavánte crossed the Rio das Mortes, the lands into which they moved were empty of people, probably because the previous native inhabitants had been destroyed. Gradually they occupied a region bounded on the north by the tributaries of the Xingu River and on the east and south by the Rio das Mortes. With the exception of skirmishes with the Boróro to the south and the Karajá on the Araguaia there is no record of fighting with other Indians. Until the twentieth century few whites, whose nearest towns were along the Araguaia or far to the west around Cuiabá, attempted to penetrate the unmapped region into which the Xavánte migrated.

Among the Xavánte who lived beyond the Rio das Mortes rejection of contact with Western society was so firmly entrenched that it was nearly a hundred years before peaceful contact was again made. It seems likely that one of the motives for this rejection was fear of contagious disease. They also rejected contact with neighboring indigenous groups that might transmit sickness. According to Maybury-Lewis (1967, 12), "The Shavante despise the Karajá and fear them, for they say that they infect them with sickness, either by contagion or sorcery, depending on the mood of the informant." A Karajá told Hermano Ribeiro da Silva (1935, 227) that bands of Xerénte used to pass by Ilha do Bananal on their way to visit their Xavánte relatives until, he said, the Xavánte chiefs told the Xerénte not to return since, because of their association with the whites, the Xerénte were bearers of disease.

There is a Xavánte legend that indicates a belief that sickness could be caused by articles belonging to whites or other strange creatures. This legend tells how long ago, when many of the bravest men in a village had died, a group of women resolved to raid the whites (Giaccaria and Heide 1975, 173–80). After days of searching the women came across a village and painted themselves to attack. Then they heard strange cries, and a wind rose; this was not a white settlement but a village of evil spirits, who disappeared as the women approached. The women entered the empty houses, searching for knives or other tools. They did not find these things, so they took some weapons and other things belonging to the spirits. When the women left the village they began to fall sick. They were afraid they would die. "Perhaps it is because of these things that we took from the village; that must be it." The women started back the way they had come, and as they traveled they recovered. When the women arrived home everybody was amazed to hear of their adventures. "What were those creatures that you met?" people asked. "They have flat faces, flat white faces, and they are covered all over with feathers. . . . Those were the Tsimihöpãri monsters!"[14] As a cautionary tale, this story points to the incorporation of fear of introduced diseases into Xavánte myths and the internalization, through the identification of outsiders with monsters, of the idea that contact with them was dangerous.

The Early Twentieth Century

From the middle of the nineteenth century to the beginning of the twentieth the Xavánte were left relatively undisturbed. One reason why the Xavánte were left to themselves was because the major economic frontiers were in the Guaporé and Pantanal regions, far from their territories. During much of this period the Xavánte lived at Tsõrepré, west of the Rio das Mortes and on the flanks of the Serra do Roncador (Graham 1995, 31; Silva 1992, 365–66; see fig. 2.2). In the following decades groups that remained in this area were the first to come into permanent contact with the Brazilian national society.

In the early part of the twentieth century, the interior of Mato Grosso was regarded as one of the world's most mysterious regions. Since no one knew what was there, anything might be. The disappearance in 1925 of the English explorer Col. Percy Fawcett somewhere between the Rio das Mortes and the upper Xingu provided a field day for the international press. The story was particularly intriguing because Colonel Fawcett was looking not only for lost mines but for lost cities and civilizations. Shortly before his disappearance Fawcett wrote: "It is certain that amazing ruins of ancient cities—ruins incomparably older than those of Egypt—exist in the far interior of Mato Grosso" (quoted in Fleming 1934, 18). As Peter Fleming, an English journalist who took part in one of the expeditions to search for Fawcett, wrote:

> Less, probably, is known about the interior of Matto Grosso than about any other inhabited area of equal size in the world. It was in 1925, it is to-day, and it is likely to remain for some time, very largely virgin territory. Between the headwaters of the great northward-flowing tributaries of the Amazon—between the Araguaya and the Xingu, between the Xingu and the Tapajoz—there are huge tracts of jungle which no white man has even attempted to enter. You can believe what you like about those regions: no one has authority to contradict you. You can postulate the existence in them of prehistoric monsters, of white Indians, of ruined cities, of enormous lakes. Fawcett plumped for ruined cities. (17–18)

The Xavánte were soon faced with less quixotic invasions of their territory than those of Fawcett and his would-be rescuers. Homesteaders were pushing into Mato Grosso and settling along the Rio das Mortes. Xavánte occasionally ambushed and killed people from these isolated farms. In retaliation, settlers organized posses to invade Xavánte villages, killing the inhabitants and burning their houses (Maybury-Lewis 1967, 3–4). Gradually the Xavánte acquired a national reputation for ferocity. The Xavánte would become perceived as being so treacherous that, according to one report, "they even ran with their feet pointing the other way to mislead their trackers" (Smith 1971, 34).

The March to the West

The push into Xavánte territory during the 1930s and 1940s was designed and directed by interests at the national level. The 1930s became a turning point in Brazilian political and economic history when a strongly centralist government led by President Getúlio Vargas stressed nationalism and state intervention in the economy and society.[15] During this period, the government emphasized the need to stimulate national development and integration. Expressing this ideology, in 1938 the Vargas regime launched the March to the West, a program intended to open up and develop the central regions of Brazil, at that time viewed as unpopulated and isolated from the process of economic development of the country as a whole. In 1940 Vargas said:

> If Brazil is unified politically, it is not unified economically. In this respect, it resembles an archipelago formed by a few islands separated by empty space. The islands have already reached a high level of economic and industrial development and their political boundaries coincide with their economic boundaries. However, there continue to be vast unpopulated spaces that have not reached the necessary conditions for development because they lack a whole series of basic measures, which the government intends to carry out. . . . In this way, the March to the West carries on the campaigns of the builders of our nation, the *bandeirantes* and pioneers, bringing them up to date. . . . We need to promote this effort in every way and by all possible means in order to fill the demographic emptiness of our territory and make our economic frontiers coincide with our political frontiers. This is our imperialism. (quoted in Velho 1976, 147–48)

The March to the West was first embodied in the exploratory effort called the Roncador-Xingu Expedition, which was organized by a government agency, the Fundação Brasil Central (FBC), established by decree in 1943 (Garfield 1996; Menezes 1999; Villas-Bôas and Villas-Bôas 1994). The mandate of the FBC was to build infrastructure that would make possible the subsequent settlement and economic development of regions in Central Brazil. This infrastructure was to include roads, schools, hospitals, airfields, and so on. Specific goals were set, such as the settlement of two hundred families a year along the Rio das Mortes and in the Serra do Roncador region. The plan was that cattle raising would become the basis of the regional economy (Garfield 1996, 82–91).

The presence of the Xavánte in this very region obviously presented a barrier that had to be overcome if the FBC plans were to be carried out. The intensification of efforts by the SPI, the Indian Protection Service,[16] in the early 1940s to establish contact with the Xavánte was in part due to the fact that the planned route of the Roncador-Xingu Expedition lay through territories known to be occupied by the Xavánte. This is made clear in a 1944 SPI document of instruction, addressed to agent Francisco Meireles, who would coordinate the "pacification" effort (SPI 1944). This document stresses the urgency of establishing peaceful contact with the Xavánte, since the Roncador-Xingu Expedition would soon cross the Rio das Mortes and necessarily would move through Xavánte territory. The FBC would collaborate with the SPI by contributing financial resources in order to accelerate the attraction process, as the FBC had no intention of delaying its own activities to await "the pacification of these Indians, who are still aloof and hostile, since we know that this process takes a long time to produce results" (1–2).

The difficulty of establishing peaceful contact with the Xavánte had been demonstrated in the 1930s by the deaths of two Salesian priests, Sacilotti and Fuchs, who were killed making the attempt. Officials of the SPI attributed their deaths to imprudence. But the SPI was soon to suffer its own tragedy when Inspector Genésio Pimentel Barbosa, with a team, set up an "attraction post" on the Rio das Mortes. The men of the expedition began to cut trails, along which they left, according to standard SPI procedure, gifts of knives, ax heads, pots, and other manufactured goods for the Indians. The attraction team was encouraged when the Indians began to take the gifts, but unexpectedly, in 1941, Pimentel Barbosa, with several of his men, was ambushed and killed by a party of Xavánte.

In the 1940s the Xavánte were well aware that the outside world was once again moving in on them. As Ravagnani (1978, 162–63) writes, they were "corralled, with no possibility for further migration, surrounded by cattle ranches, with their territory invaded from all sides, with powerful motor launches on their rivers, their lands traversed by expeditions, their villages subjected to surprise attacks and their houses ransacked and robbed." The power and determination of the outsiders were evident. From newspapers and magazines published in these years we learn that flights over Xavánte villages were common. In 1940 Getúlio Vargas, on a trip to the Araguaia region to promote the March to the West, directed his plane to fly over uncontacted villages of the "extremely ferocious" Xavánte Indians (Garfield 1997, 749). The Brazilian press, with sensational reports of daring forays into Xavánte territory, sought to satisfy the public's curiosity. The publisher of *O Cruzeiro,* Brazil's foremost weekly newsmagazine, sent a light plane with a reporter and photographer to buzz Xavánte villages at low altitude (Morais 1994, 419–20; Nasser 1944). As the plane plunged toward the houses, men armed with bows and arrows and clubs rushed out to shoot at the invaders. The plane flew so low that a thrown club hit and damaged its tail. It spread panic that an elder from Etéñitépa would recall in the 1990s: "[The plane] gave us great fear! Great fear! People were running all around, hiding their children, not knowing what to do . . . great fear!" (Sereburã et al. 1998, 117; see figs. 3.4 and 3.5).

The progress of an SPI pacification team organized in 1944 under the leadership of Francisco Meireles and installed at an attraction post on the banks of the Rio das Mortes was slow. But within two years Meireles could report that the first Xavánte had accepted gifts, and a few months later they were visiting the post.[17] The group that first accepted contact was headed by a man, born at Tsõrepré, called Apöwẽ. According to the Xavánte view of events, as explained by Warodi, Apöwẽ's eldest son, it was Apöwẽ, rather than Meireles, who "initiated and controlled the contact," saving the whites from other Xavánte who were still angry (Graham 1995, 33–34). The decision was not unanimous; while Apöwẽ's group accepted contact, others remained hostile. Reporters present around the time of contact agree that the Xavánte, once they had become accustomed to trade, often initiated the barter by offering arrows in exchange (Fonseca 1948). Apöwẽ's people, who were closest to the SPI team, became wealthy in trade goods. By 1949

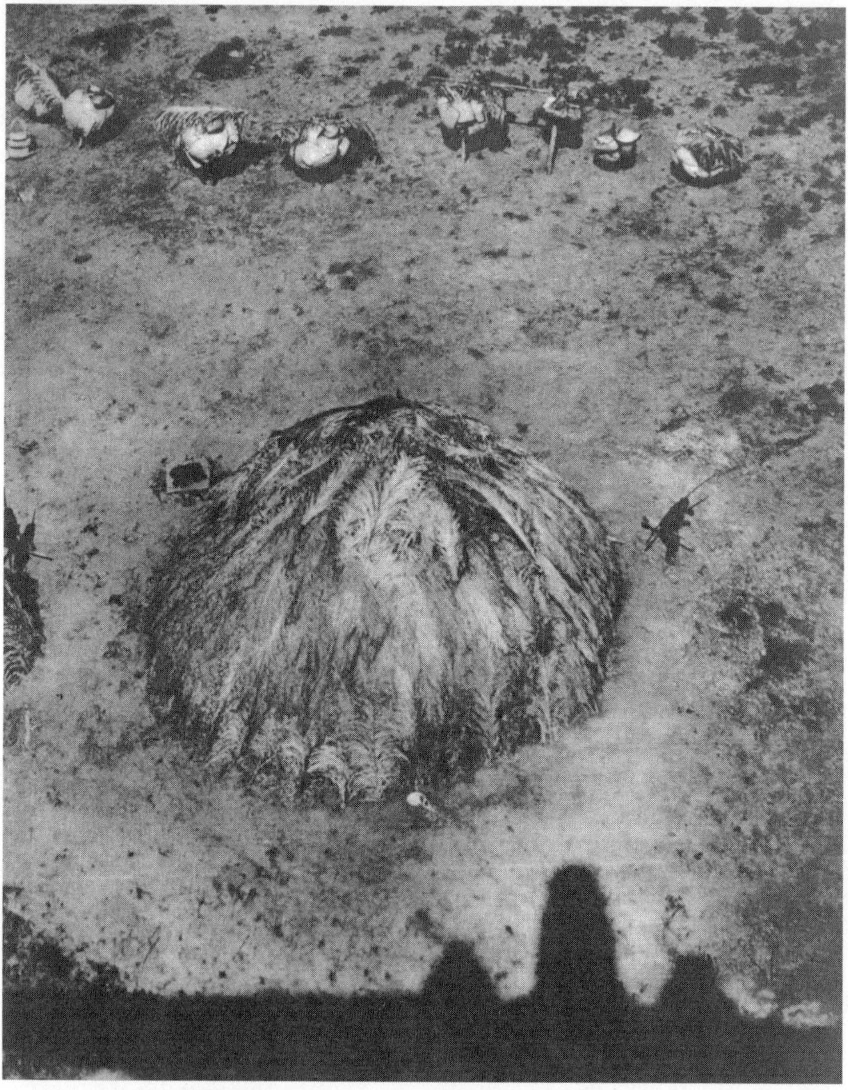

Fig. 3.4. Aerial photograph of an uncontacted Xavánte village, published in *O Cruzeiro* in 1944. Men have rushed out of their houses and are shooting arrows at the plane, whose shadow can be seen in the right bottom corner. These flights are still remembered by some Xavánte. (Reproduced from *O Cruzeiro,* 24 June 1944, 55.)

Fig. 3.5. Uncontacted Xavánte village at Aroboñipopa. Near the center of the village there is a ring of posts where people hung storage baskets or bunches of drying corn. (Reproduced from *O Cruzeiro,* 24 June 1944, 59.)

Fig. 3.6. A member of the SPI team that contacted the Etéñitépa Xavánte in 1946 exchanging trade goods for arrows with a Xavánte. (Courtesy of Museu do Índio, Rio de Janeiro.)

Meireles had distributed some four thousand machetes and two thousand ax heads, as well as a variety of other tools and utensils (Souza 1953a; Souza 1953b, 103; see fig. 3.6).

Although Meireles was able to establish contact with one Xavánte group in 1946, it would be several more years before all the Xavánte came into permanent contact. Weakened by disease and worn out from their confrontations with settlers, various Xavánte groups entered into peaceful relations with different representatives of the Brazilian national society during the 1950s and 1960s. Many Xavánte groups took refuge at Catholic Salesian missions or SPI posts, where they often arrived sick and hungry.[18]

The Immediate Postcontact Period

The experience of the Xavánte in the period that followed the onset of contact in the 1940s and 1950s was highly heterogeneous. In part, this heterogeneity is due to the fact that some groups made contact with the government agency (SPI) and others with religious missions. In addition, some groups remained in their own territories, while others were forced to move away from their traditional lands.

Anthropologist Aracy Lopes da Silva (1986, 31–44), taking into consideration the social and geographic diversity of postcontact experiences, has proposed a division of present-day Xavánte communities into three large groupings (see fig. 2.1). The first consists of the Xavánte with whom Meireles made contact in 1946 and their descendents, who are now settled on the Pimentel Barbosa and Areões reservations. In the 1950s this group split; those who accepted Apöwẽ's leadership moved in 1956 to the vicinity of the SPI attraction post at São Domingos, where they remained until around 1964. Some of the dissidents, after a number of moves, eventually settled at an SPI post near Xavantina that took the name Areões, while others moved to São Marcos. The second consists of two Salesian missions. In the 1950s the Salesian mission for the Boróro at Merure took in a number of Xavánte refugee groups. Since the Xavánte were traditional enemies of the Boróro, the Salesians set up two new missions, one at Sangradouro and the other at São Marcos, both on the upper reaches of the Rio das Mortes. During the 1960s São Marcos took in more Xavánte, including an entire group from Marãiwasede in the north that had lost its lands to a large ranching conglomerate from São Paulo. The Xavánte at São Marcos and Sangradouro have been exposed to missionary influence. Most members of the third group trace their origins to the precontact village of Parabubu located on the Couto Magalhães River near the spot where it enters the Culuene, a Xingu tributary. In the 1950s, due to conflicts with settlers, this village broke into smaller groups, which were forced further west and outside their traditional range. Most of them were eventually settled at two SPI posts that were set up for them, one on the Batovi River and the other at Simões Lopes. These Xavánte were proselytized by American Protestant missionaries of the South American Indian Mission.

The 1960s represented a low point in the postcontact experience of most Xavánte groups. Many seem to have suffered severely from epidemic disease (see chapter 5), and Xavánte population numbers fell to their lowest point. According to Silva (1992, 372–74), the 1960s were a period during which the Xavánte absorbed the impacts of contact, were exposed to the everyday routine of the missions and government posts, and became accustomed to using products of Western manufacture. Above all, it was a pause when they sought the refuge of government and religious institutions against the systematic and growing pressure that they had been enduring for at least the previous thirty years.

The decade of the 1960s was not only the one that immediately followed permanent contact of Xavánte with outsiders; it was also the decade during which the Brazilian government initiated a series of programs intended to develop the region of eastern Mato Grosso. As we will see in chapter 6, these programs involved highway construction, intensified migration, and colonization and agricultural projects. In a way, the new phase of economic development in Mato Grosso constituted the fulfillment of the policies and programs, including the March to the West, which were promoted by the government starting in the 1930s.

Xavánte reservations were demarcated during the late 1960s and the early 1970s in the midst of these rapid and profound economic transformations. Not surprisingly, this was a period of intense disputes and widespread speculation over land. Powerful economic interests, some from companies based in southeastern Brazil, were involved in these activities. The territory occupied by Xavánte groups became limited to patches isolated from one another as the land around the missions and SPI posts was bought up and occupied by homesteads and ranches, some of the latter occupying thousands of hectares. Throughout the 1970s and 1980s the Xavánte were deeply involved in efforts to win back territory, not included in the demarcations, on which they claimed rights (Garfield 1996, 331–96; Graham 1995, 37–42; Menezes 1982; Silva 1992, 374–76).

The Etéñitépa Xavánte in the Late Twentieth Century

The Etéñitépa Xavánte, on which this study is based, can be traced directly to the group led by Apöwẽ that the SPI contacted in 1946. Unlike others, the Etéñitépa group remained in the region that the Xavánte had occupied since they crossed the Rio das Mortes in the nineteenth century. Also, they were not subjected to the influence of missionaries. According to Graham (1995, 37), "these Xavante can be characterized as the most traditional of the contemporary Xavante communities" (see also Silva 1986, 35).

From the 1950s to the 1970s Apöwẽ's group relocated several times. For several years after making peaceful contact in 1946, it lived at the site of the present-day village, known as Etéñitépa (see fig. 2.2). In

1956, it settled close to the SPI post at São Domingos on the banks of the Rio das Mortes. Around 1960 there was a split, when some people moved to a place called Ötõ. Later part of this group moved back to São Domingos, while others went to São Marcos. Around 1964, Apöwẽ's group moved to a site known as Barreira Vermelha, which was on the opposite bank of the Rio das Mortes and upstream from São Domingos. In 1972, after the demarcation of the Pimentel Barbosa reservation, the Xavánte moved again, this time returning to Etéñitépa, where they made the village they continue to occupy (see Graham 1995, 34–35, 254; and Maybury-Lewis 1967, 21–30). From 1972 to 1980 the entire group lived at Etéñitépa. In 1980, after the Xavánte won more territory for the reservation, Etéñitépa split. The dissident faction founded a village at the site of an abandoned ranch, calling it Caçula. Another faction left in 1983 to found Tanguro (Garfield 1996, 539–40; Graham 1995, 51–53). Finally, in 1996, Pe'azarupré was established. All these splits came about because of political disputes and factional rivalries.

The 1950s and 1960s: The Years by the River

A few years after contact Apöwẽ's group built its village within walking distance of the SPI post at São Domingos on the Rio das Mortes. The post, once a seemingly inexhaustible source of trade goods, was drastically reduced in importance once the Xavánte were no longer to be feared. Nevertheless, a visiting SPI inspector chastised Ismael Leitão, the São Domingos agent, for his "obsolete and paternalistic" method of administering the Indians, as he had allowed them to continue their "hunting and gathering routine" and distributed medicine, tools, and clothing without obliging the Xavánte to work for them (Velloso 1957).

In 1957–58 anthropologist David Maybury-Lewis lived with and studied this group (Maybury-Lewis 1967, 1988). At that time, "the Shavante rarely spent more than a week or two in their base village, usually on the occasion of an important festival. The influence of the Indian agent seemed to be directly proportional to his supply of gifts. He was careful to give the lion's share of these to the chief and members of his lineage." However, when Maybury-Lewis returned in 1962 with a team of geneticists and physical anthropologists (see Neel et al. 1964) he found the situation altered. "In 1958 the village had comprised seventeen huts, plus a bachelors' hut. In 1962 there were only ten, and a bachelors' hut. There had been a considerable decrease in the population due to epidemics and internecine warfare" (Maybury-Lewis 1967, 28). Much of this strife was probably provoked by jealousy over the distribution of trade goods (Maybury-Lewis 1988, 177). Accusations of sorcery also increased with the rise in deaths from disease, and this led to killings (Maybury-Lewis 1967, 187–88; see also chapter 5). Moreover, the Xavánte now wore clothes and had become increasingly dependent on the Western goods that the post provided.

At this time the SPI only controlled the land, and therefore could only protect the Indians, in the immediate vicinity of the posts. Proposed demarcation of protected land for the Xavánte failed to be carried out because of lack of funds and personnel (Garfield 1996, 256–69). Agents of the SPI were instructed to patrol the post area to prevent invasions, and to oppose titling of invaded land, but their reports and appeals to headquarters often brought no results. Ismael Leitão reported indignantly that the government of Mato Grosso had sold off the very land on which the São Domingos post had been located since the 1940s (Garfield 1996, 262).

In 1966, the lands of the Xavánte of Marãiwasede, the only group that had not yet been "pacified," were purchased by corporate investors from São Paulo for a large cattle ranch. All the Marãiwasede Xavánte, many of whom were relatives of the Etéñitépa Xavánte (Maybury-Lewis 1967, 29, 93), were rounded up and taken in a Brazilian Air Force plane to the Salesian mission at São Marcos. In the next year almost a third of these refugees died in a measles epidemic (Garfield 1996, 294). Since 1985, the Marãiwasede survivors, who left São Marcos, have settled in a "temporary" village (Água Branca) at the southern end of the Pimentel Barbosa reservation, as they attempt to regain their traditional lands (see fig. 2.2).

In 1967, the military government, in power since 1964, ordered an investigation of the SPI that found evidence of widespread corruption and violations of indigenous rights. In the same year a fire, which some held to be suspicious, destroyed the archives of the SPI (Garfield 1996, 335–43; Ramos 1998, 154–57). A new Indian agency, the National Indian Foundation, known as FUNAI, was established with wide powers but was placed under tight military control. For the next twenty years most of the presidents of FUNAI were military men. Although a number of SPI employees were charged with crimes or dismissed, others were eventually exonerated and reemployed by FUNAI.

The 1970s and 1980s: Of Rice and Land

Largely because of years of neglect, which allowed investors to buy up and register title to land within traditional Xavánte territory, when FUNAI attempted to demarcate reservations for Xavánte groups settled in different areas of their precontact territory, as well as those who had been displaced in the early contact years, the results pleased neither the Indians nor the landowners. Landowners claimed that the reserves were taking away "productive lands." The Indians complained that the land demarcated was inadequate for their subsistence and did not include traditional areas of settlement. In fact, during General Bandeira de Melo's administration of FUNAI in the early 1970s, the agency became notorious for issuing "negative certificates" attesting that no Indians were present in areas targeted for development and where prospective landowners were eager to register titles. The ability to issue certificates had the effect of encouraging graft and corruption among FUNAI officials.

Pimentel Barbosa reservation was demarcated in 1972 with an area of 224,000 ha. Since that time the Xavánte of Pimentel Barbosa have worked almost constantly to regain lands of traditional occupation outside the reservation boundaries, including the site of Tsõrepre, their "mother village" (see fig. 2.2). They have applied to this process both the skills gained in the factional politics of their own society and their growing experience of white society. As Garfield, referring to the land struggles of the 1970s and 1980s, wrote:

> With unrelenting force, the Xavante would pressure the federal government to expel invaders from their lands, amplify their reservations, and push for their socioeconomic development. In short, with similar pomp and media coverage, the Xavante would invert the pilgrimage of Vargas which catapulted them to national attention. . . . With Brasilia as their Mecca, the "true sense of Brazilianness" for the Xavante lay in the March to the East. (1996, 396)

In 1976, when Flowers arrived at Pimentel Barbosa for fourteen months of fieldwork (see Flowers 1983a), interaction between the Xavánte leadership and owners or administrators of the ranches that surrounded the reservation was characterized by surface tolerance with undercurrents of suspicion and resentment. Two years before, one Xavánte leader had connived with local FUNAI officials to persuade the community to agree to the exchange of 62,000 ha of reservation land for cattle, sewing machines, and the promise to build a bridge. The land was subsequently parceled out and sold. When the cattle failed to arrive, a leadership struggle developed between Surupredu, who favored ceding land for cattle, and Warodi, Apöwẽ's eldest son, who opposed it. The Xavánte also disputed the southern boundary of the reservation, where they claimed that the FUNAI cartographers who surveyed it had switched the name of two rivers, thus demarcating a smaller area and allowing the land left outside to be sold. When Flowers left the reservation in 1977 organized teams of Xavánte men were cutting a trail to mark what the community held to be the correct boundary of the reservation.

Some Indians were growing rice for sale but had difficulty getting it to market in the one truck that belonged to the community. A number of men periodically left the reservation to work on neighboring farms. As we describe in chapter 6, the Xavánte Rice Project was just beginning, and people were looking forward eagerly to the arrival of the tractors and harvesters with which they were already familiar from observing them at work on neighboring farms. Eastern Mato Grosso was developing at a frantic pace, with the government providing fiscal incentives for agribusiness and cattle ranching as well as private colonization projects.

Several young boys, sons of prominent men in the community and designated by the elders for future leadership roles, were studying in Brazilian cities, where they lived with middle-class families. When these boys came of age and returned to the village in the next decade, they would guide the community in different directions from those taken by their fathers.

When Apöwẽ died in 1978 at the age of around seventy-five, and Ismael Leitão, who was Meireles's assistant when the Xavánte were contacted by the Rio das Mortes, retired and left Etéñitépa, the last links with the paternalism of the old SPI days were severed. Warodi, who succeeded as chief, aggressively pursued the reclamation of Pimentel Barbosa territory. He was not the only Xavánte leader to do so. Xavánte who were driven west in the 1950s had returned to seek their traditional territory on the Culuene and were fighting a war of attrition with the huge ranch that occupied most of the land they claimed (Garfield 1996, 397–425).

At this time Mario Juruna, from São Marcos, was attracting media attention with his tape recorder, which he used to record interviews with FUNAI officials in order to hold them accountable for their statements (see Juruna et al. 1982). In 1980 Juruna was invited to attend the Russell Tribunal on indigenous rights held in Rotterdam, and the Brazilian government garnered negative publicity when FUNAI refused to grant him a passport to travel. In 1982, Juruna was elected to Congress as a delegate from Rio de Janeiro and from his seat made fiery speeches accusing the government of oppression and wrongdoing.

Outside support for Indian causes was also developing. In 1972 the Catholic bishops of Brazil, inspired by liberation theology, created the Conselho Indigenista Missionário, or CIMI (Missionary Council for Indigenous Affairs). Throughout the 1970s and 1980s CIMI organized assemblies at which indigenous leaders shared their accounts of struggle. Other Indian advocacy groups appeared at this time, as supporting the Indian cause became one way to criticize the military government, which was losing national support.

Leaders from Pimentel Barbosa and other reservations made innumerable trips to Brasília, trips that became increasingly adversarial as FUNAI inaction continued (Vale 1980a, 1980b). In May 1980 a group of Xavánte leaders stormed into the office of the president of FUNAI and refused to leave, aggressively demanding that "he do something about guaranteeing their lands and that he root out corruption in FUNAI itself" (Maybury-Lewis 1983, 55). This incident was given wide coverage in the Brazilian press (fig. 3.7). At last the Etéñitépa Xavánte took the decision to expel the encroaching ranchers, with or without the support of FUNAI. At Pimentel Barbosa, men painted and armed with war clubs and bows and arrows surrounded the headquarters of one of the ranches before dawn. A young ranch hand described what happened.

> At dawn the house was surrounded. One of them [Xavante] summoned the foreman to tell him that we had until noon to be gone. Then he gave a signal, a call, and several Xavante immediately stormed the house. . . . Then they helped us pack the pickup with all we could carry. . . . Just before noon they sent us on our way. They'd always been friendly to us, but Jesus, they were angry then! We didn't even try to resist. As we were leaving we saw the women arriving with their baskets, to carry off whatever they could manage. (quoted in Graham 1995, 38)

Fig. 3.7. Xavánte holding the president of FUNAI hostage. The cartoon refers to one of the many occasions on which the Xavánte challenged FUNAI's authority. At the right is a representation of Xavánte leader Mário Juruna, with his famous tape recorder. The president's adviser reassures him, "Calm down. The U.S. cavalry is on its way." (Courtesy of *Jornal de Brasília,* 7 May 1980.)

The takeover was accomplished without violence, but it sent a clear message of Xavánte determination. In 1979, following a meeting at Etéñitépá with leaders from other Xavánte reserves, the people of Pimentel Barbosa set about demarcating their reservation in accordance with their demands. Finally FUNAI, in 1980, officially demarcated Pimentel Barbosa according to its present boundaries, with an area of 328,966 ha (CEDI 1991, 503). The government eventually indemnified the owners of the ranches that were expropriated.

Between 1980 and 1985 the Xavánte Rice Project was at its height (see chapter 6 for more details). The proliferation of new villages on Xavánte reservations at this time was in part the result of conditions created by the project. A new village could claim a "project" with a tractor, a truck, and salaries for community leaders (Graham 1987). But if FUNAI thought that the project would keep Xavánte leaders home on the reservations tending to their rice fields they were mistaken. Xavánte leaders were conspicuous in Brasilia, making constant demands of the FUNAI administration, which was going through a critical period of instability as the military regime weakened. Between 1983 and 1986 FUNAI had eight presidents, each lasting only a few months (Ramos 1998, 252–53). As the project drew to an end, FUNAI abandoned broken machinery, forcing the communities to harvest their rice crop by hand or pay for outside assistance. Instead of supporting the communities, the agency resorted to buying off Xavánte leaders with individual handouts. "In July 1987 FUNAI dispensed Cr $300,000 (US $6,676) to each leader to buy miscellaneous items such as soap, cloth, and cooking oil. There would, however, be no 'project' in 1988" (Graham 1995, 61).

In 1985, the military regime was replaced with an elected civilian government. In 1988, a constitutional assembly was convened in Brasilia to draw up a new federal constitution. During the assembly the presence of an effective Indian lobby, made up largely of Kayapó and Xavánte, formally painted and befeathered, kept the issue of indigenous rights in the public eye. According to Ramos (1998, 82, 257–59) the effectiveness of this lobby was largely due to the efforts of the União das Nações Indígenas (UNI), a pan-Indian organization formed in 1980 by Ailton Krenák and other young leaders. The new constitution brought changes to the status of Brazilian Indians (Souza 1994). Among the most important was that "Indigenous societies are no longer considered to be disappearing cultures whose inevitable and desirable destiny would be their incorporation into the so-called 'national communion'" (Gaiger 1989, 8, cited in Ramos 1998, 259). Not only was the distinctiveness of Indian societies recognized, but the new constitutional rights freed them to organize in associations that could receive funding directly from national and international organizations (Graham 2000; Ramos 1998, 259–60; Souza 1994). In 1988 the Associação dos Xavantes de Pimentel Barbosa was founded.

In 1988 Warodi died. He was the Pimentel Barbosa chief who had led the struggle for land, saw the rise and demise of the FUNAI rice project, and exhorted his people to "continue forever Xavánte" (Graham 1995).

The late 1980s were years when the deforestation of Amazonia held international attention. In 1988 Chico Mendes, a political leader who defended the collective rights of rubber tappers to forest resources, was murdered. The Kayapó gained wide press coverage in 1989 when they organized a meeting of Amazonian Indian groups at Altamira to protest the proposal to build, with World Bank funding, a series of dams on the Xingu River that would flood indigenous lands. In 1988 the rock singer Sting visited the Kayapó and set up the Rainforest Foundation to lobby for demarcation of Kayapó land under the banner of environmental protection. The Kayapó shrewdly understood that visual aspects of their culture could be effective political weapons. Western environmentalists tended to identify Kayapó efforts to achieve self-determination through control of land and resources as "native environmentalism" (see Conklin and Graham 1995; Conklin 1997; Ramos 1998, 267–83; and Turner 2000). However, the Kayapó have valuable gold and timber on their land, and the ways in which some Kayapó leaders have allowed the exploitation of these resources for their own enrichment, and with little regard for the environmental consequences, have left the Kayapó open to charges of "environmental hypocrisy."

In the 1980s the Xavánte took a different road, seeking to achieve political control and material advantages for their communities by exploiting their relationship with FUNAI, leading to charges that they profited unduly from FUNAI favoritism (Ramos 1998, 263–65). After the demise of the Xavánte Rice Project, support from FUNAI was no longer an option. As we show in chapter 6, the Xavánte of Pimentel Barbosa have returned to a more "traditional" subsistence economy, but they still have basic needs for tools and transportation as well as education and health care. In the 1990s they began to explore new sources of funding to meet the material needs of their growing population.

The 1990s: Looking Beyond the Reservation

In the 1990s most of the relations of the Etéñitépá Xavánte with the outside world were mediated through the leadership of the young men who were educated in Brazilian cities during the 1970s and 1980s and returned to marry in the community.

In 1991, Milton, Warodi's half brother, who in 1985 had succeeded him as chief, left Etéñitépa with his faction and joined Caçula when he was accused of selling reservation cattle to his own advantage. Suptó, who then took over as chief, was educated in Ribeirão Preto, São Paulo, as were Paulo, and José Paulo, two other young men who became influential and now earn FUNAI salaries for their work in the community (Pereira 1997a).

Two others boys, Cipassé and Jurandir, went to high school in Goiânia. Cipassé was greatly influenced by his contacts with Ailton Krenák, the coordinator for UNI, and participated in the plans of the Centro de Pesquisa Indígena (CPI, Center for Indigenous Research), which was set up in Goiânia in collaboration with the Catholic University of Goiás, and the state agricultural agency. "A primary CPI objective was to provide Brazil's indigenous peoples with Western knowledge and skills in applied biological sciences so that they could use new technologies in their reserves" (Graham 2000, 55). The CPI gained financial support from a number of international agencies and built a research center, staffed by professional consultants, on the outskirts of Goiânia. The Goiânia center closed in 1992 when the CPI turned to implementing programs on reserves in other regions of the country.

When the Associação dos Xavantes de Pimentel Barbosa was established in 1988, Cipassé became its first president. In 1988 he also married Severiá, a university-educated Karajá woman from Goiânia, and they became a husband and wife team. They made their home first in Goiânia, later in Nova Xavantina, a town some one hundred miles from the reservation, although they made periodic visits to the village and usually maintained a house there. The Jabiru Project, with which Cipassé and his wife were closely associated, was a product of CPI research. It was comprised of two projects believed to be complimentary. One project was to restore the ecological integrity of the reservation, which had been damaged by deforestation for pasture and monocropping, by planting native tree species whose fruit could be harvested and processed to make marketable products. The other was to develop a plan for game management and conservation. At first the plan included the idea that the Xavánte would raise peccaries in semicaptivity, but this was quickly rejected by the community. These projects appealed to international funding agencies, not only because they had a strong conservation component but because they appeared to be "grassroots" programs that would lead to self-sufficiency (Santos 1991). Both projects were funded through the Associação dos Xavantes de Pimentel Barbosa, the native fruit program by the Inter-American Foundation, and the game management plan by the World Wildlife Fund (WWF).

The native fruit program turned out to be a fiasco, in part, as Graham

(2000) points out, because most of the funding money was spent on building a processing factory and other infrastructure at Nova Xavantina, while no one acquired the technical expertise to actually process, package, and market the fruit. Moreover, when Xavánte women after a short time became aware that most of the work involved was falling on them with no immediate return, their early enthusiasm for the project faded. After 1994, international funding came to an end and was not renewed. The processing plant was never used and joined the rusting machinery left over from the FUNAI rice project as material reminders of misguided attempts to promote economic development based on modern technology without providing the community with the technical knowledge to sustain it.

The wildlife conservation program, supported by the WWF, has taken a different course. The community has supported it since its inception in 1991 because Xavánte hunters were becoming increasing concerned about dwindling game and wanted to find ways to manage the resource that would allow them to maintain the supply (see chapter 6). They realized the importance of research to ascertain which animals were actually endangered by overhunting. They have cooperated with the research efforts of population biologists, with whom they have developed a plan to rotate hunting areas to allow game to recover and to intensify fishing in the dry season (Leeuwenberg 1997; Leeuwenberg and Robinson 2000). As Graham points out,

> The wildlife management project had two advantages over other collaborations with donors that emerged under the auspices of Project Jabiru. First, the idea resonated powerfully within the community and was rooted in community experience and expertise. And second, there was no rush to implement a plan by importing an elaborate theoretical framework or building physical infrastructure that no one knew how to operate. (2000, 64–65)

Although the community at first enthusiastically supported Cipassé's ideas, it withdrew support when he was perceived to be taking undue advantage of his role as indigenous mediator. While outsiders believed that he spoke for the village, community members increasingly regarded him with suspicion, believing that he was amassing personal wealth at community expense. Cipassé was an innovator, but his position between two different cultures led to problems that he was unable to overcome.

> His vision broke new ground, ushering in an era of interaction with outsiders in which the Xavante would play a more decisive role. He established a precedent for project collaboration in which Xavante representatives would design and implement activities. He set a precedent for community members to assume administrative responsibilities such as grant writing, budgeting and accounting. Without him, Project Jabiru would likely never have occurred. However, in the process of building it up, Cipassé acquired more power than the society sanctioned. (Graham 2000, 55)

In 1996 Cipassé's family left Etéñitépa to found a new village (Pe'azarupré), and in 1997 he lost control of the Associação dos Xavantes de Pimentel Barbosa when José Paulo was elected president.

In the late 1990s, community leaders continue to forge relationships with individuals and institutions at the national and international levels. These young men are not only literate and fluent speakers of Portuguese; they are also well traveled. They are committed to upholding and fostering pride in cultural traditions within the community and also in making aspects of Xavánte culture better known to outsiders. Paraphrasing Ramos (1998, 176), they have been turning culture and ethnicity into political capital and are molding these concepts to serve their own goals. The release in 1994, by Quilombo Musica and Warner Music Brasil, of *Etenhiritipá,* a compact disc of Xavánte singing, led to a recording session at Etéñitépa with the rock group Sepultura, which gained wide publicity (Almudena 1995; Anonymous 1995; Vianna 1996). The Etéñitépa Xavánte also participated in assembling a collection of their material culture for the American Museum of Natural History (Williamson 1995). In 1992 the Etéñitépa community hosted for two weeks a group of German teenagers whose visit was sponsored by an environmentalist magazine. The young people, who lived in two *ri uptabi* houses that were built for them, declared that they had enjoyed their unusual summer camp experience, although they admitted to being terrified when the Indians took them to the Rio das Mortes on a fishing expedition and left them alone for hours with the mosquitoes (Pappiani 1992).

In 1997, the Etéñitépa Xavánte, with the assistance of a pro-Indian organization, the Núcleo de Cultura Indígena (NCI), produced a book, *Wamrêmé Za'ra: Nossa Palavra* (Sereburã et al. 1998), to commemorate in their own way the fifty years that had passed since contact by the Rio das Mortes. Young community members who are literate in both the

Xavánte and Portuguese languages recorded and translated Xavánte myths and memories of the elders from the time of contact (Sereburã et al. 1998). At the same time an exhibit in São Paulo included photographs of the Xavánte "pacification" in 1946 taken from the files of the Museu do Índio in Rio de Janeiro. These historical photographs, along with recent photographs taken in the community, were included in the book. The NCI also made a video that attempts to present the history of contact from the viewpoint of the participants (Pereira 1997b).

Caimi Waiassé, a young man from Etéñitépa, is studying video making in São Paulo with filmmaker Vincent Carelli, whose project, Video in the Villages teaches young Indians the elements of video making so they can record the customs and traditions of their cultures (Oricchio 1997). In 1998 Caimi took his video *One Must Be Curious,* about his observations of life at Etéñitépa, to New York, where it was shown at the Museum of the American Indian.

Community members also participated in writing a textbook to be used in schools about Xavánte culture and history called *Para Sempre A'uwê* (Leeuwenberg and Salimon 1999). Most recently, some Etéñitépá Xavánte have built in a São Paulo park a replica of a Xavánte traditional house and are holding workshops at which schoolchildren, teenagers, and interested adults can learn more about Xavánte life (Alves 2000). At the opening Suptó told reporters, "We hope that they will become aware that Indians exist, and so they will respect our children and grandchildren, because our ancestors were not respected" (Carvalho 2000).

At a commemoration of the five hundredth anniversary of the discovery of Brazil in April 2000, a group from Pimentel Barbosa danced for the presidents of Brazil and Portugal. At the same time Suptó took advantage of the opportunity to give the presidents a basket containing videos, books, CDs, and a letter complaining of lack of government support for Indian land claims (Anonymous 2000).

Despite the diversity of the connections developed by the Etéñitépa Xavánte in recent years, the community continues to lack basic educational and health services (see chapters 7 and 8). At present no one leaves the community to work for wages. For day-to-day needs the principal outside resources are FUNAI salaries earned by the community leaders and social security pensions, which elderly people collect. Although obvious economic differences between the leaders and their families and other community members have not appeared, leaders have opportunities for education and travel that are not available to others, which may indicate that new forms of social differentiation are emerging.

A recent concern is the proposed Tocantins-Araguaia inland waterway, designed for barge shipment of soybeans and other agricultural products to ports on the Amazon, thus lowering shipping costs to Europe (see chapter 6). According to proposals, part of the waterway will follow the Rio das Mortes, which forms the eastern boundary of both the Pimentel Barbosa and Areões reservations. The water traffic and pollution that the waterway would bring threaten the reserves by disrupting game, affecting fishing, and encouraging illegal trespassing on Xavánte land. The Xavánte have campaigned against the project, rejecting proposed compensation, and in 1997 lawyers working on behalf of the two reservations obtained a federal court injunction halting the Rio das Mortes portion of the project (Graham 2000).

Conclusion

This chapter makes clear that the Xavánte have been interacting with the outside world for a long time and that this interaction has been closely connected with the expansion of economic and demographic frontiers in Central Brazil over the past three centuries.

Throughout the colonial period the Portuguese Crown maintained policies whose primary intention was to bring the Xavánte and other Indians in Central Brazil under the control of the state so that regional economic resources might be exploited. In the eighteenth century, this policy assumed two major configurations: one was offensive war to break up and destroy native groups; the other was the construction of *aldeias* where thousands of Indians were settled, forcing them to become agricultural laborers. The eventual response of those Xavánte groups that survived warfare and rejected settlement was migration in the direction of less inhabited regions and determined defense of their isolation.

Starting in the 1930s, frontier expansion again impacted the Xavánte. In this period, the conquest and occupation of eastern Mato Grosso became important goals of the government. As a century and a half earlier, these goals were important enough to justify expenditures in human and economic resources to bring the Xavánte, who were seen as obstacles to colonization and economic development, under control of the state. Again, the Xavánte experienced epidemics, population decline, sedentarization, and loss of territory. This time migration was no longer an option. In the decades following contact, the Xavánte were forced to cope with new forms of interethnic relations and attempts to make them serve the interests of the Brazilian national society. Like the colonial governors of the past, the bureaucrats who designed the rice project of the 1980s attempted to incorporate the Xavánte into the regional economy by making them agricultural laborers. The project failed, and at present the Xavánte are taking the initiative in trying to find new ways to deal with the society that surrounds them.

One important conclusion that we can draw from this chapter is that there are evident parallels in the ways through which the surrounding national society in different historical periods has attempted to subjugate not only the Xavánte but other indigenous societies occupying economically important geographic space. Strategies for restraining Xavánte mobility and circumscribing them physically and socially were central elements of the policies implemented by the colonial government in the eighteenth century as well as by the Brazilian state in the twentieth. Even though encapsulation seems have prevailed in the second half of the twentieth century, the Xavánte example shows that interethnic relations are not shaped entirely by the will of the state but by the interplay of often opposing interests and strategies.

Chapter 4

Biological Variability and Continuity

In previous chapters we have placed the Xavánte, especially the group living at Etéñitépa, in their geographical, social, and historical settings. In this chapter we will compare biological data from the Xavánte at the morphological, biochemical (protein), and molecular (DNA) levels with similar data available for other South American Indian groups in order to evaluate similarities and differences. We will be referring to data derived from field studies conducted at Etéñitépa in the 1990s by our research team as well as data obtained in previous research on the Xavánte. Our aim is to situate the Xavánte within the wider context of human biological variation in lowland South America.

James Neel, Francisco Salzano, and their collaborators carried out the first in-depth bioanthropological research on the Xavánte in 1962. Their investigations were made at the village of São Domingos, where the Xavánte group that now lives at Etéñitépa lived from 1956 to around 1963. The research team included physicians, geneticists, and an anthropologist, and the results of their investigations were published in an article entitled "Studies on the Xavánte Indians of the Brazilian Mato Grosso" (Neel et al. 1964). This detailed report is almost ninety pages in length; it covers a wide range of topics and includes analyses of demographic, genetic and medical data.[1]

A few years later, the team that worked at São Domingos expanded its research to include two other Xavánte groups from the villages of Simões Lopes and São Marcos.[2] The work with the Xavánte was the first of a long series of investigations by James Neel, Francisco Salzano, and collaborators among a number of different Amazonian groups, including the Kayapó, the Makiritáre, and the Yanomámi. The field methodology used in these studies in the 1960s, 1970s, and 1980s follows to a large extent the model initially applied in the study of the Xavánte (Neel 1970, 1994; Salzano and Callegari-Jacques 1988).

In recent decades the technical resources available to geneticists and biological anthropologists for studying biological variability have been greatly augmented by new discoveries and technical advances. When the study of human biological variability became a topic of widespread interest in the nineteenth century it was limited to the analysis of morphological characteristics (Comas 1966). The problem with this approach was that, since these traits are multifactorial, it was difficult to separate which portion was due to inheritance and which to environment. With the discovery of the ABO blood groups by Karl Landsteiner in 1900, new possibilities for research in human biology opened up. A number of other blood groups were discovered in the first half of the twentieth century, further expanding the possibilities for investigating human variability at the population level (see Mourant et al. 1976 and Tills et al. 1983). In the mid-1950s, analyses at the protein level were greatly enhanced by the development of starch gel electrophoresis, which allowed for a better characterization of human polymorphisms (Roychoudhury and Nei 1988). Finally, a new era began in the 1970s, with the possibility for direct study of the genetic material per se, that is, deoxyribonucleic acid, or DNA.

When the Xavánte were studied in the 1960s, the primary emphasis was on collecting data related to morphological features (anthropometry, dermatoglyphics, and other measurable physical characteristics), proteins, and blood groups. More recent research on the Xavánte, including that carried out by our group, has used technology based on molecular genetics. In reviewing the data derived from genetic research on the Xavánte over the last few decades we are, in a way, following the history of how technological advances revolutionized research on human biological variability in the second half of the twentieth century.

Morphology

During the 1960s field studies the morphological characteristics of the Xavánte were examined in several ways, namely: (1) standardized photographs, afterward evaluated by equally standardized procedures; (2) anthropometric measurements; (3) skin, eye, and hair color; and (4) finger and palm prints as well as direct observation of toe patterns. The São Domingos data were reported in Neel et al. 1964, while a somewhat more restricted set of characteristics was investigated and presented in Niswander et al. 1967 for the Simões Lopes Xavánte. In the field some peculiarities of Xavánte morphology were already apparent: relatively high stature (for Native Americans), marked dolichocephaly (long head), great facial breadth (resulting in a bizygomatic: head width index higher than 100 in males), and low nasal index (long, narrow nose).

From the 1970s to the 1990s, investigators continued to use the morphological data collected among the Xavánte in the 1960s by Neel and his associates, contrasting their measurements with those from a series of other South American Indian populations. For instance, Marcellino et al. (1978) compared them with the Kayapó, Guaraní Kaiwá, Kaingáng, Tenetehára, and Yanomámi in relation to six measurements and found that the contribution of the shape component of variance to interpopulation variation was about three times as important as that of size.

Neves et al. (1985) made a wider comparison, which included eleven other indigenous groups in Brazil besides the Xavánte. A summary of their results is given in table 4.1. As we can see, the Xavánte averages are

TABLE 4.1. Morphological Measurements from the Xavánte Compared to Those from Eleven Other Indigenous Populations in Brazil

Characteristic	Xavánte Average	South American Indians	
		Range	Average
Males			
Number studied	76	1,129	1,129
Stature	169.2	153.3–173.7	160.6
Sitting height	87.5	77.5–86.0	82.3
Head length	19.8	17.9–19.3	18.7
Head breadth	14.9	14.6–15.7	15.0
Bizygomatic breadth	15.0	13.0–14.7	14.0
Bigonial breadth	11.3	9.7–12.6	10.6
Face height	13.1	11.3–13.1	11.8
Nasal height	6.0	4.6 5.8	4 9
Nasal breadth	4.2	3.6–4.2	4.0
Females			
Number studied	77	855	855
Stature	155.5	144.3–153.9	148.0
Sitting height	81.0	74.4–79.6	76.8
Head length	18.6	17.3–18.5	17.8
Head breadth	14.0	14.0–14.7	14.3
Bizygomatic breadth	13.7	12.4–13.5	13.1
Bigonial breadth	10.4	9.1–10.1	9.6
Face height	11.9	10.1–11.8	10.8
Nasal height	5.4	4.2–5.6	4.6
Nasal breadth	3.7	3.3–3.8	3.6

Source: Neel et al. 1964 for Xavánte data; Neves et al. 1985, where the primary publications are listed.

Note: The female data are more limited, including samples from eight other indigenous groups from Brazil.

(with the one exception of head breadth) always above the averages found in the other groups, and in many instances (four cases among the males, six cases among the females) their values fall outside the range observed for the other populations. When a principal-components analysis was applied to these data, the distinctive position occupied by the Xavánte was clearly apparent (see fig. 4.1 for the male results); essentially the same pattern was found in the females.

Da Rocha et al. (1974) compared relationships among the Xavánte, Kayapó, Kaingáng, and Yanomámi for twelve anthropometric measurements, obtained using Mahalanobis's generalized distance, with the frequencies in the same groups of fourteen alleles from six loci expressed on blood, evaluated through Cavalli-Sforza and Edwards's distances. They were impressed by the high correspondence observed between the two data matrices.

Do morphological traits, as evaluated through anthropometry, influence mating choice? This question was considered by Stark et al. (1990), using data from 336 couples from ten Xavánte, Kayapó, and Kaingáng villages. Fifteen measurements were considered, and the correlation coefficients between husband and wife were generally positive. None of the negative correlations reached a statistically significant level ($p < 0.01$).

Xavánte measurements were also used by Rodriguez-Delfin et al. (1994), together with those from the Kayapó, Kaingáng, and Tikúna, to examine whether a negative correlation could be found between their variances (twelve to sixteen characteristics) and those obtained from seven to eleven blood polymorphic loci. Other authors, in other populations, had observed this correlation and interpreted it in the context of a developmental homeostasis hypothesis (namely, the heterozygotes clustering around the population mean would be better buffered against eventual perturbations during development, which would lead to the indicated negative correlation). But no evidence for such a relationship was found in the Brazilian Indian data.

The standardized photographs obtained during the 1960s Xavánte research were used in an attempt to place facial variability on a semiquantitative basis, employing a method of paired comparisons developed by F. Keiter. The results indicated that the amount of facial variability found among the São Domingos and Simões Lopes Xavánte was not very different from that observed in the urban population of Hamburg, Germany (Neel et al. 1964; Niswander et al. 1967). Similar results were obtained using the coefficients of variation of anthropomet-

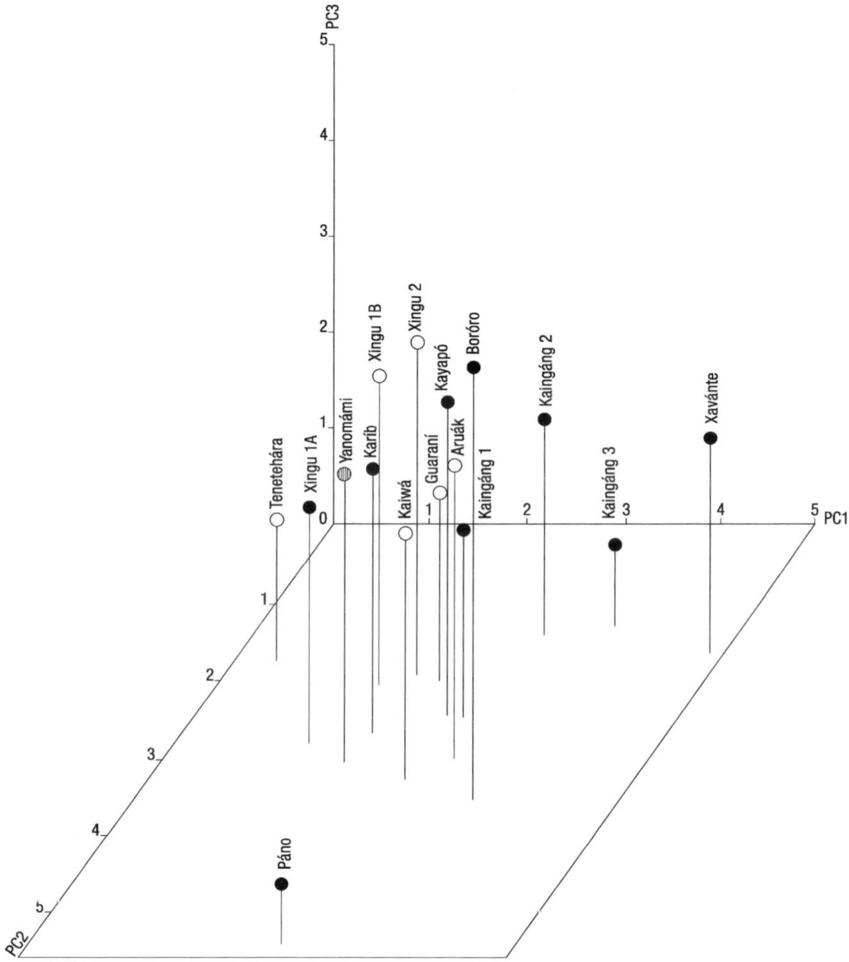

Fig. 4.1. Principal-components analysis of male Brazilian Indian anthropometric data. The circles indicate the languages spoken by individuals of the different indigenous populations: Filled circles, Jê-Páno-Karíb; open circles, Andean Equatorial; hatched circles, Macro Chibchan. Note the peripheral position of the Xavánte. (Reprinted from Neves et al. 1985, from the *American Journal of Physical Anthropology,* Wiley-Liss, Inc., a subsidiary of John Wiley & Sons, Inc.)

ric traits. Although they tended to be smaller among the Xavánte, the differences were not striking.

The dermatoglyphic data have been analyzed less frequently. The most remarkable finding, stressed by Neel et al. (1964), was the high frequency (41 percent) of arches in the toes, compared to that observed in the fingers (3 percent), in both males and females. Although differences in the same direction had been observed in European and Asian populations, they were not so striking. The correlation coefficient for the number of triradii present in fingers and toes of the Xavánte (0.35), however, was of the same order of magnitude as that observed in Hamburg (0.42). As for the palmar data, a high frequency of thenar patterns and a high ratio of thenar to hypothenar patterns were noted; just the opposite occurs in Europeans.

Table 4.2 contrasts the values for the number of triradii in the fingers observed in São Domingos and Simões Lopes with those found in fifteen other South American Indian groups. The intervillage differences are small, and the Xavánte numbers are not markedly different from those observed in other populations.

Protein Blood Genetic Variability

The number of protein genetic systems now available for study is large, and data on about thirty-eight of them are available for the Xavánte. Table 4.3 presents information about them. Two main methods have been

TABLE 4.2. Finger Pattern Intensity Index (Triradii Number) from the Xavánte, Compared to Data from Fifteen Other Indigenous Populations in South America

Sex and Population	Number Studied	Average	Range
Males			
Xavánte			
São Domingos	37	14.3	—
Simões Lopes	91	15.5	—
South American Indians	764	14.1	11.3–17.1
Females			
Xavánte			
São Domingos	39	14.6	—
Simões Lopes	71	15.3	—
South American Indians	404	13.3	10.2–16.5
Males plus Females[a]			
Xavánte			
São Domingos	76	14.4	—
Simões Lopes	162	15.4	—
South American Indians	1,019	13.8	10.8–16.9

Source: Coope and Roberts 1971, where the primary publications are listed.

[a]For some series only the combined males plus females data are available, while in others information on just one sex is furnished.

TABLE 4.3. Information about the Protein Systems Considered

System according to Type	Detection Method[a]	Chromosome Region Location[b]
Blood Groups		
ABO	1	9q 34
MNSs	1	4q 28–q 31
P	1	Chrom. 6
Rh	1	1p 36.2–p 34
Lutheran	1	19p 13.1–q 13.11
Kell	1	7q 33
ABH secretion	1	19p 13.1–q 13.11
Le(a) secretion	1	19p 13.1–q 13.11
Lewis	1	19p 13.1–q 13.11
Duffy	1	1q 12–q 21
Kidd	1	18q 11–q 12
Diego	1	17q 21–q 22
Wright	1	No data
Nonenzymatic Proteins		
Hemoglobin	2	11p 15.5/16p 13.53–p 13.11
Albumin	2	4q 11–q13
Ceruloplasmin	2	3q 21–q24
Group-specific component	2	4q 12
Gamma-globulin	1	14q 32.33
Haptoglobin	2	16q 22.1
Km	1	2p 12
Transferrin	2	3q 21
Enzymes		
Acid phosphatase	2	2p 23
Adenylate kinase	2	9q 34.1
Carbonic anhydrases		
Type 1	2	8q 22
Type 2	2	8q 22
Serum cholinesterases		
Type 1	2	3q 25.2
Type 2	2	16p 11–q 23
Esterase A	2	11q 13–q 22
Esterase D	2	13q 14.11
Glyoxalase 1	2	6p 21.3–p21.2
Glucose-6-phosphate dehydrogenase	2	Xq 28
Lactate dehydrogenases		
Type A	2	11p 15–p 14
Type B	2	12p 12.2–p 12.1
Peptidase A	2	18q 23
Peptidase B	2	12q 21
Peptidase C	2	1q 42
Phosphogluconate dehydrogenase	2	1p 36.2–p 36.13
Phosphoglucomutase 1	2	1p 22.1

Source: McKusick 1988.

[a]Method 1 is immunological; method 2 is electrophoretic (molecule separation according to its electric charge).

[b]The first number indicates the chromosome; p = short arm; q = long arm. The numbers following the letter indicate regions and subregions of that arm.

used for their characterization (immunological and electrophoretic), and the markers can be classified in three broad categories (blood groups, non-enzymatic proteins, and enzymatic proteins). They are spread widely on the genome, occurring in sixteen different autosomes (1–4, 6–9, 11–14, 16–19) and in the X chromosome. Therefore, 73 percent of the autosomes and one of the two sex chromosomes are represented in the total set.

Information about the degree of this variability is given in table 4.4. Of the thirty-eight systems considered, nineteen (ABO, Lutheran, Kell, Wright, Hemoglobin, Albumin, Ceruloplasmin, Adenylate kinase, Carbonic anhydrases 1 and 2, Serum cholinesterase type 1, Esterase A, Glucose-6-phosphate dehydrogenase, Lactate dehydrogenases A and B, Peptidases A, B, and C, and Phosphogluconate dehydrogenase) generally do not show variation among South American Indians. The Xavánte results confirmed this finding. However, Weitkamp et al. (1973), using three starch gel testing procedures, were able to identify a rare albumin variant in two of 214 Xavánte individuals from São Marcos. The variant is similar to albumin Pushtoon in the pH 5.0 system but clearly different from it in pHs 6.0 and 6.9. It is also different from the normal (A) type at pH 5.0. Unfortunately, the further studies necessary for its complete characterization at the molecular level were not performed.

Considering now the remaining nineteen systems listed in table 4.4, several comparisons can be made: (1) the frequencies found in the two surveys conducted at São Domingos and Etéñitépa with an interval of twenty-eight years (1962 and 1990); (2) eventual differences among the three Xavánte villages; and (3) the place occupied by the Xavánte when data from all other South American Indians are considered.

Comparison between the 1962 and 1990 surveys had to be restricted to the systems studied at both of these periods that showed variation, namely, MNSs, P, Rh, Duffy, Haptoglobin, and GC (Salzano et al. 1997b). For four of them (MNSs, Duffy, Haptoglobin, and GC) no significant differences were found, and the unusually high *GC*2* frequency observed in the first survey was confirmed. In the second sample we found a rare variant associated with the *GC*2* allele. The double bands showed a lower electrophoretic mobility than those determined by *GC*1S,* and they could therefore be classified in the GC1C group of variants (Cleve and Constans 1988).

The P difference may have occurred due to the use of a particularly strong reagent in the early determinations, but in relation to Rh sampling problems may have been involved. In contrast to the previous

TABLE 4.4. Protein Genetic Data from the Xavánte, Compared to Frequencies (in percentages) of South American Indians in General

System and Allele or Haplotype	Xavánte Frequencies						South American Indians, Average
	São Domingos (a)	Etéñitépa (b)	Average (a and b)	Simões Lopes	São Marcos	Xavánte, Average (all)	
Blood groups							
ABO							
Number studied	78	85	163	171	289	623	63,014
*ABO*0*	100	100	100	100	100	100	98
MNSs							
Number studied	79	59	138	171	287	596	24,374
*L*MS*	33	29	31	39	40	37	22
*L*Ms*	38	43	40	48	34	40	51
*L*NS*	8	16	12	1	12	9	6
*L*Ns*	21	12	17	12	14	14	21
P							
Number studied	78	85	163	171	289	623	29,335
*P*1*	66	51	58	65	63	62	45
Rh							
Number studied	79	34	113	171	289	573	28,735
*RH*RZ*	3	0	2	7	4	4	6
*RH*R1*	55	56	56	63	57	58	56
*RH*R2*	39	16	32	25	35	32	34
*RH*R0*	3	28	10	0	4	4	3
*RH*r*	0	0	0	5	0	2	1
Lutheran							
Number studied	78	—	78	—	—	78	9,863
*LU*A*	0	—	0	—	—	0	<1
Kell							
Number studied	78	85	163	171	288	622	31,057
*KELL*k*	100	100	100	100	100	100	>99
ABH secretion							
Number studied	55	—	55	168	114	337	4,477
*SEC*Se*	34	—	34	34	41	36	87
Secretion Lewis(a)							
Number studied	—	—	—	168	113	281	3,639
*SEC*Le(a)*	—	—	—	61	67	63	45
Lewis							
Number studied	—	—	—	171	167	338	15,498
*LEWIS*Le*	—	—	—	53	60	56	51
Duffy							
Number studied	79	85	164	171	289	624	18,868
*FY*A*	50	45	47	59	53	53	68
Kidd							
Number studied	—	—	—	171	274	445	16,824
*JK*A*	—	—	—	45	40	42	44
Diego							
Number studied	78	78	156	171	289	618	26,554
*DI*A*	17	19	18	20	15	17	11

(*continues*)

TABLE 4.4.—*Continued*

	Xavánte Frequencies						
System and Allele or Haplotype	São Domingos (a)	Etéñitépa (b)	Average (a and b)	Simões Lopes	São Marcos	Xavánte, Average (all)	South American Indians, Average
Wright							
Number studied	78	—	78	171	288	537	9,144
*WR*A*	0	—	0	0	0	0	<1
Nonenzymatic proteins							
Hemoglobin							
Number studied	75	83	158	153	222	533	30,671
*HB*A*	100	100	100	100	100	100	>99
Number studied	75	83	158	153	222	533	8,718
*HB*A2*	100	100	100	100	100	100	99
Albumin							
Number studied	74	83	157	171	214	542	17,471
*ALB*A*	100	100	100	100	99	>99	98
*ALB*XAV*	0	0	0	0	1	<1	<1
Ceruloplasmin							
Number studied	NA	83	NA	NA	NA	540	14,405
*CP*B*	NA	100	NA	NA	NA	99	99
*CP*ACAY1*	NA	0	NA	NA	NA	1	1
Group-specific component							
Number studied	78	80	158	171	275	603	12,480
*GC*1*	33	25	29	32	30	31	80
*GC*2*	67	74	70	68	70	69	20
*GC*VAR*	0	1	<1	0	0	<1	0
GC subtypes (isoelectric focusing)							
Number studied	—	80	80	—	—	80	3,723
*GC*1F*	—	11	11	—	—	11	27
*GC*1S*	—	14	14	—	—	14	50
*GC*2*	—	74	74	—	—	74	23
*GC*VAR*	—	1	1	—	—	1	<1
Gamma globulin							
Number studied	NA	NA	NA	NA	NA	453	17,144
*GM*ag*	NA	NA	NA	NA	NA	76	73
*GM*axg*	NA	NA	NA	NA	NA	24	24
*GM*fb*	NA	NA	NA	NA	NA	<1	1
Haptoglobin							
Number studied	67	83	150	171	269	204	30,154
*HP*1*	45	55	51	47	48	48	63
Km							
Number studied	—	—	—	143	248	431	17,902
*KM*1*	—	—	—	63	37	46	57
Transferrin							
Number studied	78	54	132	171	272	575	22,724
*TF*C*	100	100	100	100	100	100	>99

TABLE 4.4.—*Continued*

System and Allele or Haplotype	Xavánte Frequencies						South American Indians, Average
	São Domingos (a)	Etéñitépa (b)	Average (a and b)	Simões Lopes	São Marcos	Xavánte, Average (all)	
TF subtypes (isoelectric focusing)							
Number studied	—	54	54	—	—	54	711
*TF*C1*	—	85	85	—	—	85	91
*TF*C2*	—	13	13	—	—	13	4
*TF*C3*	—	1	1	—	—	1	<1
*TF*C4*	—	1	1	—	—	1	2
Enzymes							
Acid phosphatase							
Number studied	—	83	83	154	222	459	19,635
ACP*A	—	22	22	18	20	20	12
Adenylate kinase							
Number studied	—	83	83	—	—	83	17,681
*AK*1*	—	100	100	—	—	100	>99
Carbonic anhydrases							
Type 1							
Number studied	—	—	—	—	86	86	7,372
*CA1*1*	—	—	—	—	100	100	100
Type 2							
Number studied	—	82	82	—	—	82	10,080
*CA2*1*	—	100	100	—	—	100	>99
Serum cholinesterases							
Type 1							
Number studied	—	90	90	—	285	375	6,680
*CHE1*U*	—	100	100	—	100	100	>99
Type 2							
Number studied	—	94	94	—	—	94	2,055
CHE2(C5+)[a]	—	20	20	—	—	20	12
Esterase A							
Number studied	—	81	81	100	86	267	8,464
*ESA*1*	—	100	100	100	100	100	99
Esterase D							
Number studied	—	82	82	—	—	82	13,933
*ESD*1*	—	63	63	—	—	63	72
Glyoxalase 1							
Number studied	—	83	83	—	—	83	7,174
*GLO1*1*	—	22	22	—	—	22	26
Glucose-6-phosphate dehydrogenase							
Number studied[b]	—	33	33	46	42	121	9,219
*G6PD*B*	—	100	100	100	100	100	99
Lactate dehydrogenases							
Type A							
Number studied	—	—	—	6	78	84	9,920
*LDHA*1*	—	—	—	100	100	100	>99

(*continues*)

TABLE 4.4.—*Continued*

System and Allele or Haplotype	Xavánte Frequencies: São Domingos (a)	Eténitépa (b)	Average (a and b)	Simões Lopes	São Marcos	Xavánte, Average (all)	South American Indians, Average
Type B							
Number studied	—	—	—	6	78	84	9,920
*LDHB*1*	—	—	—	100	100	100	>99
Peptidase A							
Number studied	—	76	76	—	—	76	11,890
*PEPA*1*	—	100	100	—	—	100	>99
Peptidase B							
Number studied	—	82	82	—	—	82	12,411
*PEPB*1*	—	100	100	—	—	100	>99
Peptidase C							
Number studied	—	82	82	—	—	82	6,835
*PEPC*1*	—	100	100	—	—	100	99
Phosphogluconate dehydrogenase							
Number studied	—	83	83	116	88	264	20,246
*PGD*A*	—	100	100	100	100	100	99
Phosphoglucomutase 1							
Number studied	—	82	82	—	—	82	1,838
*PGM1*1A*	—	21	21	—	—	21	33
*PGM1*1B*	—	64	64	—	—	64	50
*PGM1*2A*	—	5	5	—	—	5	4
*PGM1*2B*	—	10	10	—	—	10	13

Source: Neel et al. 1964; Gershowitz et al. 1967; Shreffler and Steinberg 1967; Tashian et al. 1967; Salzano and Tondo 1968; Weitkamp et al. 1973; Tanis et al. 1974; Neel 1978; Gershowitz and Neel 1978; Alcântara et al. 1995; and Salzano et al. 1997b for the Xavánte data. The South American Indians averages were obtained from our data bank, available on request.

Abbreviations: NA—results not available by village.

[a] Phenotype frequency only, since the genetics of this system seems not to be simple.

[b] Males only.

investigation, four cDe individuals were observed in 1990, while none had been found in 1962. This led to a high estimate of *RH*RO*, with the complementary reduction of the *RH*R2* frequency. Everything considered, the admittedly limited information obtained suggests that the social, economic, and environmental transformations that have occurred in Eténitépa over the last thirty years were not sufficient to significantly alter the gene pool of its population. As we have seen in chapter 2, marriage practices at Eténitépa have not changed over recent decades; the Xavánte there continue to marry within their own group, thus limiting gene inflow.

For the intervillage comparisons, nine variable systems could be compared (MNSs, P, Rh, Duffy, Diego, Acid phosphatase, Haptoglobin, Albumin, and GC), and for three of them (MNSs, Rh, and Duffy) significant differences were found (table 4.4). The São Domingos and Etéñitépa Xavánte show low prevalences of *L*MS, RH*RZ, RH*R1,* and *FY*A* but high frequencies of *L*Ns* and *RH*RO* or *RH*r,* compared to the other two groups. Multivariate analysis (application of the D_A distances and the UPGMA method to sixteen systems studied in the three populations) furnished a dendrogram with a clear separation of Simões Lopes from the two other communities. This was confirmed using alternative statistical procedures. Because of their geographical proximity, we might have expected a closer relationship between São Marcos and Simões Lopes, but due to the shifting system of village formation detailed in chapter 2 it is difficult to establish firm hypotheses about the genetic similarities and differences among Xavánte populations.

The final comparison that can be made with the results listed in table 4.4 relates to the genetic relationships between the Xavánte and the other South American Indian populations. The table presents information on thirty-eight systems, but nineteen of them show little variation among Amerindian populations and are not informative for this purpose. In about half of the remaining nineteen clear differences (above 10 percent) can be found between the Xavánte average and the South American Indian averages. They occur in the MNSs, P, ABH Secretion, Le(a) Secretion, Duffy, GC, Haptoglobin, KM, and Phosphoglucomutase 1 systems. The differences found in the ABH Secretion and GC comparisons are particularly marked. In relation to the first the Xavánte show an unusually low frequency of secretors (36 percent only of *SEC*Se,* against an average of 87 percent, obtained from 4,477 individuals of other groups). The opposite occurs in relation to *GC*2:* 69 to 74 percent among the Xavánte, contrasted with 20 to 23 percent in other South American Indian populations (depending on whether classical or isoelectric focusing electrophoretic techniques are used). Moreover, the complementary difference seems to be much more marked for *GC*1S* (14 vs. 50 percent) than for *GC*1F* (11 vs. 27 percent).

A previous multivariate analysis of these genetic relationships, involving twenty-five genetic systems and thirty-three South American Indian populations, yielded somewhat inconclusive results, suggesting that the systems available did not discriminate sufficiently for this purpose (Salzano et al. 1997b). However, the results of a more limited comparison, including twelve loci, of three other Jê and one Tupí-Guaraní group, disclosed a clear separation of the Xavánte from the other groups, emphasizing their distinctiveness (Salzano et al. 1997a).

Analysis at the DNA Level

The amount of data concerning DNA polymorphisms (common variants) among South American Indians is much less than that available for protein markers. Therefore, most of the analyses to be presented below derive from results obtained by researchers from the Department of Genetics of the Universidade Federal do Rio Grande do Sul, Porto Alegre, and their associates, who systematically surveyed the Xavánte plus four other groups (Gavião, Suruí, Waiwái, and Zoró) for these systems. Although the sample is small, since these societies speak languages of three distinct stocks (Xavánte speak Jê; Gavião, Suruí, and Zoró speak Tupí; Waiwái speak Karíb), it includes widely distributed groups.

Table 4.5 furnishes information about the sixteen systems considered. Fourteen of these are distributed along ten autosomes (1, 2, 5, 6, 8, 11, 12, 15, 17, 19). The Y chromosome was investigated, while in the mitochondrial DNA the first hypervariable segment of the control region (HVS–I) was subjected to sequence analysis.

The variability surveyed involved: (1) single nucleotide variation, tested by the restriction fragment length polymorphism method (which verifies whether a given enzyme cuts or not a certain point of the DNA) or by oligonucleotide hybridization (DNA segments amplified by the polymerase chain reaction, or PCR, are blotted on nylon membranes after electrophoresis in agarose gels and hybridized with ^{32}P-labeled short, 15–50 nucleotides long probes) or by direct sequencing; (2) presence or absence of single DNA insertions or deletions; and (3) evaluation of variability in short or large repeats distributed in tandem. Results from the determination of the constitution of several regions can be combined in arrangements of their distribution in *cis,* called haplotypes.

Xavánte prevalences involving the alleles or haplotypes of these sixteen systems, compared to similar data from other South American Indians, are presented in table 4.6. Due to the high variability of these systems, no less than 179 comparisons could be made. They depend greatly on the type and amount of this variability, but it is important to emphasize that over one-third (36 percent) of the Xavánte frequencies fall outside the range observed in the other groups.

What are the most distinctive Xavánte features in this array? They show especially different characteristics in the low-density lipoprotein receptor (LDLR) and mtDNA systems. Thus, low prevalences in relation to the averages found for LDLR in the other groups occur in alleles *LDLR*H+* (54 vs. 77 percent) and *LDLR*M+* (11 vs. 49 percent), while the opposite is true for *LDLR*A+* (73 vs. 45 percent) and

TABLE 4.5. Information about the DNA Systems Considered

System according to Type	Identification Characteristics	Chromosome Region Location
Nuclear DNA		
Autosomes		
Apolipoprotein B	Haplotypes, 3 RFLPs, 1 insertion/deletion	2p 23–2p ter
Apolipoprotein E	Point mutation, RFLP	19q 13.1
Beta globin	Haplotypes, 5 RFLPs	11p 12.5–11p 12.8
Coagulation factor F13A1	STR, repeat unit: 4bp	6p 24–6p 25
Cytochrome P4501A1	Haplotypes, 2 RFLPs	15q 22–q ter
D1S80	VNTR, repeat unit: 16bp	1p 26–1p 35
Dopamine D2 receptor	Haplotypes, 3 RFLPs, 1CA dinucleotide STRP	11q 22–11q 23
Dopamine D4 receptor	VNTR, repeat unit: 48 bp	11p 15.5
Dopamine transporter protein (SLC6A3)	VNTR, repeat unit: 40 bp	5p 15.3
Human lymphocyte antigens (HLA)	Oligonucleotide hybridization, 6 loci	6p 21.3
Lipoprotein lipase	Haplotypes, 2 RFLPs	8p 22
Low-density lipoprotein receptor	Haplotypes, 5 RFLPs	19p 13.2–p 13.1
T lymphocyte CD4	STR, repeat unit: 5bp	12p 12–12p ter
Tumor suppressor TP53	Haplotypes, 2RFLPs, 1 insertion/deletion	17p 13.105–p12
Y Chromosome		
DYS19, alphoid heteroduplexes (α h)	Haplotypes, repeat variation	p. centromere
Mitochondrial DNA		
Hypervariable segment I (HVS–I)	Nucleotide variability by sequence analysis (SNPs)	First 360 base positions

Source: McKusick 1988.

Abbreviations: RFLP = restriction fragment length polymorphism; SNP = single nucleotide polymorphism; STR = short tandem repeat; STRP = short tandem repeat polymorphism; VNTR = variable number of tandem repeats; bp = base pairs. For anonymous DNA sequences, the convention is to use D (=DNA) followed by 1–22, X or Y, to indicate the chromosome location, then S for a unique segment or another alternative, and finally a serial number (like D1S80). As for the chromosome region location, p = short arm; q = long arm. These letters are followed by numbers, which indicate the regions and subregions of that arm; ter = terminal.

TABLE 4.6. DNA Genetic Data from the Xavánte, Compared to Frequencies (in percentages) of Other South American Indians

System and Allele or Haplotype	Xavánte Frequency	South American Indians	
		Range	Average
Apolipoprotein B			
Number studied	31	109	109
Alleles			
*5'βSP*24*	28	17–28	22
*5'βSP*27*	72	64–83	74
*5'βSP*29*	0	0–9	4
*XbaI*X+*	39	14–26	20
*MspI*M+*	81	85–97	90
*EcoRI*E+*	97	92–98	94
Haplotypes			
1(27+++)	8	0–2	1
2(24+++)	21	2–12	7
3(24++−)	0	0–2	1
4(24+−+)	0	4–9	7
5(27−++)	44	60–77	68
6(24−++)	2	0–17	7
7(29−++)	0	0–2	1
8(29−+−)	0	0–7	3
9(27−+−)	6	0–2	2
10(27−−+)	15	0–0	0
11(24−−+)	4	0–4	1
12(27+−+)	0	0–4	2
13(27+−−)	0	0–4	1
14(29+−+)	0	0–2	<1
Apolipoprotein E			
Number studied	31	593	593
*APOE*2*	0	0–4	<1
*APOE*3*	98	51–96	79
*APOE*4*	2	2–47	21
Beta-globin			
Number studied	30	126	126
1(−−−−−)	0	0–60	1
2(+−−−−)	60	82–93	88
3(−−−−+)	0	0–2	<1
5(−+−++)	15	0–4	1
6(−+−−+)	18	3–12	8
7(−++−−)	5	0–5	1
11(−−−++)	0	0–2	<1
16(−+−−−)	2	0–0	0
Coagulation factor F13A1			
Number studied	32	111	111
*F13A1*4*	5	0–25	9
*F13A1*5*	25	20–57	44
*F13A1*6*	45	10–50	34

TABLE 4.6.—*Continued*

System and Allele or Haplotype	Xavánte Frequency	South American Indians	
		Range	Average
*F13A1*7*	16	0–17	4
*F13A1*8*	9	0–14	4
*F13A1*9*	0	0–14	3
Cytochrome P4501A1			
Number studied	21	194	194
Alleles			
*CYP1A1*Val*	97	54–81	69
*CYP1A1*m2*	95	72–96	84
Haplotypes			
*1(*Ile/*m1)*	3	4–27	14
*2(*Ile/*m2)*	3	8–42	19
*3(*Val/*m1)*	0	0–5	2
*4(Val/*m2)*	94	54–75	64
D1S80			
Number studied	25	293	293
*D1S80*14*	0	0–2	<1
*D1S80*17*	0	0–2	<1
*D1S80*18*	44	19–35	31
*D1S80*21*	0	0–4	1
*D1S80*22*	0	0–1	<1
*D1S80*23*	0	0–3	1
*D1S80*24*	14	6–37	16
*D1S80*25*	16	0–17	7
*D1S80*26*	4	0–6	2
*D1S80*27*	0	0–4	1
*D1S80*28*	0	0–27	7
*D1S80*29*	0	0–6	1
*D1S80*30*	10	2–56	26
*D1S80*31*	12	0 16	4
*D1S80*35*	0	0–4	<1
*D1S80*36*	0	0–1	<1
Dopamine D2 receptor			
Number studied	28	272	272
1(B1/D2/16/A1)	84	20–82	52
2(B1/D2/15/A1)	0	0–1	<1
3(B1/D2/14/A1)	0	0–1	<1
4(B1/D2/16/A2)	0	0–2	<1
5(B1/D2/14/A2)	0	0–7	1
6(B2/D1/15/A1)	0	0–7	1
7(B2/D1/16/A2)	0	0–2	<1
8(B2/D1/15/A2)	0	0–4	1
9(B2/D1/14/A2)	0	0–4	1
10(B2/D1/13/A2)	0	0–2	<1
11(B2/D2/14/A1)	0	0–1	<1

(*continues*)

TABLE 4.6.—*Continued*

System and Allele or Haplotype	Xavánte Frequency	South American Indians Range	South American Indians Average
12(B2/D2/13/A1)	0	0–1	<1
13(B2/D2/16/A2)	0	0–28	7
14(B2/D2/15/A2)	0	0–4	1
15(B2/D2/14/A2)	16	14–38	26
16(B2/D2/13/A2)	0	0–12	6
Others	0	0–6	2
Dopamine D4 receptor			
Number studied	28	305	305
*DRD4*2*	0	0–14	3
*DRD4*4*	32	14–52	30
*DRD4*5*	5	0–5	1
*DRD4*6*	4	0–18	5
*DRD4*7*	43	23–78	60
*DRD4*8*	18	0–1	<1
Dopamine transporter protein			
Number studied	28	107	107
*SLC6A3*10*	100	100–100	100
Human lymphocyte antigens			
Number studied	74	203	203
Alleles			
*DPB1*0101*	0	0–<1	<1
*DPB1*0201*	0	7–16	11
*DPB1*0202*	0	0–<1	<1
*DPB1*0301*	3	3–6	4
*DPB1*0401*	5	2–9	6
*DPB1*0402*	80	36–62	53
*DPB1*0501*	7	10–16	12
*DPB1*1101*	0	0–3	1
*DPB1*1301*	0	5–9	7
*DPB1*1401*	1	1–6	4
*DPB1*1601*	0	0–<1	<1
*DPB1*1701*	0	0–1	<1
*DPB1*2701*	0	0–4	1
Blank	11	0–3	1
*DRB1*0101*	0	0–<1	<1
*DRB1*0102*	0	0–<1	<1
*DRB1*0103*	0	0–<1	<1
*DRB1*1501*	0	<1–9	4
*DRB1*1502*	0	0–3	1
*DRB1*1602*	30	0–2	1
*DRB1*0301*	0	0–3	2
*DRB1*0403*	0	0–3	2
*DRB1*0404*	7	3–11	6

TABLE 4.6.—*Continued*

System and Allele or Haplotype	Xavánte Frequency	South American Indians Range	South American Indians Average
*DRB1*0405*	0	0–<1	<1
*DRB1*0407*	7	0–6	4
*DRB1*0411*	0	6–9	8
*DRB1*0417*	0	6–9	9
*DRB1*1104*	0	0–4	1
*DRB1*1301*	0	0–1	1
*DRB1*1303*	0	0–<1	<1
*DRB1*1401*	0	0–<1	<1
*DRB1*1402*	30	9–26	15
*DRB1*1406*	0	17–27	23
*DRB1*0701*	0	1–9	4
*DRB1*0802*	26	13–22	18
*DRB1*0901*	0	0–1	<1
*DRB1*1001*	0	0–<1	<1
Blank	0	0–2	1
*DQA1*0101/4*	0	0–2	1
*DQA1*0102*	0	<1–9	4
*DQA1*0103*	0	1–3	2
*DQA1*0201*	0	1–9	4
*DQA1*03*	15	23–31	28
*DQA1*0401*	26	13–21	17
*DQA1*0501*	61	41–46	42
Blank	0	2–3	2
*DQB1*0501*	0	0–2	1
*DQB1*0503*	0	0–<1	<1
*DQB1*0601*	0	0–3	1
*DQB1*0602*	0	<1–9	4
*DQB1*0603/7*	0	0–1	1
*DQB1*0201*	0	1 13	6
*DQB1*0301*	61	36–48	41
*DQB1*0302*	15	23–32	26
*DQB1*0303*	0	0–1	<1
*DQB1*0402*	26	13–20	17
Blank	0	0–7	4
Lipoprotein lipase			
Number studied	25	111	111
Alleles			
*LPL*H+*	98	83–100	89
*LPL*P+*	76	53–98	73
Haplotypes			
1(++)	76	54–98	74
2(+−)	22	2–32	16
3(−−)	2	0–16	10

(*continues*)

TABLE 4.6.—*Continued*

System and Allele or Haplotype	Xavánte Frequency	South American Indians Range	South American Indians Average
Low-density lipoprotein receptor			
Number studied	28	103	103
Alleles			
*LDLR*T+*	25	15–17	16
*LDLR*H+*	54	65–90	77
*LDLR*A+*	73	36–58	45
*LDLR*M+*	11	36–60	49
*LDLR*N+*	88	40–64	50
Haplotypes			
1(− + − + −)	7	14–50	30
2(− − + − +)	47	0–35	18
3(− + − − +)	0	2–23	13
4(− + + − +)	21	8–24	16
5(+ + − − +)	14	0–2	<1
6(+ + − + −)	4	4–17	13
7(+ + − − −)	2	0–0	0
8(− − + + −)	0	0–9	2
9(− − − − +)	0	0–4	1
10(− + + + −)	0	0–4	2
11(− − − + −)	0	0–2	1
12(+ − − + −)	0	0–2	<1
13(+ + + − +)	5	0–6	2
T lymphocyte CD4			
Number studied	32	111	111
*CD4*8*	81	56–77	63
*CD4*9*	0	0–5	1
*CD4*13*	19	23–44	35
Tumor suppressor TP53			
Number studied	25	89	89
Alleles			
*TP53*D−*	100	98–100	99
*TP53*B−*	28	7–25	16
*TP53*M−*	0	0–7	2
Haplotypes			
1(− − −)	28	7–25	15
2(− + +)	72	75–93	82
3(− + −)	0	0–7	2
4(+ − −)	0	0–2	<1
DYS19, alphoid heteroduplexes (α h)			
Number studied	5	32	32
IIA haplotype	100	70–100	92
Mitochondrial DNA			
Number studied	25	216	216
*mtDNA*16093C*	36	0–36	3

TABLE 4.6.—*Continued*

System and Allele or Haplotype	Xavánte Frequency	South American Indians	
		Range	Average
*mtDNA*16111T*	16	0–86	22
*mtDNA*16175G*	7	0–22	2
*mtDNA*16189C*	84	15–79	36
*mtDNA*16217C*	84	0–54	17
*mtDNA*16223T*	16	54–100	79
*mtDNA*16241G*	36	0–8	1
*mtDNA*16290T*	16	11–81	26
*mtDNA*16319A*	16	11–81	26
*mtDNA*16362C*	16	22–100	65

Source: Cerna et al. 1993b; Bevilaqua et al. 1995; Heidrich et al. 1995; F. Santos et al. 1995; Ward et al. 1996; Hutz et al. 1997, 2000; Bortolini et al. 1998; Kaufman et al. 1999; Andrade et al. 2000; Mattevi et al. 2000; Bogdawa et al. 2000; Gaspar et al. 2000; Kvitko et al. 2000.

Note: When alleles and haplotypes are indicated in a given system and no other information is furnished, the order of the alleles in the haplotypes is that of their previous listing.

*LDLR*N+* (88 vs. 50 percent). This is reflected on the LDLR haplotypes *1*(−+−+−) (7 vs. 30 percent) and *2*(−−+−+) (47 vs. 18 percent). As for mtDNA, the corresponding values are low prevalences for mutations *16223T* (16 vs. 79 percent) and *16362C* (16 vs. 65 percent), with the opposite occurring for *16093C* (36 vs. 3 percent), *16189C* (84 vs. 36 percent), *16217C* (84 vs. 17 percent), and *16241G* (36 vs. 1 percent).

In other systems there are not so many differences. In HLA a low frequency is found for *DRB1*1406* (0 vs. 23 percent), but high for *DPB1*0402* (80 vs. 53 percent), *DRB1*1602* (30 vs. 1 percent), and *DQB1*0301* (61 vs. 41 percent). In Cytochrome P450 1A1 the values for *CYP1A1*Val* are 97 vs. 69 percent, reflected in haplotype *4*(**Val/*m2*) (94 vs. 64 percent). Finally, there are marked differences for the frequencies of haplotype *5*(*27*−++) of Apolipoprotein B (44 vs. 68 percent) and haplotype *2*(+−−−−) of beta-globin (60 vs. 88 percent).

Genetic Diversity and Population Relationships

How do the Xavánte compare to other South American Indian groups in terms of genetic diversity? Are these estimates much influenced by the genetic system considered? Do the population relationships obtained using protein, nuclear DNA, and mitochondrial DNA show the same pattern?

Table 4.7 presents information related to the two first questions.

Due to the widely different amount of information available for other groups, we concentrated our attention on the five populations mentioned in the previous section. Due to their recent history of population intermixture, it would be expected that the Waiwái would show the most marked degree of variability (Callegari-Jacques et al. 1996). When we consider the isolated DNA systems listed in table 4.7, however, they do not depart markedly from the other four populations. In the global evaluation they are first in the rank for nDNA and mtDNA, but the difference is most clearly marked in the latter system. They occupy an intermediate position in the two comparisons involving protein markers.

What about the Xavánte? They show marked variability in the beta-globin, F13A1, DRD4, and TP53 systems, but when all nuclear DNA

TABLE 4.7. Measures of Diversity in the Xavánte and Four Other Indigenous Populations of Brazil

System and Measures of Diversity	Populations					Source
	Gavião	Suruí	Waiwái	Xavánte	Zoró	
Individual indices						
Beta-globin						
Haplotypic diversity	22	31	22	58	12	Bevilaqua et al. 1995
Coagulation factor F13A1						
Average heterozygosity	70	58	68	72	50	Bogdawa et al. 2000
D1S80						
Average heterozygosity	67	71	78	75	59	Hutz et al. 1997
Dopamine D2 receptor						
Average heterozygosity	71	31	60	27	75	Hutz et al. 2000
Dopamine D4 receptor						
Average heterozygosity	45	44	65	70	47	Hutz et al. 2000
Low-density lipoprotein receptor						
Average heterozygosity	79	80	71	72	84	Mattevi et al. 2000
Tumor suppressor TP53						
Average heterozygosity	32	38	33	41	13	Gaspar et al. 2000
Mitochondrial DNA						
Gene diversity	87	ND	ND	68	78	Ward et al. 1996
Nucleotide diversity	112	ND	ND	84	112	Ward et al. 1996
Global evaluation						
Protein (23 loci)	18	13	17	17	17	Hutz et al. 1999
Protein (8 most polymorphic loci)	48	37	47	48	50	Hutz et al. 1999
Nuclear DNA (8 systems)	54	41	57	56	48	Hutz et al. 1999
Mitochondrial DNA (HVS–I)	12	5	17	8	12	Hutz et al. 1999

Note: ND = not determined.

systems are considered the amount of variation observed is high, but slightly lower than that of the Waiwái, and they present an intermediate position when the protein and mtDNA markers are considered. These considerations, plus the fact that nDNA estimates of diversity are generally much higher than those obtained using the other types of systems, indicate the necessity, for any type of intergroup comparison, of making comprehensive evaluations that include as many types of markers as possible.

The population relationships obtained with these same sets of characteristics are shown in figure 4.2. The pattern that most closely agrees with the history, socioeconomic, and cultural data, as well as the geographical location of these groups, is that depicted by the nuclear DNA alleles. In this case the three Tupí-Mondé groups cluster together, with the Suruí somewhat separated. Then we have the Xavánte and Waiwái cluster, with the Xavánte branch being the longest. The Xavánte mtDNA pattern, on the other hand, clearly deviates from the others, while in the protein trees the Xavánte cluster with the Zoró, also in a peripheral position.

The dendrograms shown in figure 4.2 exemplify the complexities faced when interpopulation relationships are investigated using different sets of markers. Comparisons involving these same populations, plus five others, were also made by Bortolini et al. (1998), using seventeen protein loci and the HVS-I mtDNA markers. The results are clearly parallel to those already discussed, and in this case the protein results most closely show patterns in accordance with linguistics (e.g., the Xavánte cluster with the Kayapó, and both speak Jê languages). The unraveling of human evolutionary history is difficult, especially due to the occurrence of cultural/biological interactions. But it is this same difficulty that makes the approach so fascinating.

Biological Variability and Continuity at Etéñitépa

When all the data presented here are considered together some specific points are worth emphasizing. The Xavánte show distinctive morphological (anthropometric, dermatoglyphic) features that set them apart from other South American Indian populations. This distinctiveness is also apparent when proteins, and especially the mtDNA results, are considered. In relation to the latter, they show only four lineages (against seven observed in the Gavião and eight among the Zoró), and the

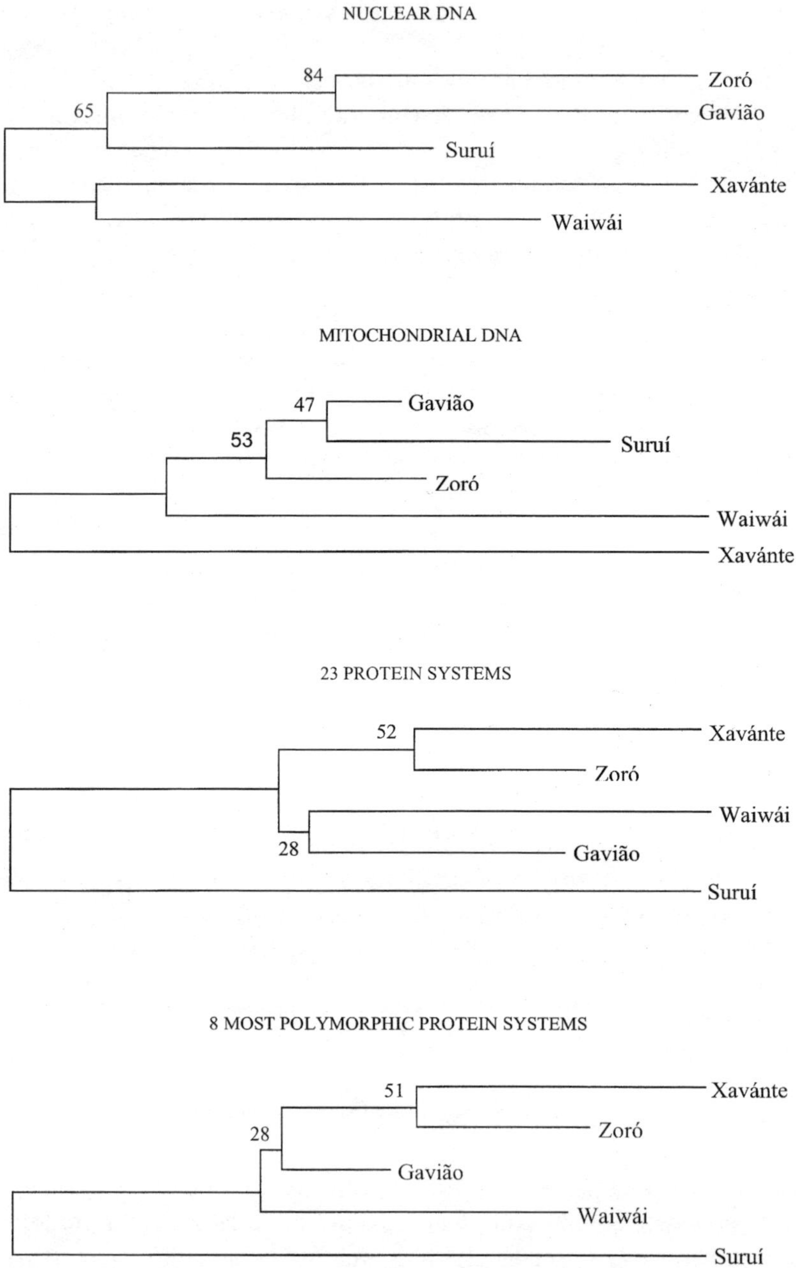

Fig. 4.2. Dendrograms showing the population relationships observed among the Xavánte and four other indigenous populations. (Reprinted from Hutz et al. 1999, by permission of Kluwer Academic/ Plenum Publishers.)

estimate of nucleotide diversity is only 67 percent of that found in five other Amerindian groups (Ward et al. 1996). This eventual mtDNA restriction of variability may be due to a sampling effect. Estimates of morphological diversity indicated that the Xavánte show results equivalent to those of a cosmopolitan city, and the diversity in nuclear DNA and proteins point to high, not low, values.

One important finding that has come out of research on biological variability among the Xavánte of Etéñitépa is the demonstration of their genetic continuity. When we compare frequency markers for protein obtained during the study of São Domingo in 1962 with those collected at Etéñitépa in 1990, almost thirty years later, we find no statistically significant differences. This genetic continuity of the Xavánte contrasts vividly with the profound demographic, ecological, and epidemiological transformations that they have undergone in recent decades, transformations that we will examine in the chapters that follow.

Chapter 5

Demographic Crisis and Recovery

Xavánte population dynamics over the last three hundred years have been affected by the expansion of Western society into Central Brazil. As we have seen in chapter 3, historical sources show that as far back as the eighteenth century the Xavánte were experiencing displacements and the impact of epidemics. The historical experience of the Xavánte over the centuries is closely tied to demographic processes such as changes in settlement pattern, population decline, recovery, recombination, and migratory movements, which, in one way or another, were products of the interaction between Xavánte and expanding colonial frontiers.

In this chapter we will analyze the demographic experience of the Xavánte that has resulted from their interaction with non-Indians. We are the first to recognize that we will only be able to recover fragments of this experience. This is because, due to lack of data for the Xavánte, it is impossible to carry out even the simplest demographic analysis of the events that they experienced before the twentieth century. Contact in the 1940s was only the latest and most enduring of several encounters between the Xavánte and Western society over the last three centuries. It was distinctive because the Xavánte no longer had the option to withdraw, since Central Brazil was becoming increasingly populated by non-Indians.

The data that we will analyze in this chapter relate to the experience of the Etéñitépa Xavánte over a fifty-year period, from the onset of permanent contact in the 1940s to the 1990s. The time around the middle of the twentieth century was a period of crisis for the group, as they suffered the effects of loss of territory, epidemic diseases, and social disruption. We document this crisis and show how the Xavánte lived through it by analyzing mortality and fertility data from retrospective reproductive histories, genealogies, and censuses collected at several different periods from the 1950s on. We also consider the interplay between political disputes and death from epidemics and its impact on specific aspects of Xavánte social organization, including marriage arrangements.

Data Sources and Collection

Unfortunately there is no adequate, systematic, and continuous system to collect demographic data on indigenous peoples in Brazil. The result is that if one attempts at present to carry out even the most basic demographic analysis, such as the construction of a population pyramid, the necessary data (information on age and sex composition) will not be available for the vast majority of indigenous groups. If such basic information is not available, neither, needless to say, are systematic data on mortality and fertility (crude death and birth rates, infant mortality rates, life expectancy at birth, total fertility rates, and so forth), which are fundamental to meaningful demographic analyses.

Even government agencies that work directly with Indians are not good sources of demographic data. Several reasons account for this. Over the years, FUNAI has not had a policy or a system to collect and disseminate demographic data. Even if such a policy were implemented, judging by present conditions it is unlikely that most FUNAI posts would be equipped to provide headquarters with the necessary data. The collection of systematic data requires the continued presence of, or at least frequent visits by, well-trained personnel. Moreover, in recent years a growing number of FUNAI posts on Indian reservations have been either irregularly staffed or shut down completely.

One would expect that the national decennial census could serve as a source of demographic information on indigenous peoples. Unfortunately, this is not the case. Although current census forms include the option of *indígena* under "color"—the 1991 census form was the first to include this option; the categories in the census of 1980 were limited to *branco* (white), *amarelo* (yellow), *preto* (black), and *pardo* or brown—Indian reservations are not systematically covered by the census. Furthermore, the census data do not allow one to derive information on specific indigenous groups from the broad category *indígena,* which lumps together all the hundreds of different groups in Brazil (Azevedo 1994, 1997; Oliveira 1994; Silva 1994).

At different times, anthropologists and other researchers have collected demographic information on specific indigenous peoples in Brazil. Geneticists have been particularly active in collecting these data. Most of the results obtained up to the end of the 1980s have been reviewed by Salzano and Callegari-Jacques (1988, 44–67), who list data on age and sex distribution for seventy-four groups. Investigations along these lines carried out in the 1990s in Brazilian Amazonia include Callegari-Jacques et al. 1996, Oliveira et al. 1988, and Salzano et al. 1990. These demographic data of genetic interest include age and sex distribution, marriage patterns, mobility, fertility, and mortality. These data are important not only to characterize the investigated groups demographically, but because they allow geneticists to investigate their population structures, which are key parameters in evolutionary studies. This is particularly true of the human species, in which, beyond biological factors (sex, age, and mobility), there are cultural phenomena that affect evolutionary processes.

In order to overcome the lack of current demographic data, a São Paulo NGO, the Instituto Socioambiental (ISA), has channeled energy and resources into the effort to collect and disseminate demographic data on indigenous peoples. The institute has published a series of reports under the general title Povos Indígenas do Brasil. The most recent one was published in 1996 (ISA 1996). Data compiled by ISA have been made available on the Internet (http://www.socioambiental.org, accessed May 2000). The demographic data made available through these reports mostly come from anthropologists and other professionals doing fieldwork. Despite the importance of this effort, the information presented does not allow in-depth demographic analysis, since for most groups the only figure published is total population size.

Therefore, for the vast majority of indigenous peoples in Brazil it is impossible to carry out meaningful demographic analyses by relying on secondary data sources, either governmental or nongovernmental. Under these conditions, most researchers have been forced to build their own data banks from scratch (see Adams and Price 1994; Black et al. 1978; Chagnon 1974; Coimbra 1989a; Early and Peters 1990; Flowers 1994a, 1994c; Neel and Weiss 1975; Salzano and Callegari-Jacques 1988, 44–67; Souza 1999; Souza and Santos 1999, 2001; and Werner 1983, among others).

The data presented in this chapter are derived from several censuses carried out at the field site and from women's reproductive histories. In 1976–77 Flowers spent fourteen months at Etéñitépa and collected these data. In 1990, when our research team did fieldwork at Etéñitépa, demographic information was again collected.[1] During our visits to Etéñitépa from 1990 to 1996 we updated our censuses.

Information made available by researchers who worked among the Xavánte in previous years was also helpful in the demographic investigation. The genealogies collected by Maybury-Lewis (1967, 317–42) provided the names and relationships among adults living at São Domingos at the time of his research in 1957–58 and 1962. James V. Neel kindly provided copies of field notes giving the names and estimated ages of people examined during the fieldwork at São Domingos in 1962 (Neel et al. 1964). This information allowed us to revise some estimates of the ages of older people, since it is clearly easier to tell if a person is one or five years old than it is to tell if he or she is thirty or thirty-five.

When anthropologists try to collect demographic information in societies like that of the Xavánte, without written records, they must rely on the statements and memories of the individuals they interview. This may lead to incorrect information. In order to circumvent this problem, the researcher should try to cross-check the data by obtaining as many independent sources of information as possible and by collecting censuses as often as is practicable.[2]

All those who have collected demographic data in preliterate societies know that one of the most difficult, yet essential, tasks is to establish, with the greatest accuracy possible, the ages of individuals in the population surveyed (see, among others, Early and Peters 1990; Early and Headland 1998; Hill and Hurtado 1996; and Howell 1979). In 1977 one of the tools Flowers used to estimate ages was an event calendar, which was especially useful in questioning parents about the birth dates of their children. As examples of these events one can mention the time when the Xavánte had their first peaceful contact with SPI employees (in 1946), when anthropologist Maybury-Lewis and his family lived among them (in 1957–58), when the group moved to its current site (Etéñitépa, in 1972), and so forth. When approximate date of birth could be ascertained for one child, the relative ages of brothers and sisters could be approximated. This is because, as we shall see later in this chapter, Xavánte women give birth during most of their reproductive years and birth intervals present some regularity.

Xavánte society presents some features that facilitated our demographic investigations. Unlike some other lowland South American groups, such as the Suruí (Coimbra 1989a), the Xavánte do not have restrictions against talking about their dead. If this had been the case, the recovery of demographic data through interviews would have been even more difficult. Age estimations were facilitated by the existence of a system of age sets. As we described in chapter 2, a group of boys, roughly between the ages of eight and thirteen, enters the bachelors' hut where the boys live together until their initiation into manhood about five or six years later. This group takes the name of one of the eight age sets, which are successively rotated, and remains identified with that age set throughout life. Women belong to the same age set as their male age mates. The Xavánte themselves use this system, saying, for instance, that such and such an event occurred when boys affiliated with a specific age set were in the bachelors' hut (see table 2.2).

As the demographic information presented here comes from interviews, most of the birth dates in our data base are approximations. It was only possible to obtain exact birth dates for those children born between 1974 (with the arrival on the reservation of a FUNAI employee who began registering births) through July 1977 (when Flowers left Etéñitépa). When the 1990 census was undertaken we were able to use those dates to make more accurate statements about the age at which childbearing begins among Xavánte women. In 1990, women born in the 1974–77 period were between thirteen and sixteen years of age, which is the age range within which Xavánte women usually give birth for the first time.

We did not attempt to collect data on adult mortality. When Flowers first collected demographic data in 1976–77 she was primarily interested in child health and nutrition (see Flowers 1983a). In connection with these interests she interviewed the mothers in order to obtain information on their reproductive histories. Although Flowers initially thought about collecting mortality data from adults, her impression was that the Xavánte were not willing to talk about the circumstances of adult death. Later on it became clear to Flowers that the 1960s were difficult years for the group, when many deaths related to political disputes took place. The Xavánte were not willing to talk about these events. There were also methodological difficulties in estimating the age at death of an adult and the time when his or her death took place. When a mother was interviewed about her offspring it was often possible to estimate a child's age at death, as well as when the death occurred, taking into consideration the child's place in the birth order and information from an event calendar. This approach could not be used for adults.

Population Changes over Time

Comparison of data collected in 1990 with information obtained in preceding decades clearly shows demographic changes in the Etéñitépa

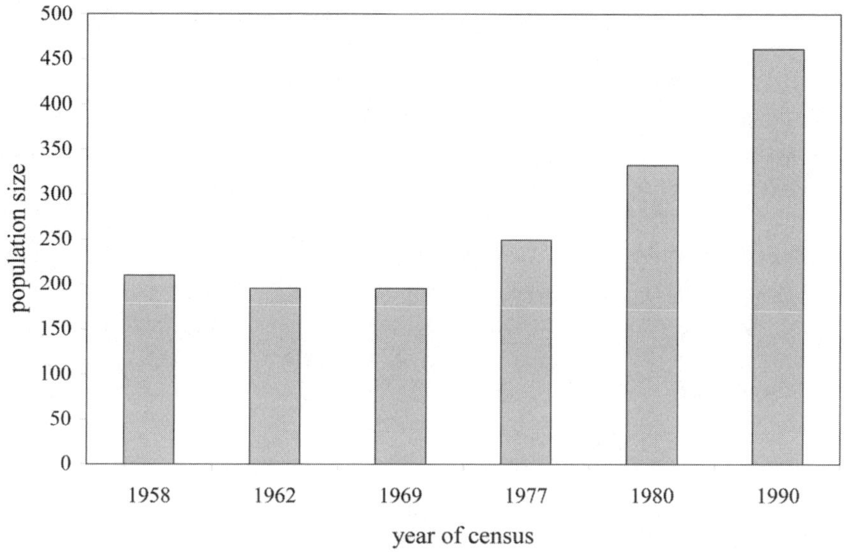

Fig. 5.1. Etéñitépa Xavánte population size at different times

Xavánte population, not only in terms of total population size, but also in age and sex composition. Figure 5.1 shows total population at six different times from 1958 to 1990.[3] From the late 1950s to the late 1960s the size of the population did not increase. Actually, it seems to have decreased somewhat. From 1969 onward there was steady growth; the population increased from around 200 in 1969 to nearly 450 individuals in 1990. Xavánte age composition has also changed over time. In 1962, Neel and his collaborators (1964, 92) observed that 39 percent of the population was under fifteen years of age. In 1990, the percentage under age fifteen had grown to 54 percent.

The detailed information needed in order to understand how population composition changed over time is available only from data collected in 1976–77, when Flowers lived among the Xavánte, and 1990, when we conducted fieldwork. In the 1970s the entire Xavánte population of the Pimentel Barbosa reservation lived in one village (Etéñitépa). During the 1980s, this village split and part of the population moved to two new villages (Caçula and Tanguro), which we will refer to as "satellite" villages. Data on total population size and age and sex composition for 1990 include the main village and the satellites.[4]

Between 1977 and 1990 the total population of the reservation nearly doubled, from 249 to 461 (table 5.1). Population pyramids for the two periods have a large base and a narrow top, which is typical of populations with high fertility and mortality (figs. 5.2 and 5.3). There are some differences in the age distributions of the two censuses. In 1977, 45.4 percent of the population was under the age of ten; in 1990, this percentage had decreased to 39.5 percent. However, the percentage of those ten to nineteen years of age increased from 18.4 percent in 1977 to 25.6 percent in 1990. Another striking difference between the two distributions is the increase in the percentage of older people (>50 years), which grew from 2.8 to 6.1 percent of the population. Probably the increase in older people occurred because during the epidemics, which subsided after the 1970s, old people were seriously affected, with higher mortality.

How can we explain the differences in the percentages of children zero to nine and adolescents ten to nineteen years of age when we compare 1977 to 1990? At first one might think that fertility has dropped, as the percentage of children zero to nine years is lower in 1990 than in 1977. This is not the case. As we will see, the fertility of Xavánte women was higher in the 1970s and 1980s than in the 1960s. Probably the major factor that explains the changes in age composition is a decline in mortality. In the 1970s and 1980s, mortality, especially that of children, dropped at Etéñitépa, allowing more children to survive beyond

TABLE 5.1. Age and Sex Distributions of the Etéñitépa Xavánte in 1977 and 1990

	1977			1990		
Age (years)	Males	Females	Total	Males	Females	Total
0–4	40	34	74 (29.7%)	53	36	89 (19.3%)
5–9	21	18	39 (15.7%)	39	54	93 (20.2%)
10–14	8	12	20 (8.0%)	34	32	66 (14.3%)
15–19	14	12	26 (10.4%)	28	24	52 (11.3%)
20–24	8	8	16 (6.4%)	17	15	32 (6.9%)
25–29	7	12	19 (7.6%)	11	15	26 (5.6%)
30–34	7	10	17 (6.8%)	14	10	24 (5.2%)
35–39	7	5	12 (4.8%)	8	10	18 (3.9%)
40–44	5	6	11 (4.4%)	7	7	14 (3.0%)
45–49	6	2	8 (3.2%)	8	11	19 (4.1%)
50–54	2	2	4 (1.6%)	4	6	10 (2.2%)
55–59	0	0	0 (0.0%)	4	4	8 (1.7%)
60+	1	2	3 (1.2%)	7	3	10 (2.2%)
Total	126	123	249 (100%)	234	227	461 (100%)

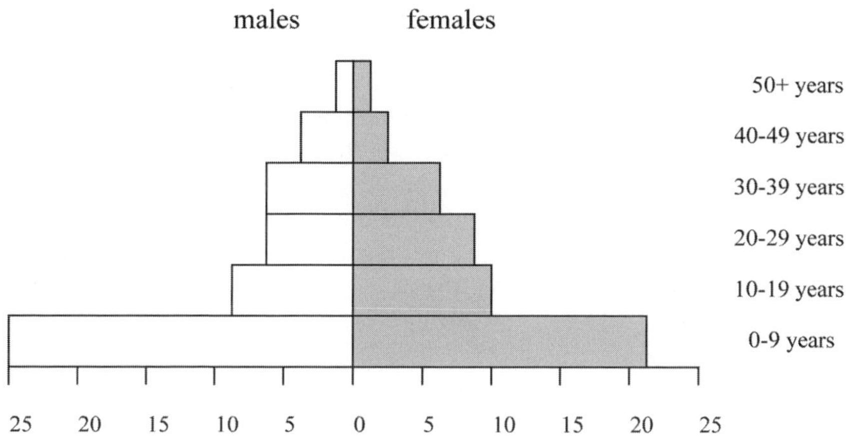

Fig. 5.2. Population pyramid, by percentage in age group, Etéñitépa, 1977

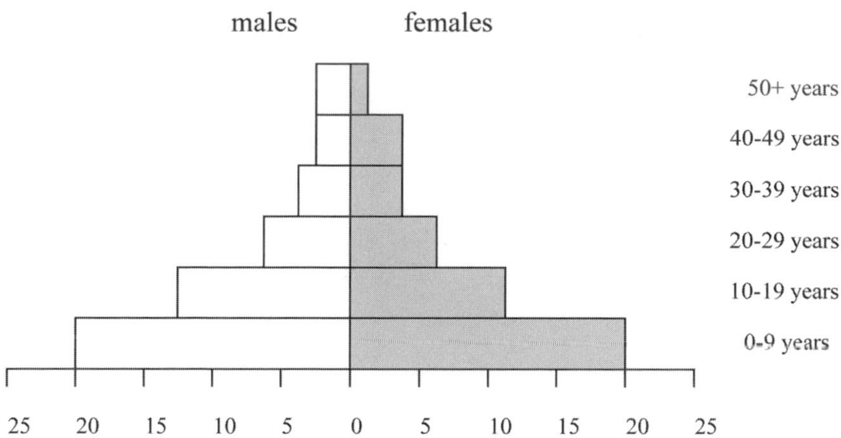

Fig. 5.3. Population pyramid, by percentage in age group, Etéñitépa, 1990

the age of ten, which increased the relative numbers in the ten to nineteen cohort.

Almost all population growth between 1977 and 1990 was due to natural increase, as there has been little migration into or out of Pimentel Barbosa. The total number of children born between 1977 and 1990 was 221, 185 in Etéñitépa and 36 in the two satellite villages.

Eighteen people enumerated in the 1977 census died before 1990 (table 5.2). These included the old chief Apöwẽ, who died in 1978 in his late seventies; one of his wives (Arewai'õ), also in her seventies; and their eldest son Warodi, who succeeded Apöwẽ as chief and died in 1988 in his sixties. The exact year of death of the other adults could not be ascertained. However, judging by their age in 1977, 2 were elderly and 3 were middle aged. Two young married men and a crippled boy in his teens also died. The remaining 7 people were under five years old in 1977. According to the mothers of 4 of these children, 2 died at around one year of age, 1 at five, and another at eight. Of these 18 people we only know for sure the cause of death of 2, both of whom died in hospitals in Brasília, Apöwẽ of pneumonia and Warodi of cancer.

Migration was a relatively minor element of population change. Six people had moved away, four from Pimentel Barbosa to other Xavánte communities. One young man had moved to Goiânia to attend school, and an older man who became mentally ill in 1977 had disappeared. In 1990, three people were missing—they were probably in one of the satellite villages but escaped the census.

Since 1977, eighteen people had moved into Pimentel Barbosa. These included seven young men who had left the reservation in the early 1970s and who had moved to Goiânia or Ribeirão Preto, where they attended school. In 1990, they had returned and were married in the village. Five people from other Xavánte reservations had arrived to settle with their Etéñitépa kin and had also married in the community. Five immigrants consisted of the white wife of a Xavánte man and their

TABLE 5.2. Age Distribution of Deaths between 1977 and 1990 at Etéñitépa

Age (years)	Number of Deaths
<1	15
1–4	11
5–9	4
10–14	2
15–19	0
20–29	2
30–39	0
40–59	3
60+	5
Total	42

four children. A few years later this family moved away from Pimentel Barbosa. Finally, one young Xavánte man had married a college-educated Karajá woman when he was at school in Goiânia, and the couple had a house in the village in 1990.

From the census alone, it is not possible to make an accurate estimate of birth and death rates of the community between 1977 and 1990. This is a population with relatively high infant and child mortality, and a certain number of children were born after the 1977 census but did not survive to be censused in 1990. To obtain information about these children, 109 women living in the main village (Etéñitépa) were interviewed. They were asked to describe the outcome of each pregnancy, including the age or stage of development at death of those no longer living in 1990. According to the women's reports, there were 212 births between 1977 and 1990. Twenty-four of these children died before 1990. These children must, of course, be counted both as births and deaths. Unfortunately, the count is incomplete for the total population of the Pimentel Barbosa reservation, for in 1990 we were only able to carry out in-depth interviews on reproductive histories with women living at Etéñitépa, and therefore we missed 18 women who were living in the satellite villages.

The population of the main village was 249 in 1977 and 385 in 1990, that is, it had a mean population of 317. The women interviewed reported 212 births between 1977 and 1990: 188 living plus 24 who were born and died between censuses. The crude birth rate for the period 1977–90 is 50.8 per 1,000.[5] For the crude death rate, we added to the 18 deaths known by comparing the two censuses the 24 deaths of children who were born and died between the censuses—giving 42 deaths over thirteen years. The crude death rate (CDR) is 10 per 1,000.[6]

From the data presented here, it is clear that birth rates are very high at Etéñitépa. For comparative purposes, the crude birth rate in Brazil in the period between 1970 and 1980 was 33.7 per 1,000 (Patarra 1995, 64). Both crude birth and crude death rates are highly influenced by the age composition of the population. Normally death rates are highest in the oldest age cohorts. At Etéñitépa, only 3 percent of the population was over fifty years of age in 1977 and 6 percent in 1990. The crude death rate in Brazil in the period between 1970 and 1980 was 9.0 per 1,000 (64), nearly the same as we found at Etéñitépa. However, there are many more old people in the Brazilian population than in the Xavánte population.

Changes in Parity and Offspring Survival

We have already seen that from the late 1950s to the late 1960s Xavánte population size did not increase but actually decreased somewhat (fig. 5.1). A further piece of evidence pointing to stressful conditions around mid-century is that at that time children under age fifteen comprised a smaller percentage of the population (39 percent in 1962) compared to 1990 (54 percent). It is likely that this is due in part to higher mortality rates in children in the 1960s. In this section we will compare information on parity and offspring survival collected in 1962 by Neel et al. (1964) with information collected in 1977 and 1990.

Table 5.3 shows the parity for thirty-one women with completed fertility. There are no women with zero parity, and only one has borne fewer than four children. These results are in accordance with the observations by Neel et al., who observed in 1962 that "there is no evidence for absolute sterility where the wife was above the age 15" (1964, 94). While the completed fertility is similar for each cohort (around seven births), the number of children surviving to age fifteen is very different. The six women born between 1910 and 1929, and who gave birth mostly before the middle of the century (between the 1920s and 1960s), saw 81 percent of their offspring survive. The oldest woman in the community in 1977 was Arewai'õ, Apöwẽ's wife, who was born around 1912. She had eight grown sons and daughters living in the village and said she had lost no children. Two other women in that cohort claimed that all their children had grown up. The situation experienced by women born between 1930 and 1939 was strikingly different. These women, the majority of whose children were born around the middle of the century (1948–75), saw only 38 percent of their live births survive. Offspring survival increased for women born between 1940 and 1949. These women had their children in the latter half of the century (1955–85) and 60 percent of their children survive. The results presented in table 5.3 strongly suggest that the period around the middle of the century was particularly stressful for the Xavánte.

Data collected from women who had not yet completed their parity also reveal changes in Xavánte demography over time. Table 5.4 compares parity by mother's age in three periods.[7] The data derive from information collected by Neel and his collaborators in 1962 (Neel et al. 1964) and from information obtained in 1977 and 1990. Overall, the mean numbers of children born to women of each age group in 1962,

TABLE 5.3. Parity of Etéñitépa Xavánte Women with Completed Fertility by Birth Cohort

Birth Cohort	Number	Years of First and Last Birth[a]	Number of Live Births													Mean Parity	Percentage of Survivors to Age 15 (%)
			0	1	2	3	4	5	6	7	8	9	10	11	12		
1910–29	6	1927–64	0	0	0	0	2	0	0	1	2	0	0	1	0	7.0	81
1930–39	8	1948–75	0	0	0	0	1	2	0	0	3	0	2	0	0	7.3	38
1940–49	17	1955–85	0	1	0	0	1	3	3	2	0	3	1	2	1	7.2	60
Total	31		0	1	0	0	4	5	3	3	5	3	3	3	1	7.2	

[a]Refers to the year when the first and last child of women in each cohort was born.

1977, and 1990 are similar. While women fifteen to nineteen years of age gave birth to 1.0 to 1.2 children, the figure for those older than forty years ranges from 6.3 to 7.1. The most striking finding in this table is the relatively small average number of children born to women of the twenty to twenty-nine age group interviewed in 1962 (2.8 live births), compared to the results of the interviews carried out in 1977 and 1990 with women in this same age group (4.2 and 3.8, respectively). This suggests that fertility of women who were in their twenties in the late 1950s and early 1960s was lower than fertility of women of comparable age in later periods.

Table 5.5, showing the comparative survival rate of children born to women in different age groups, reveals the time at which mortality in children began to increase.[8] In 1962, mothers fifteen to nineteen and twenty to twenty-nine years old had fewer surviving children (50.6 and 64.3 percent of those born) than women of similar age in 1977 (64.7 and 75.0 percent) and 1990 (82.6 and 88.5 percent). It was the observation of high rates of mortality of children born to women fifteen to twenty-nine years old in 1962 that prompted the comment in Neel et al. (1964, 95–97) that infant mortality rates were rising. The authors wrote, "There appears to be a higher mortality among the children of the younger mothers, a finding whose possible significance is augmented by the fact that the children of the younger mothers have had a shorter risk period. . . . Taken at face value, our data indicate a population at present replacing itself, but there are ominous signs that this

TABLE 5.4. Comparative Parity by Age Group of Xavánte Mothers

	Age Groups (years)			
	15–19	20–29	30–39	40+
São Domingos, 1962				
Number of women	15	9	4	7
Number of live births	18.0	25.2	22.0	44.0
Mean number of live births	1.2	2.8	5.5	6.3
Etéñitépa, 1977				
Number of women	14	19	18	12
Number of live births	17	80	101	82
Mean number of live births	1.2	4.2	5.6	6.8
Etéñitépa, 1990				
Number of women	23	25	15	28
Number of live births	23	96	99	198
Mean number of live births	1.0	3.8	6.6	7.1

situation may not persist" (97). Fifteen years later, in 1977, those women, now in their middle years (thirty years and older), had seen nearly half of their children die. By the 1970s children born to women younger than thirty years were surviving more often (64.7 and 75.0 percent) than in the preceding decade. By 1990, survival rates of children born to women in this age group had improved significantly (over 80 percent). In 1990 it was only women who were age forty and over who had as few as 55.6 percent of their offspring surviving, and they were those who were fifteen to twenty-nine years old in 1962.

A Closer Look at Fertility and Mortality Trends

The results discussed in the previous section suggest that the late 1950s and the 1960s were, demographically speaking, stressful for Etéñitépa Xavánte, with this stress affecting parity and offspring survival. Now we will turn to more detailed analyses of fertility and mortality trends using standard demographic indicators to obtain a clearer picture of the magnitude and timing of the demographic crisis. Data come from interviews with Xavánte women carried out in 1976–77 and 1990.

Our analysis is based on three time periods, that is, 1956 and before, 1957–71, and 1972–90. We chose 1957 as a cutoff point because from this time on there were important changes in patterns of

TABLE 5.5. Surviving Offspring by Age Group of Xavánte Mothers

	Age Groups (years)			
	15–19	20–29	30–39	40+
São Domingos, 1962				
Number of women	13	09	04	07
Number of live births	18.0	25.2	22.0	44.0
Number of surviving offspring	9.1	16.2	20.0	33.0
Survivors (%)	50.6	64.3	90.9	75.0
Etéñitépa, 1977				
Number of women	14	19	18	12
Number of live births	17	80	101	82
Number of surviving offspring	11	60	54	39
Survivors (%)	64.7	75.0	53.5	47.6
Etéñitépa, 1990				
Number of women	23	25	15	28
Number of live births	23	96	99	198
Number of surviving offspring	19	85	82	110
Survivors (%)	82.6	88.5	82.8	55.6

interaction between the Xavánte and non-Indians. Maybury-Lewis (1967, 27–29) shows that the second half of the 1950s was a turning point for the group. In 1958, the São Domingos Xavánte were still powerful and numerous. However, when the anthropologist returned in 1962 he observed that major changes had taken place, as there had been population decrease due to epidemics and warfare. Maybury-Lewis wrote: "The Shavante had learned by now of their comparative impotence, and they were acutely conscious of their own dwindling numbers, a topic to which they returned again and again in conversation with me, who had known their village in the days when it was numerous and strong" (27–29).

The year of 1972 was chosen as another cutoff point for our analysis because that was when the Xavánte moved their village from the margin of the Rio das Mortes to its present location (Etéñitépa). Epidemics, which had affected the Xavánte in the previous decade, receded. During the 1970s and 1980s there were major changes in their subsistence, including a shift toward a more sedentary lifestyle and even cash crop production. Sustained interaction with the outside world became routine (see chapters 3 and 6).

Trends in Infant and Child Mortality

During the reproductive interviews with Xavánte women in 1976–77 and 1990 we asked not only how many children they had but which children had died and when. From these data we computed abridged life tables for children younger than ten years of age.[9] The results of the abridged life tables clearly show that child survival to age ten was much lower from 1957 to 1971 (43 percent or 431 per 1,000) than in the two other periods. From 1927 to 1956, nearly 73 percent of the children born reached age ten and in 1972 to 1990 nearly 83 percent (table 5.6).

Children of age 0.9 years and younger were those who experienced the highest mortality rates. In the period when survival rates were the lowest (1957–71), over 20 percent ($_1q_0 = 0.203$) of the children did not survive the first year of life. In the other two periods mortality rates were also highest in children under age 1.

One interesting finding from the abridged life tables is that the lower the overall mortality the higher the concentration of deaths in very young children. Table 5.7, derived from the survival rates (1_x function), shows that 60 percent of deaths in the 1972–90 period occurred in children

TABLE 5.6. Abridged Life Tables for Xavánte Children 0–10 Years of Age, Sexes Combined, at Different Time Periods, 1927–90

	1927–56				1957–71				1972–90			
Age (years)	Enter (n_x)	Die (d_x)	Mortality (q_x)	Survivors (l_x)	Enter (n_x)	Die (d_x)	Mortality (q_x)	Survivors (l_x)	Enter (n_x)	Die (d_x)	Mortality (q_x)	Survivors (l_x)
0	56	5	0.089	1000	128	26	0.203	1000	294	30	0.102	1000
1	47	3	0.043	911	94	8	0.085	797	245	10	0.041	898
2	40	2	0.075	872	71	7	0.099	729	219	0	0.000	861
3	36	1	0.028	807	57	4	0.070	657	199	1	0.005	861
4	32	1	0.031	784	46	4	0.087	611	188	1	0.005	857
5	30	1	0.033	760	37	4	0.108	558	170	1	0.006	852
6	27	0	0.000	734	30	2	0.067	498	158	1	0.006	847
7	26	0	0.000	734	23	0	0.000	464	134	1	0.007	842
8	22	0	0.000	734	19	0	0.000	464	125	1	0.008	836
9	21	0	0.000	734	14	1	0.071	464	106	0	0.000	829
10				734				431				829

under age one. Only 18 percent of the deaths were in children age three and over. During the period when mortality reached its highest level, from 1957 to 1971, close to 40 percent of the deaths were in children age three and over. In other words, during the demographically stressful period from the late 1950s to the early 1970s Xavánte parents not only experienced the loss of a very large number of their sons and daughters but the children dying were comparatively older, making the situation, as they frequently told us, even more painful psychologically.

Epidemics of infectious diseases were almost certainly the leading cause of increased mortality rates in the years between 1957 and 1971. Unfortunately, information about the etiology of epidemics occurring during these years is meager. Other than the writings of Neel et al. (1964) and Maybury-Lewis (1967), which refer to the time prior to 1963, we were not able to locate further information on the situation of this community in the 1960s. When Maybury-Lewis (xxxi) returned to São Domingos in 1962, four years after his original fieldwork, he noted that the population had decreased, which he attributed in part to epidemics. Moreover, Neel and his collaborators collected some interesting data that strongly suggested the occurrence of epidemics in the Xavánte. In 1962, these investigators carried out antibody studies and found that a high percentage of the individuals tested positive for measles and pertussis. Although they were cautious about the interpretation of their findings, they wrote: "Our observations regarding antibodies could be evidence for serious epidemics among the inhabitants of this village in the recent past" (Neel et al. 1964, 129).

It is often the case that epidemics of infectious diseases in indigenous groups precede or immediately follow the onset of permanent contact. The Etéñitépa was the first Xavánte group to make peaceful contact with SPI agents, in 1946 (see chapter 3). Interestingly, our demographic data suggest that they began to experience very high mortality

TABLE 5.7. Distribution of Deaths in Xavánte Children under Age 10 at Three Time Periods according to Age Group

Age Groups (years)	1927–56 (%)	1957–71 (%)	1972–90 (%)
0–0.9	33	36	60
1–2.9	39	25	22
3–4.9	18	17	5
5–10	10	22	13

starting in the second decade of contact (from 1956 to 1971), rather than in the years immediately following contact. One reason why mortality might only have increased some time after contact may be because it was not until 1956 that the Xavánte settled in a village close to the SPI post, at São Domingos (Graham 1995, 34; Maybury-Lewis, 1967, 27). Before this time they had lived at Etéñitépa, the site of their present village and, although the Xavánte would come often to the post, they were still relatively independent. From the 1950s onward, the trend was for the Xavánte to be in much more frequent contact with outsiders, which might have facilitated the introduction of contagious diseases. Referring to the situation at São Domingos in 1962, Maybury-Lewis wrote that "Its inhabitants still went out on trek, but they were equally inclined to go on begging trips to the nearest Brazilian cities" (28).

In the most recent time period (1972–90) mortality levels in children ten years of age and younger are much lower than between 1957 and 1971 (table 5.6). The 1960s was a period of crisis for the Xavánte, when epidemics and social disruption coincided. Under these conditions, it is likely that there were breakdowns in food production. In the 1960s, the Xavánte were living by the Rio das Mortes, where malaria was endemic. The site where their present village is located, to which they moved in 1972, is also the place where the Xavánte lived in the 1940s and early 1950s. The Xavánte say that they moved back to Etéñitépa because they consider it to be a "healthy" place where agricultural land is good, which implies that their nutrition may have improved. In the 1970s and 1980s, although not on a regular basis, the Xavánte were also receiving vaccines provided by the Brazilian Ministry of Health (BCG, polio, measles, and DTP), which might have had some effect upon mortality.

It should be emphasized, however, that mortality in Xavánte children remains high, well above regional and national averages for Brazil. As indicated in the abridged life table for the 1972–90 period, 294 children were born during these years and 30 died before reaching one year of age (table 5.6), that is, a value of ${}_1q_0$ equal to 0.1020. This figure is an approximation of the infant mortality rate (IMR) for Etéñitépa Xavánte during this period (out of every 1,000 children born, 102 died during the first year of life). The IMR for Brazil in 1981 (which is approximately the midpoint for the period for which we have data for the Xavánte) was 68.7 per 1,000. For the Center-West region, which includes the state of Mato Grosso, where Xavánte reservations are located, IMR was 54.3

per 1,000. Xavánte IMR was close to the value for the Northeast region (103.3 per 1,000), which is the poorest region of Brazil and the one that showed the highest IMR in 1981 (Simões and Monteiro 1995, 155). As we demonstrate in chapters 7 and 8, the fact that mortality rates among the Xavánte remain high is related to exposure to many infectious diseases and deficient health care.

Trends in Fertility

Table 5.8 shows data on age-specific fertility rates (ASFR) and total fertility rates (TFR) for Xavánte women according to the three time periods. Total fertility rate is a widely used measure of fertility because, in addition to not being influenced by the age and sex composition of the population, its interpretation is rather easy to grasp. For example, the total fertility rate was 7.86 between 1972 and 1990, that is, a hypothetical cohort of Xavánte women would have around eight live births during their childbearing years if they were exposed to the schedule of age-specific fertility rates for that time period (ASFR of 0.050 between ten and fifteen years of age, 0.356 between fifteen and nineteen, 0.412 between twenty and twenty-four, and so forth).

TABLE 5.8. Age-Specific Fertility Rates (ASFR) for Xavánte Women, Ages 10–44, at Different Time Periods, 1942–90

Age (years)	1942–56			1957–71			1972–90		
	W/Y	B	ASFR	W/Y	B	ASFR	W/Y	B	ASFR
10–14	91	5	0.055	148	10	0.068	317	16	0.050
15–19	56	19	0.339	145	52	0.359	225	80	0.356
20–24	38	10	0.263	129	31	0.240	170	70	0.412
25–29	20	6	0.300	95	22	0.232	167	54	0.323
30–34	16	7	0.438	56	8	0.143	138	34	0.246
35–39	10	1	0.100	37	5	0.135	132	24	0.183
40–44	8	1	0.125	20	0	0.000	110	0	0
TFR		8.10			5.88			7.86	

Note: The fertility data for 1942–56 and 1957–72 were derived from the 1977 fertility survey. The data for the 1972–90 period were derived from the 1990 fertility survey. W/Y stands for women/years, that is, the number of years spent by women of this age category during the specified interval of time. B stands for number of births. ASFR stands for age-specific fertility rate, that is, the number of births to women of this age category during the specified time interval divided by the number of women/years. TFR stands for total fertility rate, that is, the sum of all the ASFRs for this time interval, multiplied by five. It represents the average number of children a woman would bear during her lifetime at the ASFR for this interval of time.

For the time period 1972–90, when sample sizes were largest, Xavánte childbearing peaked at twenty to twenty-four years, followed by a gradual decline to age forty (table 5.8). Fertility is almost as high in women fifteen to nineteen as it is in the twenty-five to twenty-nine category. While Xavánte women start childbearing very early (often around age thirteen or fourteen), the fertility of the ten to fourteen age category is very low. Childbearing also appears to cease at a relatively early age, as fertility of women forty to forty-four is zero.

The results show that Xavánte fertility reached its lowest level between 1957 and 1971 (TFR = 5.88). Therefore, the late 1950s and the 1960s was not only a period when mortality was high, as we already saw, but when fertility experienced a drop. As for the reasons why fertility dropped, this was a period marked by social disruption, as well as epidemic disease, for the Xavánte, which probably resulted in increased instability of marriage and early widowhood for some women. From the reproductive interviews we learned that eleven women, while still in their twenties, were widowed or divorced during the 1960s and there was a space without childbearing before they remarried. Even though the Xavánte are polygynous, some women may remain single after widowhood or divorce, and, while they may have one or two additional children, it could be expected that their fertility would be lower than that of women in established unions. In 1976–77, when Flowers lived at Etéñitépa, all women in their twenties were married (figure 5.4).

The fertility of Xavánte women in 1957–72 was about 25 percent lower compared to the two other periods. Some women indicated that seeing so many of their children die, including some who had passed babyhood, they "lost heart to have children."[10] Despite the extent of the crisis the group was facing, Xavánte women were still giving birth on average to close to six children. One wonders whether mortality-fertility interactions might have been strengthened during those years. In a review of fertility in so-called anthropological populations Wood (1990, 233–34) writes that one of the outstanding issues yet to be addressed in the literature concerns fertility-mortality interactions. According to him, there is growing evidence that an important regulator of fertility levels in small-scale societies is not reproduction per se but mortality. The part of Wood's argument that concerns us here relates to the interaction between child mortality and women's fertility, which he refers to as "reproductive compensation." He writes: "when a nursing child dies, lactation is terminated and the mother resumes ovulating sooner than she otherwise would

Fig. 5.4. Young couple in their section of the extended family house. Although the house has no partitions, each nuclear family has a designated space in it. (Photo by N. M. Flowers, 1977.)

have" (233). Hence, paradoxically, under a situation of high infant mortality the total fertility rate of a given group of women might become "inflated" because of decreased birth intervals associated with resumed ovulation due to early termination of lactation.[11]

Our data do not allow for conclusive statements concerning the role of child mortality in influencing Xavánte fertility levels. However, the period between 1957 and 1971 was certainly the time when this sort of interaction would be expected to occur most frequently. Probabilities of dying in the first two years of life in 1957–71 (${}_1q_0 = 0.203$ and ${}_2q_1 = 0.085$) were at least twice as high as in the two other periods (table 5.6). That is, Xavánte mothers were losing more children who were still breast-feeding. It seems reasonable to argue that, with an increased number of their babies dying and the duration of breast-feeding shortened, Xavánte women at that time had increased probabilities of becoming pregnant. During these years, Xavánte women may have had more births because of shortened periods of lactation infecundability due to early death of babies.[12] In summary, the fertility decline may not fully reveal the extent of the crisis because of the effects of reproductive compensation.

Epidemic Disease and Social Order

Not all deaths that occur during postcontact epidemics in Amazonia are directly due to sickness. When an unfamiliar and debilitating disease strikes a community people may become frightened or disoriented and the bonds that normally unite kin may break down. When women are sick they may be unable to feed or care for their children (Coimbra 1987; Neel 1982a; Neel et al. 1970; Ribeiro 1956). Nutrition may suffer if men are unable to hunt or women cannot collect plant foods (McGrath 1991), and under these conditions mortality may increase. The availability of basic medical and nursing care, even food and water, when an epidemic breaks out may make a great difference. Mota (1955) and Nutels (1968) reported mortality differences during a measles epidemic in the Xingu National Park in 1954. According to Nutels, "Of the 564 patients, 114 died. Among those who received medical care, the death rate was 9.6 percent; among those who could not be treated in time, it reached 26.8 percent" (70).

In addition to biological stress due to economic breakdown, another major indirect cause of adult deaths may be scapegoating, as increase in mortality brings about accusations of sorcery and revenge deaths (Early and Peters 1990, 80; Ferguson 1990; Ross 1984). In some Amazonian societies "natural" causes are never accepted as a reason for death (Crocker 1985, 36–37; Seeger 1981, 219–20; Wagley 1977, 171–72). Others distinguish in various ways between deaths due to natural causes and those that originate in the spirit world or sorcery worked by malign individuals (see Buchillet 1992, 216; Crocker and Crocker 1994, 170; and Langdon 1992b, 50).

Almost all Amazonian societies have spiritual specialists, individuals who claim the power to communicate with spirits and to influence human fate by curing illness or, conversely, by harming others (Langdon 1992a, 3–5). These specialists, often referred to in the literature as shamans, are not usually community or factional leaders; rather, their power comes from the ways in which they may use their spiritual knowledge for good or evil. The same individual who cures as a shaman may be suspected of being a "witch" when unexplained deaths increase, and he or she may eventually be executed by the community (Crocker and Crocker 1994, 112–13; Seeger 1981, 86–88, 174; Wagley 1977, 186–89). The concentration of suspicion on an individual believed to have shamanic powers and his or her eventual elimination from the society may have the effect of limiting social turmoil.

The Xavánte do not have the social role of shaman, as described in most other Amazonian societies (see Langdon 1992a, 3–5). Certain men have defined ritual roles inherited through lineage membership: a senior man of the Uhö lineage of the Öwawẽ clan, for example, claims the power to make magic that attracts peccaries (Flowers 1983a, 236; Giaccaria and Heide 1984, 125–26). All adult members of the society are believed to have the ability to communicate in dreams with the souls of dead kin (Graham 1990, 90–91; 1995, 99–102). "They are the dead others who, although spoken of with some trepidation, are thought of as benevolent, unlike the feared dead among other Gê groups" (Graham 1995, 100). As elders reach the final stages of the life cycle, they identify increasingly with the ancestors. Informants told Graham (100–102) that men who have played exceptional roles as chiefs or counselors join a class of immortals who are said to live in the sky. Warodi, the chief at the time of Graham's fieldwork, in his speeches and his telling of myths often spoke as one who, in life, had already joined the immortals. Even young men, after they are initiated, have the ability to "receive" songs from the ancestors, whom they hear and see in their dreams. The dreamer then teaches the new song to his age set, which performs it in public (125–28).

Graham (1993) points out that in Xavánte society the ideal of collective solidarity is constantly at odds with the ambitions of individual leaders and their factional loyalties. This is demonstrated in the discussions that take place in the *warã* (men's council), which meets in the evening to discuss the affairs of the community. "In mature men's speech, Xavánte harness their expressive resources to create a balance between factions and to counteract the omnipresent tendency toward fission. They do not succeed in all instances, and when they do not, a minority faction may leave to found a new village" (725).

Spiritual curing is also collective. It is carried out in a public ceremony in which both men and women of the community participate (Flowers, field notes; Giaccaria and Heide 1984, 238–43; Maybury-Lewis 1988, 271–72). Flowers saw it held in 1976 for Sibupá, one of Apöwẽ's sons, who was seriously ill. The sick man's brother cleared a circle in front of his house and covered it with fresh palm leaves. The wife of another brother laid down mats and placed on them blankets and baskets of bananas. At sunset they led Sibupá, looking weak and ill, out of his house and laid him on the mats. All the initiated men, ceremonially painted, some wearing tall feather headdresses, formed a great semicircle around him. One by one, the elders danced up to the sick man, knelt by him and cupping their hands sucked in their breath as if to draw out the sickness. All night the men danced and sang, willing the disease to depart.

Xavánte Factions and Sorcery

The Xavánte do not see the ability to work sorcery as the attribute of certain individuals who have the power in themselves. "They believe that Shavante are only able to inflict sickness and death on other Shavante by the power of certain rituals involving the manipulation of magical substances or implements. These substances are powders, which can be used both offensively and defensively" (Maybury-Lewis 1967, 276). Only men are suspected of working sorcery. When a prominent man becomes sick or dies, men of an opposing faction, those who presumably would have the most to gain by his death, fall under suspicion. Such an event is likely to trigger a dispute in which members of the weaker faction are killed or driven out of the village. As Maybury-Lewis (188–89) points out, a sorcery case is a political matter because "all Shavante cases are essentially cases between groups rather than disputes between individuals." Rather than blame being concentrated on a culpable individual, it falls on the entire faction.

In the 1950s, apparently, the Xavánte did not necessarily attribute sickness and death to witchcraft. "They have some understanding of contagion and are aware, for example, that if they consort with a Brazilian who has a cough, then they too are liable to catch it from him. Should they catch his cough they will not believe that he has bewitched them unless they have good reason to" (Maybury-Lewis 1967, 274). But the intense factionalism that always threatens to break down Xavánte society was exacerbated by the unusual number of deaths, often of prominent men.

In the years that followed contact in the late 1940s, Apöwẽ and the people of his Wamãri lineage held the advantage when it came to the distribution of "presents" from the Indian agent as well as from the anthropologist (Maybury-Lewis 1967, 28, 171). "The influence of the Indian agent seemed to be directly proportional to his supply of gifts. He was careful to give the lion's share of these to the chief and the members of his lineage, so even if the community as a whole did not feel warmly toward him at a particular time, he could be certain that its most influential people would not side against him" (28). Maybury-Lewis contrasts the formal distribution of game after a collective hunt, when the chief presided over a scrupulously equitable distribution of meat, with the melee that took place when the Indian agent had a steer slaughtered for the community. "The chief's household thus invariably came off best . . . the difference in the shares was such that some households would have meat sufficient for one or two days longer than others' portions would provide for" (202). In other ways the introduction of manufactured goods affected the community. Unlike artifacts that people make themselves, these cannot easily be replaced if stolen, and Xavánte therefore insisted on exclusive rights over such items as fishhooks or knives, leading to resentment and accusations of theft (182–83).

In the late 1950s and early 1960s the synergism among intense factionalism, the occurrence of epidemics, and disputes over trade goods seem to have been particularly intense. An epidemic of influenza that broke out around 1960 apparently became the catalyst for a series of sorcery accusations that, together with the disease itself, diminished the community. Although the Wamãri may have believed that their primary motive in moving to eliminate other factions was to revenge the death of kin through sorcery, by this means they also assured control over the distribution of trade goods. The details of what actually occurred are difficult to untangle.

> Some Wamãrĩ had died in an epidemic of influenza. The faction held the Topdató responsible and killed a number of them. . . . I never could determine how many Topdató were killed at this time, for the enemies of the Wamãrĩ were eager to attribute every death which had occurred since my previous visit to the ferocity of the dominant faction. The Wamãrĩ assured me, on the other hand, that no Topdató had been killed by them. They insisted that the Topdató had suffered even more severely in the epidemic than they themselves had. It seems certain, however, that some Topdató were, in fact, killed, though some may have died of influenza. (Maybury-Lewis 1967, 176)

In the early 1960s it was already becoming apparent that the disputes were bringing about visible consequences upon social composition:

> At the time of my second visit [in 1962] . . . the Wamãrĩ were still firmly established as the dominant lineage, for there were few to oppose them. The only other lineages still represented in the community were respectively the Uhe and the Dzutsi, both considerably depleted. . . . São Domingos was one of the most homogeneous Shavante villages I have stayed in, factionally speaking" (Maybury-Lewis 1967, 177).

In 1976 Flowers found that people were reluctant to talk about what had happened in those dark times. However, murmurings about sorcery recur in the village from time to time. Flowers's neighbor blamed the failure of a marriage arranged between two young people to sorcery worked by the girl's father. As recently as 1984, after a village split, Etéñitépa people accused the leader of a faction that left to found another village of trying to "poison" the Etéñitépa soccer team (Graham 1995, 104).

Effects of Demographic Crisis on Social Organization

In the final portion of this chapter we explore some relationships among political disputes, violence, and death from epidemics and investigate their impacts upon specific aspects of Xavánte demography and social dynamics. In 1976–77, Flowers collected detailed genealogical data, including information on clan and lineage affiliation. Comparison of the 1970s data with Maybury-Lewis's censuses and genealogies taken in 1957–58 and 1962 provides compelling evidence as to the relationship between demographic crisis and political disputes and the resulting long-term impacts on Xavánte social organization. The almost complete elimination of certain lineages and clans led to social imbalances that eventually affected marriage arrangements.

Table 5.9 compares the numbers of Xavánte adult men and women of the clans and lineages seen by Maybury-Lewis in the late 1950s and early 1960s with those present at Etéñitépa in 1976–77.[13] In the 1970s Flowers sought to investigate what happened to those who were missing (who had died, moved, etc.). A most remarkable finding is that no men from the Poridza'õno Tebe, Topdató Aiute'mañãri, and Topdató Wahi lineages were at Etéñitépa in the 1970s. That is, the Etéñitépa group was even more homogeneous, with regard to the composition of factions, than in the preceding decades.

TABLE 5.9. Fate of Adults Present in São Domingos in 1958–62 up to 1977 according to Clan and Lineage

Clans:	Poridza'õno					Öwawẽ					Topdató			
Lineages:	Wamãri		Tebe		Uhö		Dzutsi		Unknown		Aiute'mañãri		Wahi	
Sex:	M	F	M	F	M	F	M	F	M	F	M	F	M	F
Living in São Domingos 1958–62	23	23	2	6	7	6	9	17	—	3	3	6	4	5
Died between 1958/1962 and 1977	4	7	1	1	4	1	3	9	—	1	2	2	4	3
Moved between 1958/1962 and 1977	2	1	1	1	—	—	—	1	—	—	1	—	—	—
Unknown what happened between 1958/1962 and 1977	—	3	—	2	1	—	1	3	—	1	—	—	—	—
Living in Etéñitépa in 1977	17	12	—	2	2	5	5	4	—	1	—	4	—	2

Source: Data from Maybury-Lewis 1967, 317–42; and Flowers 1983a, 1974.

In order to reveal this pattern even more clearly, in table 5.10 we pooled the data into two groups: men and women of the politically dominant lineage (Poridza'õno Wamãri, headed by Apöwẽ) and men and women of the other six lineages.[14] From a total of twenty-three Wamãri men and twenty-three Wamãri women recorded in Maybury-Lewis's genealogies and censuses, over half (seventeen or 74 percent) of the men and (twelve or 52 percent) the women were still living at Etéñitépa in 1976–77. For non-Wamãri the reduction was much greater, particularly in the case of men. Only 28 percent of the non-Wamãri men and 42 percent of non-Wamãri women seen by Maybury-Lewis were still living at Etéñitépa in the 1970s. Even more striking is the fact that the major reported cause of the reduction in non-Wamãri men was death, which accounted for 56 percent. Only 18 percent of Wamãri men died between 1958 and 1962 and 1976 and 1977. As it is unlikely that different lineages living in the same village would suffer different rates of mortality due to epidemics, it is more probable that reduction in the non-dominant lineages was due to political disputes.

The Wamãri were unquestionably the dominant political force in Etéñitépa in the 1970s. On the basis of genealogical and census data Flowers found that, in 1977, of the twenty-eight men over the age of thirty, nineteen were Wamãri. Interestingly, although this lineage ended up having almost complete political control, members were forced to face some social consequences brought about by the imbalance between the clans. According to Xavánte marriage rules, the Wamãri, who are affiliated with the Poridza'õno clan, are supposed to take spouses from the two other clans (see chapter 2). So many men of the Öwawẽ and Topdató clans had died in the 1960s that few remained to father daughters who Wamãri men could marry.

In 1976–77 there were eleven young Wamãri men between the ages of fifteen and twenty-five who were already initiated and therefore entitled to take wives, who, according to Xavánte rules of clan exogamy, they must find among the girls of the two other clans. Flowers observed that only three of these young men had found wives. Although there were seven Öwawẽ girls between the ages of ten and twenty, three of them were married to older men and one was betrothed to a young man who was away from the village. Hence there were only three girls available for marriage but eleven young Wamãri men. Meanwhile, the situation of young men of the Öwawẽ clan was quite different. Of the nine young men of this clan between the ages of fifteen and twenty-five, seven were

TABLE 5.10. Fate of Adults Present in São Domingos in 1958–62 up to 1977 according to Lineage

	Men		Women		Both Sexes	
Lineages	Poridza'õno Wamãri	All Other Lineages	Poridza'õno Wamãri	All Other Lineages	Poridza'õno Wamãri	All Other Lineages
Living in São Domingos 1958–62	23 (100%)	25 (100%)	23 (100%)	43 (100%)	46 (100%)	68 (100%)
Died between 1958–62 and 1977	4 (18%)	14 (56%)	7 (30%)	17 (40%)	11 (24%)	31 (46%)
Moved between 1958–62 and 1977	2 (9%)	2 (8%)	1 (4%)	2 (5%)	3 (7%)	4 (6%)
Unknown what happened between 1958–62 and 1977	—	2 (8%)	3 (13%)	6 (14%)	3 (7%)	8 (12%)
Living in Etéñitépa in 1977	17 (74%)	7 (28%)	12 (52%)	18 (42%)	29 (63%)	25 (37%)

Source: Data from Maybury-Lewis 1967, 317–42; and Flowers 1983a, 174.

married, three of them polygynously. There was such an "abundance" of young Poridza'õno Wamãri girls for Öwawẽ young men to marry that four young men belonging to this clan had recently moved from other villages to Etéñitépa and married there.

Due to the scarcity of spouses, in the mid-1970s young Wamãri men were growing restive. Several of them realized that they would have to wait a long time to marry because of the lack of potential spouses or because they were betrothed to very young girls. Although the prospects of getting married soon were remote for a number of Wamãri men, none of them left the village to seek wives and only two "incestuous" marriages (as defined by rules of clan exogamy) had taken place. One of the sons of the powerful Apöwẽ, leader of the village and the Wamãri lineage, for a time considered marrying a Guarani Indian girl, the daughter of the foreman on a neighboring ranch. He gave up the idea when the girl's parents said he must leave the village if he did so and be married by the Church (Flowers 1983a, 175–79).

In the 1970s Flowers predicted that the trend would be toward an increase in the size of the Öwawẽ, as men from lineages of this clan were enjoying more access to women. She wrote: "The young Öwawẽ men were married to women at their fertility peak, and clearly were fathering more children than Wamari of their age" (1983a, 179). The comparison of 1976–77 and 1990 census data shows that, indeed, over this period the relative size of the Poridza'õno Wamãri decreased from 55 to 48 percent of the population (table 5.11). The increased presence of non-Wamãri is mainly due to the larger number of children under age fifteen, who were born from 1975 to 1990 and are the offspring of Öwawẽ men. From the 1970s to the 1990s, the percentage of the non-Wamãri children below age fifteen grew from 44 to 58 percent. This example shows that, due to the mechanism built into the kinship structure, as long as the Xavánte

TABLE 5.11. Distribution of the Xavánte Population according to Lineage Affiliation in 1976–77 and 1990

	1977–76			1990		
Age (years)	Poridza'õno Wamãri	All Other Lineages	Total	Poridza'õno Wamãri	All Other Lineages	Total
0–14	74 (56%)	59 (44%)	133	104 (42%)	144 (58%)	248
15–29	31 (51%)	30 (49%)	61	57 (52%)	53 (48%)	110
30+	31 (56%)	24 (44%)	55	60 (58%)	43 (42%)	103
Total	136 (55%)	113 (45%)	249	221 (48%)	240 (52%)	461

avoid "incest" it is impossible for a lineage-based faction to grow indefinitely and not face some adverse consequences stemming from its own political domination and demographic hegemony.

Conclusion

In this chapter we have seen that the Etéñitépa Xavánte suffered a severe demographic crisis in the decades that followed contact with the Brazilian national society in the 1940s. This crisis was the result of both higher mortality and a drop in fertility, which, for a time in the 1960s, threatened the biological survival of the group. Starting in the 1970s mortality dropped and the population began a period of rapid growth. The Etéñitépa Xavánte population nearly doubled between 1977 and 1990.

The demographic crisis of the Xavánte, which we described in detail, may be not too different from those undergone by hundreds of other New World indigenous populations since the sixteenth century, crises that resulted in demographic collapse on two continents (Cook 1998; Crosby 1972; Denevan 1992; Dobyns 1983; Hemming 1978; Larsen and Milner 1994; Thornton 1987; Verano and Ubelaker 1992; Whitehead 1988, 1993). The literature on lowland South American indigenous populations is full of accounts that report the occurrence of postcontact crisis. For many the crisis was so severe that the population decline became irreversible and resulted in biological extinction. However, this crisis has been documented, using demographic data, only for a small number of groups (Black et al. 1978; Early and Peters 1990; Hill and Hurtado 1996; Price 1994; Werner 1983).

In this chapter we have tried to show that in the Xavánte case the consequences of the demographic crisis were mediated by factors linked to the social organization of the group. During the period of epidemics, sorcery accusations became more frequent, exacerbating intragroup violence. Ethnographic evidence, as well as our demographic data, shows that this violence affected in particular men of the factions that were politically weaker. The Xavánte example shows that, although demographic crisis is almost a universal experience in the history of Amazonian indigenous groups, its effects may vary in accordance with specific social and cultural characteristics of the society undergoing the crisis. In addition, the case of the Etéñitépa Xavánte indicates that the impacts of epidemics involve more than population reduction, also influencing social dynamics, such as marriage practices, even decades after the crisis itself.

Chapter 6

Subsistence, Ecology, and the Development Trap

If one reads reports by scientists, journalists, and other travelers who crossed Central Brazil in the first half of the twentieth century it is not uncommon to come across descriptions of the *cerrado* as an extremely inhospitable environment for human life. The engineer Manuel Rodrigues Ferreira, a member of the Instituto Histórico e Geográfico de São Paulo, who took part in the Roncador-Xingu Expedition in the 1940s, wrote:

> These Indians live in a region which is hostile to animal life, since the soil is not fertile, watercourses are few, and the vegetation consists predominantly of wizened scrub. These aborigines, therefore, are subjected to all the harshness of the environment. There are no fish, and birds are rare. . . . To survive, the Xavánte struggle against the hostile nature that surrounds them. (Ferreira 1946, 72–73)

Anthropological accounts also characterized the *cerrado* as a difficult place to make a living, describing the regions inhabited by Jê-speaking peoples as "relatively arid and unproductive uplands" (Steward and Faron 1959, 363) or an "infertile plateau of sandy soil covered by scrubby xerophytic growth" (430).

In-depth anthropological research carried out among the Xavánte since the late 1950s has produced a body of information describing how indigenous peoples of Central Brazil relate to the *cerrado* in a way that sharply contrasts with descriptions such as these. According to Maybury-Lewis,

> The way in which Shavante make use of a seemingly unpromising environment to supply their needs is a feature of their life which strikes an outside observer most forcibly. . . . Shavante life was so well adapted to their environment that as late as 1958 a visitor got

> an impression of abundance and efficiency in their villages, which was in striking contrast to the feeling of poverty and inadequacy conveyed by Brazilian settlements in Central Brazil (1967, 61)

Early descriptions of the *cerrado* as unproductive have changed in recent decades as more research has been carried out. Actually, it was not until the 1970s that the ecology of Brazilian *cerrado* landscapes really started to become better known, as an increasing number of research projects in the fields of botany, zoology, and agronomy were initiated.[1] The picture that has emerged from this large array of research is the *cerrado* as a diversified environment rich in both plant and animal species.

Resources and Subsistence

The Xavánte have exploited the *cerrado* in a myriad of ways. Their traditional subsistence may be characterized as a combination of hunting, gathering, and horticulture. Emphasis placed on each one of these practices has changed over time as the Xavánte have experienced different historical and sociopolitical conditions. At present, although substantial modifications have taken place, the Etéñitépa group still largely relies on the *cerrado,* so much so that early ethnographic descriptions of subsistence practices, such as those of Giaccaria and Heide (1984) and Maybury-Lewis (1967), might sound remarkably contemporaneous to someone coming to their village today. The use of faunal and floral resources from the *cerrado* by the Etéñitépa community is probably more intensive than that of any other Xavánte group. In part this stems from the fact that Pimentel Barbosa is the largest Xavánte reservation (table 2.1), the one with the least degraded vegetation, and the one enjoying the lowest population density.

Gathering

According to Maybury-Lewis (1967, 43),

> without hunting, Shavante culture would have been very different; but without gathering, the Shavante could not have existed at all. In 1958 the São Domingos Shavante did not eat meat every day and even went without meat for a number of days at a stretch when they

> were too busy to hunt. A day never went by, however, when the wild products of the region were not available in the community.

This statement clearly illustrates the importance in the Xavánte diet of gathered wild products, especially roots and tubers.

This pattern has changed dramatically over the decades. In spite of this, the *cerrado* flora continues to serve as a vital source of diversity in Etéñitépa Xavánte subsistence. Palm nuts and hearts of palm are highly appreciated and often, though not always, eaten fresh. Hearts of palm may be dried in the sun and pounded into flour, which is boiled to make a kind of thick soup considered suitable food for small children, or made into loaves, which are baked, like maize loaves, in an earthen oven (Flowers 1983a, 238–39). Major species of palms that may provide juicy hearts and/or nuts are *butiá* (*Butia* sp.), *indaiá* (*Attalea exigua*), *babaçu* (*Orbignya phalerata*), *tucum* (*Astrocaryum* sp.), *macaúba* (*Acrocomia aculeata*), and species of *Syagrus, Acanthococcus,* and *Bactris* (table 6.1). The yellowish or bright orange mesocarp of *tucum* and *macaúba* may be eaten raw. In order to reach the oleaginous endosperm, however, the hard-shelled nuts are either boiled in water or roasted directly on the fire. Ripe kernels may also be cracked open and their tasty endosperm eaten raw. The fruits of the *buriti* palm (*Mauritia flexuosa*) are highly appreciated. These fruits are the size of a peach and are covered with small dark red scales, under which is a bright orange, strong-flavored, oily mesocarp. This is eaten raw, after removal of the scales.

While various products from palm trees are available during most of the year, a number of other plant resources included in the Xavánte diet are available only in season (table 6.1). Important seasonal fruits include *cajuí* or *cajú-do-campo* (*Anacardium humile*), *araticum* (*Annona crassiflora*), *murici* (*Byrsonima* sp.), *mangaba* (*Hancornia speciosa*), *cagaito* (*Eugenia dysenterica*), and *jatobá* (*Hymenaea stignocarpa*). *Piqui* (*Caryocar brasiliense*) is an apple-sized fruit with an oil-rich yellow pulp that is often gathered and consumed by the Xavánte when it ripens. The inner surface of the seeds is covered with small thorns that may be very treacherous to those unaccustomed to eating it. *Piqui* is an important source of vitamin A, niacin, thiamin, and riboflavin (Araújo 1995; Handro and Barradas 1971).

In addition to palm products and fruit, a large variety of edible roots and tubers are relished by the Xavánte. The distinctive underground plant organs (roots, tubers, and rhizomes) characteristic of many

TABLE 6.1. The Most Common Wild *Cerrado* Fruits Collected by the Xavánte

Plant Family and Species	Name of Fruit in Portuguese[a]	Fruit Productivity[b,c] (per tree)	Months of Fruiting[b,d] J	F	M	A	M	J	J	A	S	O	N	D
Anacardiaceae														
Anacardium humile Mart.	*cajuí*	60–600									**	**	**	**
Annonaceae														
Annona crassiflora Mart.	*araticum*	50–200		**	**	**								
Apocynaceae														
Hancornia speciosa Gomez	*mangaba*	100–500										**	**	**
Arecaceae														
Acrocomia aculeata (Jacq.) Lodd	*macaúba*	250–1500	**								**	**	**	**
Attalea exigua Drude	*catolé, indaiá*	60–120	**						**	**	**	**	**	**
Butia sp.	*coquinho azedo*	1,600–2,000									**	**	**	**
Mauritia flexuosa Linn.f.	*buriti*	2,000–6,000	**	**	**	**	**	**						**
Syagrus sp.	*gueroba, aricuri*	750–2,000							**	**	**	**		
Caryocaraceae														
Caryocar brasiliense Camb.	*piqui*	500–2,000	**	**									**	**
Hippocrateaceae														
Peritassa campestris (Cambess.) Smith	*saputá*	not available							**	**	**			
Salacia crassifolia (Mart.) Peyr.	*bacupari*	not available									**	**	**	**

Leguminosae														
Dipteryx alata Vog.	*baru*	1,000–3,000	**	**	**									
Hymenaea stignocarpa Mart. ex Hayne	*jatobá*	200–2,000				**	**	**	**					
Inga alba (Sw.) Willd.	*ingá*	not available							**	**	**			
Malpighiaceae														
Byrsonima spp.	*murici*	100–500									**	**	**	
Moraceae														
Brosimum gaudichaudii Trec.	*mama-cadela*	60–400								**	**	**	**	**
Myrtaceae														
Campomanesia pubescens (DC, O.Berg)	*gabiroba*	30–60									**	**	**	
Eugenia calycina Camb.	*pitanga*	500–2,000									**	**	**	**
Eugenia dysenterica DC.	*cagaito*	500–2,000									**	**		
Psidium firmum Berg.	*araçá*	not available										**	**	**
Rubiaceae														
Alibertia sessilis (Vell) K. Schum.	*marmelo*	100–200		**	**	**	**							
Sapotaceae														
Pouteria ramiflora (Mart.) Radlk.	*curriola*	200–800	**	**								**	**	**
Sterculiaceae														
Guazuma ulmifolia Lamb.	*mutamba*	1,000–3,000								**	**	**		

[a]There are no English names for most of these fruits.

[b]Based on Almeida 1998 and Almeida et al. 1998.

[c]Fruit productivity is influenced by the size of the tree, rainfall, and local soil conditions.

[d]The fruit season in the *cerrado* is highly dependent on the beginning of the seasonal rains. A longer dry season will postpone fruiting and reduce productivity due to its effects on flowering.

cerrado species enable them to better withstand the dry winter months and occasional burning by accumulating large quantities of starch in their enlarged roots or tubers during the rainy season (Eiten 1972). Giaccaria and Heide (1984, 79) list fourteen different kinds of wild roots that the Xavánte distinguish as edible. According to these authors, "two or three varieties of these roots are gathered and also cultivated; the small tubers are kept in baskets hung on the posts of the hut, or placed outside on poles until planting time" (79). In the appendix of his ethnography, Maybury-Lewis (1967, 334) lists the Xavánte names for seven kinds of commonly eaten *cerrado* roots. Unfortunately, neither of these authors obtained the precise botanical identification of these plants.

The Xavánte also collect a great variety of plants used for other purposes: medicine, house building, making artifacts such as mats and baskets, and manufacturing ceremonial regalia. In the 1990s, the Etéñitépa Xavánte continued to make ample use in their daily lives of articles made from raw materials gathered in the *cerrado*. The *buriti* palm is particularly important in providing fiber for baskets and mats (fig. 6.1). Xavánte women collect bunches of the young, central leaves of *buriti,* detach the segments, and let them dry in the sun. The Xavánte beat the dry leaf segments and shred them into fiber. They roll the fibers between their hands to make cordage. Bunches of *buriti* fiber, known to local Brazilians as *seda-de-buriti,* are often seen hanging from poles inside houses. Xavánte women make larger, carrying baskets by weaving the segments of the leaves of *buritirana* (*Mauritiella armata*). These baskets are taken to the gardens to carry harvested crops back to the village, and they are taken on gathering trips. In the houses they are hung on the wall to hold clothing and other household possessions. The largest are used as cradles in which babies sleep and can be carried around. When hunters butcher game they make "one-way" baskets on the spot to bring the meat back to the village. Large sleeping mats and smaller mats to hold cooked food are also woven from *buriti* fronds. Xavánte houses are roofed with *buriti* fronds combined with those of *babaçu* and other palms.

The Xavánte distinguish more than ten varieties of honey collected from hives found in the *cerrado* and gallery forest (Giaccaria and Heide 1984, 80; Maybury-Lewis 1967, 39). The direct use of insects for food has been reported by a few authors. Maybury-Lewis (39–40) and Giaccaria and Heide (1984, 80) refer to the consumption of bee larvae, winged leaf-cutting ants (*Atta* sp.), locusts, and certain beetles. Unfortu-

Fig. 6.1. Woman weaving a sleeping mat. The Xavánte do not use hammocks but sleep on large mats woven of *buriti* palm fiber, which are laid on sleeping platforms over thick layers of palm leaves. (Photo by N. M. Flowers, 1977.)

nately, other than these very general references, not much is known about how the Xavánte use insects in their diet.

Gathering is mainly, but not exclusively, a female activity. Xavánte women are responsible for most of the gathering of food items to be consumed in the household as well as fibers used in the weaving of baskets and other utensils. Men also engage in gatherings, as, for instance, when large quantities of palm fronds, poles, and wooden sticks are needed in order to repair a house or build a new one. On some occasions men, women, and children may go out together to collect larger amounts of tubers, hearts of palm, and/or wild fruits to be brought back to the village.

The Xavánte in the 1950s spent several months of the year on trek, splitting up into several smaller groups (Maybury-Lewis 1967, 59). However, they maintained a base village where they assembled at planting time and after harvest. Xavánte treks were far from being haphazard. Treks were planned and discussed in the men's council to take advantage of several resources, not all of which were foods. According to Maybury-Lewis, "It is not unusual for a community to split into bands that travel in different directions but come together again after a comparatively short interval. . . . It might be said that the territory of any particular community was that area which it was able to exploit in the course of a year's wanderings" (53).

At present, the Xavánte continue to trek, mostly during the dry season, but in ways that depart from those of previous times (Flowers et al. 1998). These changes relate not only to the fact that their land is limited by reservation boundaries but to their sedentism in a permanent village that is linked to a post maintained by FUNAI. Most gathering is now carried out within a range of 10 to 15 km from the village. In contrast to the past, when the majority of the community would go out on trek at certain periods of the year and spend weeks or months away from the base village, nowadays only a few families leave at a given time and usually for a much shorter period. Families often go to specific areas of the reservation, such as that known as Corixão or the shores and lagoons of the Rio das Mortes (see fig. 2.2). Although trekking does not appear to be central to Xavánte subsistence at present, it remains a family activity considered to be productive and enjoyable.

Hunting

Maybury-Lewis (1967, 35) pointed out that, although hunting might not be economically as important as gathering, "meat far and away transcends other forms of food in the Shavante esteem." This certainly continues to be true in the present. Several decades after permanent contact, hunting remains a central activity for the Etéñitépa Xavánte.

Peccaries are the Xavánte's favorite game. Both collared and white-lipped peccaries are found, mostly in gallery forests and marshy areas,

Fig. 6.2. Hunter who has just shot a tapir. The tapir is the largest game animal that the Xavánte hunt. On the same day four hunters shot three tapirs and two peccaries, which weighed a total of around 800 kg. (Photo by R. V. Santos, 1995.)

where they may gather in mud wallows. Groups of collared peccaries may contain fifteen to twenty individuals, though usually fewer. White-lipped peccary herds are much larger, numbering fifty individuals or more. Tapir and deer are also highly prized game (fig. 6.2). Tapirs inhabit riverine and marshy environments and stand out as the largest herbivorous ungulates in the neo-Tropics, weighing up to 250 kg. The brocket deer is mostly a gallery forest dweller, although it may be seen in areas of *cerrado* bordering the forest. The much larger marsh deer prefers open woodlands and areas of secondary growth. Table 6.2 lists the most important game species for the Xavánte and the average adult weight of the animals.

Anteaters, armadillos, and various species of midsize rodents are also often hunted for food. While the large anteater and most armadillos are easily spotted in the *cerrado,* midsize rodents like the paca are typical gallery forest dwellers and are more difficult to locate, since they have nocturnal habits and spend the daylight hours in burrows on the banks of rivers or in old tree roots. Xavánte go out at night with flashlights to shoot pacas, which often raid the gardens. Coatis, the only carnivores considered edible by the Xavánte, prefer areas of closed canopy, where they may be found in groups. Agoutis are much hunted for meat and are very widespread, being found throughout the forest and in areas with heavy undergrowth cover, like abandoned gardens. According to Leeuwenberg (1995), smaller mammals that may be occasionally killed for food include the *cuí* (*Galea* sp. and *Cavia* sp.), the *tapiti* or Brazilian rabbit (*Sylvilagus brasiliensis*), and opossums (*Didelphis* sp.).[2]

Despite its abundance in all waterside habitats and its weight (35 to 65 kg), the *capybara* (*Hydrochaeris hydrochaeris*) is not hunted for food.

TABLE 6.2. Mammal Species Most Often Hunted by the Xavánte

Species	Portuguese/English Names	Adult Weight (kg)
Tapirus terrestris	*anta*/tapir	220–250
Tayassu pecari	*queixada*/white-lipped peccary	25–45
Tayassu tajacu	*catitu*/collared peccary	17–35
Blastocerus dichotomus	*veado-galheiro*/marsh deer	30–50
Mazama americana	*veado-mateiro*/red-brocket deer	24–48
Mazama gouazoubira	*veado-catingueiro*/gray-brocket deer	11–25
Myrmecophaga tridactyla	*tamanduá-bandeira*/giant anteater	20–40
Tamandua tetradactyla	*tamanduá-mirim*/southern anteater	4–8
Priodontes maximus	*tatu-canastra*/giant armadillo	25–30
Dasypus novemcinctus	*tatu-galinha*/nine-banded armadillo	3–6
Euphractus sexcinctus	*tatu-peba*/yellow armadillo	3–6
Cabassous unicinctus	*tatu-rabo-mole*/naked-tailed armadillo	2–5
Dasypus septemcinctus	*tatuí*/seven-banded armadillo	1–2
Agouti paca	*paca*/paca	5–13
Dasyprocta agouti and D. azarae	*cutia*/agouti	3–6
Nasua nasua	*quati*/coati	3–7

Although different species of monkeys (*Cebus* sp., *Alouatta* sp., *Callicebus* sp.) and marmosets (*Callithrix* sp.) may be found in the gallery forests, at present the Xavánte do not hunt them for food.[3]

Although the Xavánte prefer to hunt in the open *cerrado,* gallery forests are source areas for species of midsize birds that the Xavánte may opportunistically hunt. Often the Xavánte shoot these birds on the way to their gardens. Birds found in the gallery forest include species of curassow (*Mitu mitu* and *Crax* sp.), *macuco* (*Crypturellus* sp.), and cuan (*Penelope* sp. and *Pipile* sp.). Various smaller birds, such as the tinamou (*Tinamus* sp.) and pigeons (*Columba* sp.), may be found in both gallery forest and *cerrado.* The muscovy duck (*Cairina moschata*) frequents lakes and river edges. The larger *ema* (*Rhea americana*) is hunted in the open *cerrado,* where it runs in groups of fifteen to twenty individuals. Another *cerrado* bird that may be hunted for food is the *seriema* (*Cariama cristata*). Parrots (*Amazona* sp.), macaws (*Ara* sp.), and toucans (*Ramphastus* sp.) may be used as food after the feathers have been carefully collected for use in ornaments. Hawks are also hunted but only for their feathers.

A few reptiles are eaten by the Xavánte but only occasionally. Among these are the large, lizardlike *teiú* (*Tupinambis teguixin*) and the tortoise (*Geochelone* sp.), which may be found in a wide range of environments, including gallery forest, open scrub, or grassland. River turtles (*Podocnemis* sp.) are caught along the main channel of the Rio das Mortes and in its many lagoons. At the end of the dry season Xavánte families may camp on the fine white beaches of the Rio das Mortes to gather turtle eggs. Caimans and snakes are not considered edible.

Although Etéñitépa Xavánte men still manufacture bows, arrows, and clubs, and most adults are skillful in using them, nowadays most hunting done is with shotguns and rifles. According to Maybury-Lewis (1967, 37), the Xavánte had already adopted firearms as their preferred hunting weapons in the late 1950s. Adult men continue to esteem highly their ceremonial clubs, which have handles beautifully decorated with basketwork and macaw feathers.

The Xavánte engage in both individual and collective hunting. The former is carried out on the initiative of any adult man and is practiced year round; the latter is a communal enterprise, decided on by the men's council on the behalf of the community, and it is particularly common during the dry season. In the mid 1970s Flowers made observations about hunting productivity and found out that, although individual hunting is the most common form, it is not the most productive. Only 21 percent of such hunts were successful, and the average share of dressed meat per hunter was 1.7 kg (Flowers 1983a, 232). On the other hand, the success rate of collective hunts (four or more men) was 67 percent, and the average share of each hunter per day was 4.7 kg in dressed meat. One of the reasons for this may be that the principal species captured, in numbers, is the white-lipped peccary. These are highly mobile and unpredictable mammals that tend to run in relatively large herds, often numbering more than fifty individuals. Thus "hunters might search [for peccaries] for days without locating a herd or, on a lucky day, kill enough animals to provide all their families with enough meat for several days of feasting" (59). If an individual hunter comes across a herd of peccaries he may kill one or two animals, but with a number of hunters present some can head off the herd, giving everyone a chance to shoot more. On two occasions during Flowers's fieldwork in 1976–77, more than twenty white-lipped peccaries were killed by Xavánte hunters, who had spotted large herds roaming close to the village (fig. 6.3). On the basis of his observations in the 1950s, Maybury-Lewis also noted that collective hunts were much more productive than the individual ones: "such [communal] hunts invariably produced huge quantities of meat—enough to feed the hunters while they were out, and to feed the entire community for three or four days after their return" (1967, 42). Meat is preserved by long roasting over a smoky fire and can be eaten for several days in this hot but dry climate (fig. 6.4).

During the dry season, collective hunts typically involve the use of fire to surround game. Driving game with fire is a central feature of Xavánte hunting strategy, and a game drive may yield considerable amounts of meat. This type of collective hunting is considered so effective and is so widely practiced by the Xavánte that Flowers noted that on August nights "the glimmer of fires can be seen far into the distance and a pall of smoke hangs over the village" (1983a, 235). Xavánte hunters return to burned fields, as they know that various mammals are attracted to the salty ashes, and deer become easy prey as they graze on burned ground, attracted by the fresh and tender grass shoots that grow back vigorously.

In 1976–77, Flowers found that fish and game were widely shared. In a small household that she observed there was only one adult man, and nearly 100 percent of the meat consumed was contributed from

Fig. 6.3. Peccaries laid out after a successful hunt. In this case a herd of peccaries ran close to the village, and all the men in the village rushed out with whatever weapons they could lay their hands on. (Photo by N. M. Flowers, 1977.)

Fig. 6.4. Peccary roasting on a barbecue made of sticks over a slow, smoky fire. When the meat is fully cooked it will be blackened with smoke on the outside but tender and appetizing inside. Game prepared in this way remains edible for several days, even in a hot climate. (Photo by N. M. Flowers, 1977.)

some other household. In a larger household 60 percent of the meat consumed was brought in by resident hunters, with the rest coming from some other household. Yet, because of widespread sharing of game, there was little difference between the two households in mean per capita per day consumption of meat and fish. According to Flowers:

> Networks for informal exchange of garden foods seem to be much more restricted than those through which game and fish are distributed throughout the village. . . . the larger household that I monitored shared [garden] food almost exclusively among themselves. On only two occasions did a visitor eat rice, and no one brought a gift of vegetable food. (1983a, 253)

A special type of collective hunt is the so-called wedding hunt (*dabatsa*). With a number of his age mates, the young Xavánte man who is about to marry goes on a hunting expedition, which may last several days, and only returns when an impressive amount of smoked and dried meat has been accumulated. Upon the return of the hunters, they deposit the chunks of smoke-blackened meat at the doorway of the boy's in-laws, from where the bride's mother's brother distributes the meat to the entire village.

At present the community truck plays an important role in Etéñitépa Xavánte hunting, as it facilitates the access of groups of hunters to various sites located at a distance from the base village. The availability of the truck allows for same day transportation of larger amounts of raw meat and whole game, much more than hunters could carry on their backs. In other words, with motor transportation Etéñitépa men spend less time away and are often able to bring back a larger amount of meat per hunter. On one occasion in February 1995 a group of three Xavánte men who had borrowed our pickup truck left the village in the morning and returned from Rio das Mortes in the late afternoon bringing back with them three tapirs and two collared peccaries, totaling some 800 to 850 kg of meat.

In the past the Xavánte would spend several months of the year on trek, moving over a wide stretch of land (Flowers 1983a, 227; Maybury-Lewis 1967, 53–59). This pattern probably diminished hunting pressure on animal populations, since, over the year, they visited a number of hunting zones relatively distant from one another.

Whether Xavánte culture traditionally included ideological and behavioral traits that might be interpreted as favoring environmental conservation is difficult to determine (see Alvard 1993, 1995). On the basis of present data we are unable to say whether they consistently practice the kind of "natural conservationism" that would involve strategies consciously intended to ensure long-term environmental sustainability, even when costly to individual hunters. One way or another, their highly mobile foraging pattern, hunting techniques, and low population density probably prevented them in the past from overexploiting environmental resources. As pointed out by Alvard (1995, 790), "a small group of hunters inhabiting a generous environment can hunt sustainably in spite of the absence of any proactive conservation on their part; this type of apparent conservation has been termed epiphenomenal."

At present, overhunting is already a major concern on some Xavánte reservations, on the smaller ones in particular. Graham (1995, 42) mentions that, at São Marcos, "to find sufficient game, hunting parties . . . often must trespass onto private lands beyond the reserve's boundaries.

This trespassing has, in some cases, led to clashes with neighboring landholders" (see also Menezes 1984, 400–408).

The Etéñitépa Xavánte are certainly not facing such a drastic situation. However, a recent investigation of hunting productivity carried out at Pimentel Barbosa has shown that, while Xavánte living on this reservation still harvest a relatively large quantity of meat from game, the possibility exists of overexploitation of some of their favorite game animals (Leeuwenberg 1995, 1997; Leeuwenberg and Robinson 2000). A study of hunting productivity at Pimentel Barbosa by the biologist Frans Leeuwenberg showed that between February 1991 and October 1993 a total of 1,703 mammals were killed by Etéñitépa hunters, including peccaries (48 percent), anteaters (20 percent), deer (14 percent), armadillos (11 percent), tapirs (4 percent), and others (2 percent) (1995, 12). Rough estimations of total meat from game in 1991, 1992, and 1993 were, respectively, 8,872, 8,307, and 15,886 kg (18).[4] Over the three years nearly 60 percent of the meat came from peccaries (30 percent) and deer (29 percent), while anteaters and tapirs accounted for most of the remainder. Leeuwenberg and Robinson (2000, 390) conclude that "a number of wildlife species important to the Xavánte might be either at low densities as a result of past hunting or their densities are declining. . . . The only two species for which there is no indication of low or declining densities are the two species of peccaries." These authors also argue that abandonment of trekking and intensified use of fire for hunting are two important factors affecting hunting sustainability at Pimentel Barbosa in recent years.

Fishing

Etéñitépa is located near a stream where only small fish are caught. More intensive fishing involves a 30 to 40 km trip to the Rio das Mortes or to one of its tributaries or lagoons. It appears that fishing became more important in Xavánte subsistence only after they acquired their first hooks and lines from Indian protection service agents back in the 1940s (Maybury-Lewis 1967, 51–52).

Xavánte fishing is seasonal, preferably taking place when the water level in the rivers and streams is falling (March–April) or during the dry season when the water level is low. One of the most prized species for the quality of its flesh is the *tucunaré* (*Cichla ocellaris,* Cichlidae), which can be found in lakes and rivers.[5] Other common cichlids include various kinds of *cará* (*Cichlasoma* sp. and *Astronotus* sp.) and *jacundá* (*Crenicichla* sp.). Several species of large neotropical catfish (Pimelodids), found only in deep rivers, are hard fighters, and the fisherman who lands one has a major prize. The most important species in this group, such as the large *surubim* (*Pseudoplatystoma fasciatum*), *piraíba* (*Brachyplatystoma* sp.), and *pirarara* (*Phractocephalus hemiliopterus*), can weigh 40 to 50 kg. Also among the pimelodids is the smaller *mandi* (*Pimelodus* sp.). The characids comprise a large number of species that the Xavánte often catch. They include various species of piranha (*Serrasalmus* sp.), *sardinha* (*Triportheus* sp.), *pirapitinga* (*Colossoma bidens*), *pacu* (*Mylossoma* sp.), and *matrinchão* (*Brycon* sp.). Other species of fish commonly caught by the Xavánte are *curimatã* (*Prochilodus nigricans,* Prochilodontidae), *pescada* (*Plagioscon* sp., Sciaenidae), *piau* (*Leporinus* sp., Anostomidae), *peixe-cachorra* (*Cynodon gibbus* and *Hydrolycus* sp., Cynodontidae), and *traíra* (*Hoplias* sp., Erythrinidae).

Fishing in the smaller streams near Etéñitépa is casual. A group of women may go off together for a day's fishing, taking their small children with them. Small boys also organize fishing parties of their own at nearby streams, bringing their day's catch of smaller fish to be roasted in the embers of the household cooking fire.

For more intensive fishing, the Xavánte travel to the margins of the Rio das Mortes, located some 40 km from Etéñitépa (fig. 6.5). The composition of parties that engage in longer fishing trips is varied and may include whole families. While in the past family groups would walk the stretch that separates the village from more productive fishing sites, nowadays this trip is usually made by truck. It should be noted that such trips are not specialized in the sense that fishing will be their only business. Men will occasionally come across other animals, a large mammal or bird, and will not miss the opportunity to add meat to their fish basket. At the same time women and children may be busy on the beach, digging up turtle eggs or collecting large quantities of palm nuts or some seasonally ripe fruit. Often these fishing trips will last a week or more. In areas where there is relatively shallow running water the Xavánte may use the fish-poison vine locally known to Brazilians as *tinguí* (*Jacquinia* sp.; Maybury-Lewis 1967, 51).

The catch is not necessarily brought to the village but may be eaten at the camp. Only when caught in large amounts will fish be placed on a framework of sticks over a smoky fire to be preserved by heat and smoke and brought back to the village. Smaller fish tend to be eaten on

Fig. 6.5. Xavánte fisherman showing his catch of *tucunaré* at the Rio das Mortes. Fishing became more important for the Xavánte after they acquired fishhooks and lines in the 1950s, and at present it is facilitated by a road across the reservation, which allows them to go to the river by truck for a day's fishing. (Photo by C. E. A. Coimbra Jr., 1996.)

the spot, as the Xavánte quickly make a fire and simply place the fish on the embers. When they return, members of longer fishing trips distribute their catch of smoked fish throughout the village. On one occasion, Flowers observed that a group of boys from the bachelor's hut who had gone on a joint fishing trip brought back a large amount of smoked fish. They then "offered a ceremonial meal to the older men, during which several of the elders made speeches praising the boys for their civic endeavor" (1983a, 238).

Agriculture

Although the Xavánte, as well as other Jê-speaking groups of Central Brazil, have been depicted in the ethnographic literature as "incipient agriculturalists," there is little reason to believe that they are hunter-gatherers only recently converted to the use of cultivation. One indication that Xavánte have been cultivating for a long time is the presence of varieties of "interlocked" soft-corn. This is considered an archaic variety of maize, which may indicate an independent trend in the early stages of maize domestication (Brieger et al. 1958). Their maize varies in kernel color from white and yellow to bright red to purplish black, with mixtures of these colors. The Xavánte esteem this variation highly, especially the bright red variety, carefully saving a stock of seed from each color variety in separate gourds stopped with beeswax (Flowers 1983a, 1983b). To keep these varieties of color and shape requires an understanding of plant selection and cultivation techniques acquired over a long period of time.

There are several examples that show that Jê groups are not as unskilled in horticulture as they were described in previous literature (Steward 1949; Steward and Faron 1959). Some indications of experimentation with plant domestication in Central Brazil are, for instance, the Nambikwára's pineapples and peanuts (Hoehne 1940; Rondon 1946) and the unique vinelike plant, *Cissus gongyloides,* cultivated by the Timbíra, Xerénte, and Kayapó (Kerr et al. 1978; Kerr and Posey 1984; Nimuendajú 1946). Other ethnobotanical studies carried out among indigenous groups of Central Brazil and southern Amazonia have revealed that a large number of species are gathered in the wild and "domesticated" by planting them, for instance, within areas of secondary growth in abandoned gardens (Baleé 1994; Frikel 1978; Posey 1988).

In the past cultivated crops provided the Xavánte with the bulk of their food supply only at specific times during the year. As noted by Maybury-Lewis (1967, 44), in 1958, after having gathered the maize harvest and eating part of it, in March the Xavánte went on trek, subsisting on wild tubers and palm shoots, while they saved the rest of the maize for the middle of the dry season (June–July) when they assembled in the village. Upon their return to the base village, the harvest of beans and pumpkins was also available. According to Maybury-Lewis (47), these were the only crops the Xavánte grew at this time.

The Xavánte plant their gardens in clearings made in the gallery forest where soil conditions and the availability of water allow for the cultivation of subsistence crops. Open *cerrado* per se is never used for swidden cultivation. Xavánte men clear garden plots by cutting down the forest in the early months of the dry season, thus allowing enough time for the felled trees and foliage to dry thoroughly and burn properly before the first rains, which are expected to fall in mid-September. They use only steel axes and large bush knives. They burn the dry vegetation, leaving the stumps standing and the larger tree trunks lying on the ground. A couple of days after the fire has subsided, men return to the fields to gather up unburned branches and sticks. These are brought together in piles and set on fire a second time. Ashes from the burned plant material will provide the essential nutrients needed for the proper growth of their crops.

Both men and women plant, but women do most of the weeding. Clearing and planting are busy times for the Xavánte, so busy that families may stay overnight in their gardens. Gardens are planted, weeded, and harvested by individual families, and the produce for subsistence or sale is regarded as belonging to the family that clears and plants the garden. According to Graham (1995, 43–44), "each adult woman has one or more of her own gardens, which are cleared by her husband and are located adjacent to those of her sisters and mother. Often accompanied by her small children, she makes several trips to the gardens per week." In the 1990s the Xavánte were consuming most, if not all, of the production of their gardens.

The Xavánte plant various crops in separate patches of the gardens. Pumpkins are often grown between rows of maize. Beans are planted in the same plots after the maize harvest. Banana plants usually grow around the perimeter of the garden. Rice is usually planted in separate gardens. Other cultivars grown by the Xavánte are sweet manioc, sweet potatoes, yams, and watermelons. Papayas and manioc are not only grown in the gardens but may be found around the village. Large and leafy mango trees are abundant in the village. Nonfood crops include gourds of different kinds, annatto (*urucum*), and cotton. Rice, though not maize, is sometimes planted in second- and even third-year plots. Bananas and manioc are the main crops grown in gardens after the third year (Flowers 1983a; Flowers et al. 1982).[6]

The maize crop ripens in January and February. Rice harvesting is an activity that continues for several months, for the Xavánte plant rice in different gardens from early October until December. Men do the cutting

Fig. 6.6. Man in his rice field. He is piling a stack of harvested rice for storage. Rice, which is an introduced crop, became an important staple in the Xavánte diet in the 1970s. (Photo by N. M. Flowers, 1977.)

and stacking of rice with some help from the women, while women thresh rice in the fields and bring it to their houses little by little as needed (Flowers 1983a, 1983b; see figs. 6.6 and 6.7). After two to three years Xavánte gardens are left fallow. However, this does not mean that plots are completely abandoned. Manioc, sweet potatoes and yams, for instance, may continue to yield in such fields for more than two years. Papaya and banana trees also continue to bear fruit and even propagate in "abandoned" gardens for a longer period of time. In addition, several colonizing plants may root in abandoned gardens. Among these is the leguminous *ingá* (*Inga* spp.), which bears fruits that are highly appreciated for their sweetness. Other spontaneously colonizing plants may have medicinal importance, yield edible nuts, or constitute a source of fibers to be used in the manufacturing of various utensils.

Domestic Animals

At Etéñitépa, as in other Xavánte villages, many animals that have been captured in the wild are kept as pets for varying lengths of time,

Fig. 6.7. Woman hulling rice in a mortar made from a tree trunk. (Photo by N. M. Flowers, 1977.)

although they cannot be said to be domesticated. Among those that Flowers saw during her stay in the village in 1976–77 were parakeets, parrots, macaws, baby peccaries, a baby ocelot, monkeys, and an *ema.* The only one that seemed to have an affectionate relationship with a family was a pet coati, which followed the women of the household to the fields, played nearby, ate lunch with them, and followed them home again in the evening.

There are numerous dogs, smooth coated, underfed, and bearing a family resemblance to one another. Presumably the Xavánte have had dogs for a long time, probably before the contact of the 1940s. They are kept as watchdogs and for hunting. Although they are not impressive looking, some are esteemed for their bravery in attacking peccaries. In 1976–77 Flowers saw one dog that was being carefully nursed back to health by his master after his back had been lacerated by a peccary's tusks. Nowadays, some chickens are kept that seem to live mostly by hanging around when rice is hulled to catch the few grains that fall.

The reservation has a herd of cattle that formerly belonged to the FUNAI post and now belongs to the Indians. A few head may be sold to finance a community necessity, like getting the truck repaired, or one may be butchered and distributed for meat. Cows' milk is not used at present. The Xavánte take little interest in caring for cattle and have hired a Brazilian cowboy to do the work.

Changes in Subsistence from the 1950s through the 1980s

As we have seen in chapter 3, the "pacification" and subsequent settlement of the Xavánte was a high priority for the Brazilian government in the 1940s. For this reason, the Indian post at São Domingos was a large-scale operation set up in order to attract the Indians and foster their sedentarization, thus bringing them under the control of the state. Agriculture was thought to be the means by which the Xavánte would be persuaded to give up their nomadic ways and become contained within a limited territory. As pointed out by Maybury-Lewis:

> The Indian Protection Service tried very hard to persuade the Shavante to cultivate various other crops, particularly manioc (cassava), rice, and bananas. The reasons for this are easy to see. While the Shavante continued their nomadic existence they made uneconomic use of their lands. They occupied tracts of territory that could

> support a larger farming population. In the course of time it was certain that this area of Mato Grosso would be "opened up" to settlement and that the Shavante would have to face the problem of how to provide for themselves on a fraction of their former territory. If they could be induced to take up farming, it would have the advantage of enabling them to adapt themselves slowly to the inevitable, and would also make the task of administering them very much easier. (1967, 48)

Even after the Xavánte built their base village near the government Indian post at São Domingos in 1956, they continued for several years to plant crops at their old garden sites and to carry harvested garden foods to their new base village. Maybury-Lewis (1967, 48) wrote, "They used to spend only three weeks, or at most a month of the year at their plantations." Basílio da Silva Barros, who was a young worker at the post at that time, told Flowers in 1976–77 that after a few seasons of planting at the gardens of their previous village the Xavánte ceased to plant at all and for about three years depended entirely on hunting, fishing, and gathering, plus handouts of manioc flour and rice from the post. At São Domingos, which was located on the Rio das Mortes, the fishing was excellent, and when the Xavánte acquired fishhooks they took enthusiastically to fishing with hook and line (52).

The late 1950s and early 1960s were times of great change for the Xavánte. Maybury-Lewis found in 1962 that they had become diminished in numbers, and had lost much of their former autonomy (1967, 27–29). At the same time the government was no longer interested in courting the Xavánte, as they had ceased to be a threat to the settlement of the region. As time went on, the Indian service cut back the number of workers at the post, and the flow of presents began to dry up. In the mid-1960s the Indian post was so impoverished that the entire group moved to a site known as Barreira Vermelha, which was on the opposite bank of the Rio das Mortes some distance upstream from São Domingos (see fig. 2.2).

In 1972 the Xavánte moved again, this time to the village they continue to inhabit (Etéñitépa). From observations made during her 1976–77 fieldwork, Flowers wrote that after they moved to Etéñitépa rice became increasingly important in Xavánte subsistence. Although hunting, fishing, and gathering remained basic activities, they now went on trek for only a few weeks during the dry season (1983a, 226).

The Xavánte Rice Project

The Xavánte of Etéñitépa experienced major changes in their ecology and subsistence between the 1970s and the 1990s. These changes were closely associated with socioeconomic and political processes taking place at the national level in Brazil during that period. Programs of regional development were implemented by the Brazilian military government from the mid-1960s on with the intention of integrating into the national economy regions regarded by the state as "relatively isolated," "underpopulated," and "underdeveloped." This policy was largely motivated by the geopolitical concern to fill what were considered to be "empty spaces." A slogan that was widely used in government propaganda of the time was "to integrate [territory] in order not to give it up" (integrar para não entregar). The Amazon was one of the main targets of these policies, which involved highway construction and promotion of migration and colonization projects.[7]

Eastern Mato Grosso, where Xavánte territories are situated, was one important target of this policy (Brandford and Glock 1985; Hees et al. 1987; Menezes 1982; Oliveira 1981). Various colonization programs, financed with private capital, were set up with wide government support in the form of credits and fiscal incentives. As a result of this frontier expansion, the population of Barra do Garças, the most important town in eastern Mato Grosso, increased more than tenfold between the mid-1960s and the end of the 1980s, from 14,000 to 150,000 inhabitants (Menezes 1982, 64). Eastern Mato Grosso, previously devoted to extensive cattle raising and subsistence agriculture, within a few years became an economic center based on large scale production of export crops, initially rice and later soybeans. The occupation of this frontier involved the formation of various agricultural cooperatives, which attracted primarily smallholders from the south of the country who migrated in search of land. There were also investments in large-scale monoculture and cattle ranching. As time passed, in Mato Grosso as in other parts of Amazonia, a process of land concentration took place, and many of the smallholding colonization projects failed. At present, large-scale enterprises dominate the region. The period from the 1970s through the 1980s saw many land disputes, among which were the attempts of the Xavánte to regain some of the land lost to white colonists in the 1950s and 1960s.

Since their territory is situated in an economic and demographic frontier region, Xavánte lands acquired great prominence in the 1970s. The anthropologist Claudia Menezes vividly expresses the degree to which the programs implemented by FUNAI were in tune with national policies.

> FUNAI's efforts . . . have been to align itself with the planning and financial offices, which were transformed by the post-1964 official economic policies into agencies of regional development. . . . This becomes clear through the orientation adopted in relation to the Xavánte, which is characterized by initiatives aiming to hitch the Indians to the dominant regional economic system, in order to turn them into large-scale agricultural producers. (1982, 74)

In the late 1970s FUNAI bureaucrats in Brasília drew up a plan called the Integrated Development Plan for the Xavánte Nation, which became known as the Xavánte Project. Its objective, in line with government policies and the economic development taking place in eastern Mato Grosso, was to turn the Xavánte into large-scale mechanized rice producers.[8] The Xavánte subsistence strategy, which was based on extensive exploitation of the *cerrado* habitat and did not involve surplus production, was held to be incompatible with the economic processes taking place in the region, which were directed toward intensive agriculture producing commodities for the national market and export. Xavánte economy was to be radically transformed into an economy involving massive investments in seeds, fertilizers, pesticides, and machinery with high maintenance costs. The Xavánte Project included plans for improving the infrastructure of the posts by building schools and small infirmaries.

The intention of the Xavánte Project, through the incorporation of the Xavánte into the capital-intensive economy that was beginning to dominate eastern Mato Grosso, was to make the Xavánte economically independent, freeing the federal government from the expense of supporting them. During the 1940s and 1950s, when the Xavánte came into permanent contact, the government saw agriculture as a way to draw the Xavánte into its sphere of influence and open up land for colonization. The intent was to make the Xavánte less mobile in order to exercise more direct control over them. Ironically, in the 1970s and 1980s, also through agriculture, the government planned to lighten the burden that the Xavánte represented, making them "independent." The FUNAI official Col. Ivan Zanoni gave statements to the press expressing the aims of the project in grandiose terms.

> In the near future . . . the rice produced by the Indians will be exported via the Rio das Mortes . . . [and] after the completion of the Tucuruí hydroelectric project and the construction of further dams on the Araguaia river . . . it will be possible to export rice to Belém and foreign countries from a port to be built at Xavantina. (Anonymous 1980)

The Xavánte Project, beyond its strict economic dimension, had clear political aims. The constant presence of Xavánte in Brasília, which is not far from their reservations, lobbying for return of the land they had lost to large landowners in the 1950s and 1960s, created tremendous political pressure. Large contingents of Xavánte would arrive in Brasília, often in paint and feathers, and, carrying clubs and bows and arrows, would burst into the offices of high-ranking FUNAI officials to press their demands. Because Brazil was still under military rule in the 1970s, and all the top officials of FUNAI were colonels appointed by the military government, these actions were daring and attracted the attention of the media and the support of the political opposition. The Xavánte became well known in Brazil for their ability to exert pressure on FUNAI. In this context, the Xavánte Project was seen as a means of placating the Xavánte and persuading them to remain on their reservations (Garfield 1996, 481–553; Ramos 1998, 264–65; Silva 1992, 376–78).

The Xavánte Project was probably more ambitious in scope and scale than any other project designed by the government agency for the economic development of native reservations. Anthropologists who witnessed the various phases of the project on the reservations referred to it as "colossal" (Graham 1995, 44) and involving "very high investments" (Silva 1992, 376). Maybury-Lewis, who visited several Xavánte reservations at a time when the project was in full swing, wrote: "The Shavante Project is undoubtedly the most ambitious development project which has been undertaken in recent years by FUNAI on behalf of any single group of Indians" (1985, 77).

The Xavánte Project was implemented on the different reservations with varied levels of intensity and consequences that were specific to each. The reports of anthropologists who were with the Xavánte in the late 1970s and early 1980s show that the Xavánte were impressed by the project, and enthusiastic about the prospects that it seemed to present. At São Marcos, one of the reservations where the project reached its peak investment in infrastructure, a rice-drying machine was installed with the processing capacity of three hundred sacks per hour and a warehouse was built for seventy thousand sacks of rice. A number of tractors and a harvester were parked near the village (Menezes 1982, 81; see also Menezes 1984 and Queiroz 1980). The use of open *cerrado* land (which the Xavánte traditionally do not use for horticulture) to plant rice, the flow of money and machinery, the coming and going of tractors plowing the land, the large harvesters, the piles of hundreds of bags of rice, the loaded trucks—all these were novel attractions for the Xavánte.

It did not take very long before the project showed signs that it was not going to make the Xavánte economically autonomous. On the contrary, it was evident that, in order to continue to operate, the program would require a constant subsidy from FUNAI. The project did not give the Xavánte the necessary technical preparation to independently manage an enterprise requiring knowledge of finance, marketing, and large-scale mechanized agriculture.

The conclusion that most observers have reached concerning the Xavánte Project is that it was not only an economic disaster but it brought unanticipated social consequences for the communities involved. There was an increase in factionalism, with resulting splits among communities and the appearance of a growing number of relatively small villages (Graham 1987; Graham 1995, 50–55; Silva 1992, 377–78). The way in which the project was set up encouraged emergent leaders to found their own villages with the aim of receiving direct benefits from FUNAI, which might include, among other things, salaries, pickup trucks, and agricultural tools. The number of villages officially recognized by FUNAI jumped from sixteen in 1980 to thirty-five in 1985. Five of these thirty-five had fewer than thirty inhabitants (Graham 1995, 50–55). At Pimentel Barbosa, two new villages (Caçula and Tanguro) appeared during these years.[9]

Subsistence Change: Time Allocation and Food Consumption Data

In 1976–77 Flowers made a detailed study of Etéñitépa Xavánte subsistence. Nearly twenty years later, in 1994, we undertook a restudy of the same community in order to collect information comparable to that obtained in the 1970s. The comparison of these two sets of data provides qualitative and quantitative information that allows us to identify the directions of change. The data reflect the emergence and failure of the Xavánte Project and show how Xavánte subsistence was affected.

Flowers collected time allocation data at Etéñitépa over a period of twelve months. In 1994 a second time allocation study was carried out in the same village, using the same methodology, by Sílvia Gugelmin, a graduate student in public health at ENSP, under the supervision of the senior researchers.[10] To study food consumption in 1976–77, Flowers investigated households at different periods of the year, collecting qualitative and quantitative dietary data. In 1994, Gugelmin collected qualitative food consumption data (Flowers 1983a, 1983b; Gugelmin 1995; Santos et al. 1997).

From analysis of the data, several points stand out. In the first place, it is apparent that, in the 1990s, as in the 1970s, horticulture is the primary subsistence activity and time spent on garden work is greater than on any other single kind of subsistence work. In the second place, the relative importance of subsistence activities remains the same in both periods: for males, the most time is devoted to horticulture, fishing is second, and hunting is third; for females, the rank order is horticulture, gathering, and fishing. The most striking difference between the two periods is that in the 1990s Xavánte adults of both sexes were spending less time on horticulture and more time on hunting, fishing, and gathering than in 1976–77. Men, in fact, have doubled the time they spend on hunting and fishing. Detailed information on time allocated to subsistence is given in table 6.3.

The results by season of time allocation for horticultural activities versus hunting, fishing, and gathering are shown in figure 6.8. Data are presented for the sexes combined due to small sample size. During the wet season, when gardening work is most intense, the Xavánte were spending more time on hunting, fishing, and gathering in 1994 than in 1976–77. This pattern also holds for the dry season.

The results also show that in the 1990s Xavánte men and women were spending less time on subsistence activities of all kinds than they were in the 1970s and devoting more time to such "personal" activities as resting, sleeping, and conversation (table 6.4). The data also reveal well-defined gender differences. In the 1990s as in the 1970s, men spent more time off the reservation than women did, while women spent more time on household-related tasks. However, the data also show

TABLE 6.3. Time Allocation Observations of Subsistence Activities by Etéñitépa Xavánte Adults, in 1976–77 and 1994, according to Sex

	Men		Women		Both Sexes	
Activities	*N*	%	*N*	%	*N*	%
1976–77						
Hunting	7	14.0	—	0.0	7	7.5
Fishing	8	16.0	3	7.0	11	12.0
Gathering	—	0.0	7	16.3	7	7.5
Horticulture	35	70.0	33	76.7	68	73.1
Total	50	100.0	43	100.0	93	100.0
1994						
Hunting	9	25.7	—	0.0	9	14.8
Fishing	12	34.3	2	7.7	14	22.9
Gathering	—	0.0	7	26.9	7	11.5
Horticulture	14	40.0	17	65.4	31	50.8
Total	35	100.0	26	100.0	61	100.0

Source: Adapted from Santos et al. 1997.

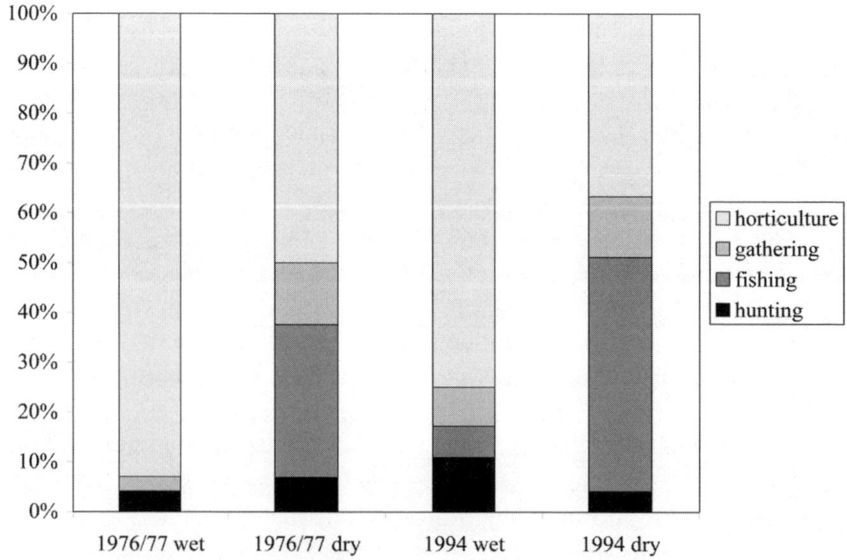

Fig. 6.8. Percentage of subsistence labor time allocated by season to different activities by Etéñitépa Xavánte adults in 1976–77 and 1994

that the amount of time devoted to activities outside the reservation did not increase over time.

Comparisons of the frequencies of consumption of food items according to their origin (horticulture, hunting/fishing/gathering, and purchased) in 1976–77 and 1994 are provided in figure 6.9. There was a sharp decrease in the frequencies of cultivated food items in the Xavánte diet and a pronounced increase in those obtained from hunting/fishing/gathering. Purchased food items also increased but only slightly.

More detailed information on the changes in food consumption according to food items and by season is provided in table 6.5. Several specific changes are notable. First, in the 1990s the Xavánte were eating rice much less often and manioc more often than in the 1970s. Second, there was a dramatic increase in gathered plant products in the Xavánte diet. Third, the relative frequencies of game and fish changed very little. Fourth, purchased foods were more numerous in the Xavánte diet in 1994. Fifth, rice was eaten more in the wet season because the rice harvest begins in February, while the consumption of manioc increased in the dry season.

TABLE 6.4. Time Allocation Observations of Etéñitépa Xavánte Adults in 1976–77 and 1994, according to Major Categories and Sex

	1976–77			1994		
Activities[a]	Men	Women	Both	Men	Women	Both
Subsistence	50	43	93	36	25	61
	(18.4)	(14.6)	(16.4)	(15.5)	(10.2)	(12.8)
Domestic	21	92	113	20	78	98
	(7.7)	(31.2)	(19.9)	(8.6)	(31.8)	(20.5)
Personal	151	140	291	131	133	264
	(55.5)	(47.5)	(51.3)	(56.5)	(54.3)	(55.3)
Outside the	44	16	60	38	7	45
reservation	(16.2)	(5.4)	(10.6)	(16.4)	(2.9)	(9.4)
Others	6	4	10	7	2	9
	(2.2)	(1.4)	(1.7)	(3.0)	(0.8)	(1.9)
Total	272	295	567	232	245	477
	(100%)	(100%)	(100%)	(100%)	(100%)	(100%)

Source: Adapted from Santos et al. 1997.

[a]Subsistence activities include horticulture, hunting, fishing, and gathering. Domestic activities include food preparation, housecleaning, child care, manufacturing, and maintenance. Personal activities include resting, sleeping, hygiene, conversation, and so on. Outside the reservation activities include wage labor, visiting authorities, attending school, and seeking health care.

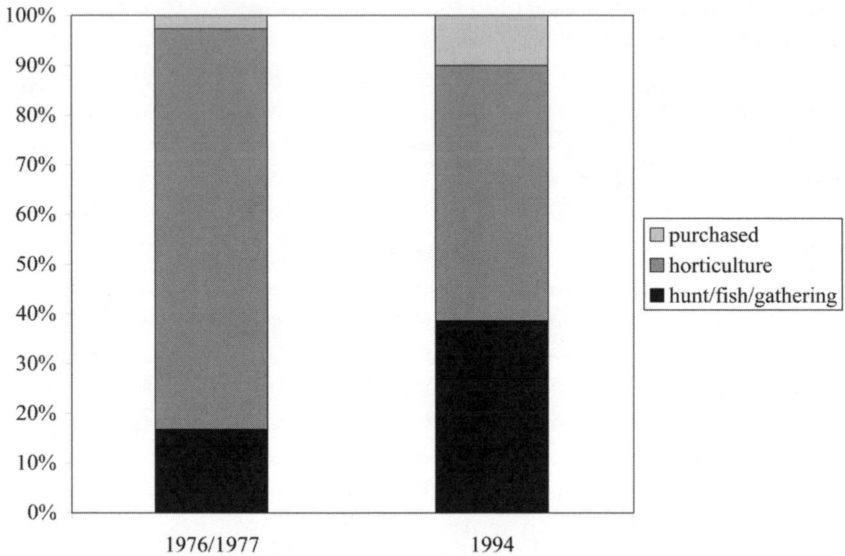

Fig. 6.9. Composition of the Etéñitépa Xavánte diet in 1976–77 and 1994, according to origin of the foods

The direction of these changes in subsistence and diet—from intensive work in horticulture to more leisure with less garden work but more hunting, fishing, and gathering, from a diet largely based on rice to a more varied diet including more wild but also more purchased foods—requires explaining. The history of the Xavánte Project at Etéñitépa provides much of the explanation.

Etéñitépa and the Xavánte Project

Flowers characterized the subsistence of the Etéñitépa Xavánte in the mid-1970s as follows.

> The Xavánte . . . are becoming progressively more involved in the local market economy, both through labor on farms of the region and by growing rice for sale. . . . When Xavánte men finished harvesting their rice . . . a number of men then left the reservation to help with the harvest on Brazilian farms, usually staying only a week or two. . . . Rice cultivation on these farms was mechanized, and the Xavánte were impressed with the labor-saving capability of

farm machinery as they observed it. . . . The Xavánte spent most of the money that they earned on clothing, tools, cigarettes, and such small food luxuries as coffee, sugar and treats of candy or fruit for their children. . . . In the season previous to my visit . . . the Xavánte had sold 300 sacks of rice. But in 1976 only three families grew rice for sale. . . . Toward the end of my stay there was much talk of a development project, to be implemented by FUNAI, for the Xavánte to increase rice production. (1983a, 256–58)

When the time allocation and food consumption data were collected in 1976–77, the Xavánte Project was not yet officially implemented at Etéñitépa. This would happen two years later. However, as Flowers's account shows, the Xavánte were already involved with rice cultivation for the market. In fact, the "seed" of the Xavánte Project was already germinating. At Areões, which is the closest reservation to Pimentel Barbosa, a pilot rice cultivation project was launched in the mid-1970s (Garfield 1996, 492; Menezes 1982, 75). In 1980 the Xavánte at Areões harvested seven thousand sacks of rice (Anonymous 1980). Although Etéñitépa Xavánte involvement in rice cultivation, as shown by our time allocation and food consumption data, was already considerable in 1976–77, they probably devoted even more time to it in the early 1980s, when the Xavánte Project reached its peak in this community.

TABLE 6.5. Observations of Consumption of Specific Foods by the Etéñitépa Xavánte in 1976–77 and 1994, Age and Sex Combined

	1976–77		1994	
Food	*N*	%	*N*	%
Cultivated				
Rice	144	60.3	200	25.6
Manioc	33	13.8	179	22.9
Beans and pumpkins	11	4.6	12	1.5
Hunting, fishing and gathering				
Game/fish	23	9.6	72	9.2
Fruits, nuts, roots, and tubers	13	5.4	222	28.5
Purchased				
Cookies, candies, coffee, powdered milk and juice, and pasta	5	2.1	76	9.7
Others	8	3.3	19	2.4
Total	239	100.0	780	100.0

Source: Adapted from Santos et al. 1997.

However, the involvement of the Etéñitépa Xavánte with the project never reached the same intensity as on other Xavánte reservations. Maybury-Lewis (1985) observed the effects of the project on three reservations: Pimentel Barbosa, Areões, and São Marcos. The Areões Xavánte were deeply involved in rice production. The Indians' houses and those of FUNAI employees were clustered together, surrounded by storage barns and sheds for agricultural machinery. There was a constant coming and going of tractors and trucks carrying groups of men between the village and the rice fields. The situation that he found at Pimentel Barbosa was quite different, with much less effect on the daily routine of the villagers. In Pimentel Barbosa,

> the Project seemed a thing apart from the everyday life of the community. The FUNAI has provided tractors and machinery for extensive rice cultivation, but the land used for this is 60 km away from the village. . . . The villagers complained about this and about the lack of transport to ferry them to and from, but did not consider relocating either the fields or their community. Instead, they sent off the young men's age grade to work in the rice fields and the rest of them went about their business in their village on the savannah. The physical separation of the rice plantation was matched by the lack of penetration of other aspects of the Project into the life of the community. (78)

In the 1990s, what we saw at Etéñitépa were the ruins of the Xavánte Project. Rusty pieces of agricultural machinery were abandoned in the vicinity of the village. Large-scale rice production was already a memory. Rice was again planted only in family gardens and used for domestic consumption.

New Directions at Etéñitépa in the 1990s

The Xavánte Project was a manifestation of the development ideology that held sway in Brazil during the 1970s and 1980s. In the 1990s there was a visible withdrawal of the government presence from Indian reservations. This coincided with broad political realignments at the federal level. Administrative downsizing, decentralization, reduced state involvement in the economy, and economic liberalism were major trends in Brazil in the 1990s. The assistance programs administrated by

FUNAI were seriously affected by these rearrangements. With its budget slashed, Indian posts were abandoned and what little assistance in education and health FUNAI had previously provided was no longer available.

Alcida Ramos (1998, 215) refers to the decade as "the pragmatic nineties." The pragmatism to which she refers is related to the initiatives of numerous indigenous groups during those years, epitomized by the Kayapó case (see also Fisher 1994; Schmink and Wood 1992, 253–75; and Turner 1995). For example, many groups, such as the Rondônia Suruí, the Nambikwára, the Parakanã and the Kayapó, became involved with commercializing nonrenewable natural resources, such as timber and gold, on their reservations (Coimbra 1989a; Price 1989; Schmink and Wood 1992, 253–75). The Xavánte did not take this route. Even if they should wish to do so, the vegetation on their reservations is not rich in valuable hardwoods and their rivers do not yield gold.

In the 1990s the Etéñitépa Xavánte economy shifted in the direction of a subsistence in which more time is spent on hunting, gathering, and fishing than in the 1970s. This is an unexpected direction if we consider that it is generally held that once an indigenous group comes into permanent contact the inevitable result is abandonment of traditional subsistence activities. Beyond the failure of the Xavánte Project, what other factors were involved in this turn of affairs?

The importance of ecological factors cannot be disregarded in looking at the shift in subsistence activities of the Etéñitépa group. As we saw in chapter 2, Pimentel Barbosa is the largest Xavánte reservation and the one in which the area/population ratio is the lowest. In the early 1980s the Etéñitépa Xavánte regained part of the land they had previously lost to encroaching ranchers. It is very likely that resources at Pimentel Barbosa are less degraded compared to other Xavánte reservations. Therefore, sufficient natural resources were available to permit a return to a subsistence involving more dependence on hunting, fishing, and gathering.

Realistically speaking, the subsistence pattern that we found at Etéñitépa in the 1990s was also made possible by an inflow of money from salaries and social security pensions. At the time of the project, some Xavánte became FUNAI employees with the title of tractor driver, administrative assistant, or head of the Indian post, and they still maintain these salaried positions. At Etéñitépa there is also a growing number of old people receiving pensions from the social security system. While it is difficult to estimate the amount of money that comes into the community through salaries and pensions, these resources make it possible for the Xavánte to acquire the manufactured goods to which they have become accustomed and which are now necessary to their subsistence.

At the beginning of the month, when the "retirees" receive their pensions, which amount to about U.S.$90 to U.S.$100 a month, there are always many trips to towns near the reservation for shopping. On several occasions, our pickup truck was borrowed for such a trip. Other sources of income are the rental of pastures on reservation land to neighboring ranchers and the occasional sale of a head from their cattle herd. At Etéñitépa reciprocity is still highly valued, so these resources are spread around the village. In the 1990s the Etéñitépa Xavánte were no longer participating in the regional economy as agricultural producers or laborers on nearby farms. Rather, their participation was mainly as consumers. A reservation that is still relatively productive, combined with a reliable (although small on a per capita basis) inflow of cash, explains why the shift in subsistence at Etéñitépa became possible. As Marshall Sahlins (1999, xi) has recently written, "The real problem [that modernization] poses for the people, is not the unlivable contradiction between the money economy and the traditional way of life. The big problems come when they cannot find enough money to support their traditional way of life."

The direction of change at Etéñitépa appears to run against the tide because, instead of becoming more dependent on the market for their subsistence, the Xavánte are increasingly using the natural resources available on their reservation. However, the connections of the Etéñitépa group in the most recent period are in fact wider than ever before. As Graham has pointed out, in the 1990s Etéñitépa was transformed from "one of the Xavánte communities least connected with the outside into the avant garde of Xavánte communities" (1995, 62).

With the failure of the mechanized rice cultivation project in the mid-1980s, as well as structural changes in the provision of public services to indigenous peoples, the various Xavánte communities began to seek alternatives to their historic overreliance on FUNAI. As we saw in chapter 3, one result was that the Xavánte became involved with national and international NGOs, especially some supporting indigenous culture and environmental programs. This realignment paralleled what was going on in other regions of Amazonia, where there was a growing presence of NGOs promoting environmental protection programs in connection with grassroots development.[11]

In the 1970s and 1980s the immediate concern for the Xavánte was to expand the boundaries of their reservation. They were relatively successful in this, as Pimentel Barbosa had its area increased by nearly 50 percent. In the 1990s the Etéñitépa Xavánte recognized that it was unlikely that their reservation would be significantly expanded in the future. Because of this the Xavánte are increasingly concerned with the preservation of their reservation against external threats. In the 1990s they have directed their efforts toward blocking a government-sponsored project to construct a commercial inland waterway in Central Brazil. This waterway has the potential to drastically alter the ecology of the Rio das Mortes, which borders the reservation and is a fundamental ecological resource for the Xavánte. In order to protect their reservation, the Xavánte have established far-reaching alliances and have framed their appeals in terms of wider interests.

Threats to the *Cerrado* Ecosystem and Xavánte Life

Although the *cerrado* is generally considered to be a more "robust" habitat than the tropical forest, the recent period of expanded use of the *cerrado* for cattle raising and industrialized agriculture has irrevocably altered the ecosystem. Some of the negative effects, according to Verdesio (1994) are increased soil erosion, water contamination by agricultural chemicals, soil compaction by agricultural machinery, and the destruction of habitats through deforestation. Grain monoculture and planting large areas with a single type of imported pasture grass impoverishes the ecosystem and may lead to weed invasion and mass attacks of insect pests. An additional source of environmental degradation due to the pace of economic development in recent decades is a rapid increase in population, much of it concentrated in new towns with inadequate systems for water supply and waste disposal.

By the early 1990s approximately 37 percent of the vast region of Brazil originally covered by *cerrado* vegetation had suffered total replacement of its vegetational cover (Dias 1994, 609).[12] About half was planted pasture. The other half was part annual crops, mostly rice, corn, and soybeans; part perennial crops, like coffee and pine and eucalyptus trees; and the rest dams, artificial ponds, and urban developments. The effects on neighboring regions, through soil erosion, water pollution, and the spread of weeds, extended far beyond the 73.3 million km^2 that were directly affected.

Most of the remaining 63 percent of the *cerrado* region is used for extensive cattle raising on native pasture, but since this land is in private hands, and agronomists believe that two thirds of the *cerrado* is suited to agriculture, it seems inevitable that replacement of *cerrado* vegetation with crops and urban development will eventually double from its present level (Dias 1994, 612). By this time there will be very little of the original *cerrado* vegetation left except in protected areas, of which Indian reservations are among the most important. Mato Grosso, for example, has 4,964,000 ha of *cerrado* in thirty-three Indian reservations but only 478,000 in two parks and three scientific reserves. Moreover, many of the parks are preserved areas only on paper.

The effectiveness of a reservation for environmental conservation depends, of course, on area and population density. As long as population density is low, indigenous groups may get their subsistence from the land with minimal alteration of the overall ecosystem. As population density increases, this may no longer be true. Terborgh (1975, 369–80, quoted in Dias 1994, 622) feels that 300,000 ha is the minimum effective area for ecosystemic preservation in tropical regions. However, of Xavánte reservations only Pimentel Barbosa, at 328,966 ha, falls into this category.

Conklin and Graham (1995) have pointed out that while conservationists and indigenous peoples may have apparently compatible aims there is much ground for misunderstanding, since "Environmentalists' primary goal is to promote sustainable systems of natural resource management. Indigenous people ultimately seek self-determination and control over resources. The degree to which these two sets of priorities coincide is debatable" (703). We would suggest that for the Etéñitépa Xavánte these goals may be tending to converge as they become increasingly aware of the importance of preserving the natural resources on which they depend for subsistence. They know that parts of the reservation were degraded when they were occupied by ranchers, who cleared the *cerrado* and planted pasture, and that the scarcity of game is in part because the reservation is surrounded by large farms and cattle ranches. Images from the LANDSAT satellite taken in 1990 show that about 80 percent of the *cerrado* habitat surrounding the reservation has been converted to farmland or planted pasture. The reservation is probably one of the largest demarcated areas of relatively undisturbed *cerrado* vegetation in Brazil (Leeuwenberg and Robinson 2000, 377–79).

The Game Management Plan

The Etéñitépa Xavánte are well aware that the territory from which they gain their subsistence is now defined by reservation boundaries. Also, in the wake of experiences such as that with the Jabiru Project (see chapter 3 and Graham 2000), both environmentalist NGOs and indigenous organizations may have gained more realistic views of what they may expect from collaboration. Thus,

> rather than attributing resource conservation to an environmental ethic presumably embedded in cultural traditions, researchers are coming to understand that limiting factors such as slow rates of population growth, low-density settlements, nomadism, and the use of traditional technologies more often must be credited with preventing the overexploitation of resources such as wildlife (Stearman 2000, 233)

The Etéñitépa Xavánte have been sufficiently concerned by the apparent decline in game resources to collaborate with Western scientists in research intended to result in a game management plan. According to Leeuwenberg and Robinson (2000, 393):

> The Wara resolved, during a meeting in March, 1992, to focus on intensifying traditional family hunts and to use distant hunting grounds. In addition, they decided to use fire in a more traditional way. They expressed little interest in western management techniques such as imposing bans, seasonal limits, or bag limits on any species. They rejected the suggestion of a hunting ban on pampas and marsh deer. In other words, novel ways to manage game were rejected in favor of more traditional approaches.

It is, of course, not certain whether these measures, which hark back to the days before trekking was limited to the reservation, will be effective in stemming the present depletion of game.

The Inland Waterway Project

The Etéñitépa Xavánte have also sought the support of environmental organizations in their determined opposition to the proposed construction of a commercial inland waterway (the Hidrovia Project), which would open the Rio das Mortes, the Araguaia, and the Tocantins to shipping. This waterway would mainly be used by barge convoys carrying soybeans and other products of industrial agriculture to European ports and, as Graham (1999, 17) points out, "The project's primary beneficiaries would be agribusiness corporations and shipping and construction companies involved in its development. In fact, the Hidrovia would represent a considerable public subsidy to expand soy monocultures throughout the cerrado." The Xavánte of Pimentel Barbosa and Areões would be particularly affected because their reservations border on the Rio das Mortes (see fig. 2.1). The *cerrado* ecosystem would be threatened by increased soil erosion and flooding, and the effect on wildlife would be devastating (Switkes 1998).

In October, 1996, the Xavánte of Pimental Barbosa and Areões held a meeting, after which they sent a letter of protest to President Fernando Henrique Cardoso, which reads in part:

> Mr. President: In the month of August, we became aware of the Rio das Mortes Hidrovia. We have a lot of trouble and we are very concerned regarding this huge project. We are sending this letter to stop the Hidrovia with our arguments and our traditional rights to these lands and rivers. . . . If this Hidrovia is to be constructed it will be the end of our animals. It will cause great damage to the environment. It will ruin an important part of the food of our people, fish and turtles. It will do away with the homes of many animals, dolphins, river otters, cayman, and others. . . . We want all work on the Rio das Mortes to be eliminated. We don't want buoys, we don't want signs, we don't want the river—the lakes, the fish, the turtles—to be ruined. We don't want the Hidrovia, Mr. President. We hope that all this will be resolved with great wisdom and honesty. We are ready to fight united as Xavántes in any way we can. We ask for help from all friends of the Xavánte people to stop the Hidrovia. (Cachoeira Declaration of the Xavánte Reserves 1996)[13]

With the assistance of the lawyers from the Instituto Socioambiental, who cited the lack of a valid environmental impact statement, the Xavánte obtained a court order in 1997 halting work on the Hidrovia. Attempts to persuade the Xavánte to change their minds by offering compensation of various kinds have been refused. So far the Xavánte and their supporters have been able to hold off the Hidrovia threat,[14] in part because of the blunders of its proponents, but we do not know if they will be able to force it to be shelved indefinitely.

Conclusion

In this chapter we have focused on Xavánte ecology and subsistence and how it changed after they came into permanent contact with the Brazilian national society in the 1940s. We showed how their economy makes use of the natural resources available in the *cerrado* environment. We also demonstrated how the Xavánte economy was profoundly affected by development policies targeting the Amazon region that were implemented, beginning in the late 1960s, by the Brazilian government. More recently the Etéñitépa Xavánte have been drawn into the worldwide environmentalist movement. A common denominator in these different processes is the tension between Xavánte ways of living and use of resources and the broader political and economic interests that impinge on them.

The Hidrovia, which was already being touted in the 1980s by government officials as the means by which the anticipated production of the Xavánte Project would be exported, is the most recent, but certainly not the last, chapter in the struggle that will continue as economic and demographic frontiers continue to expand and consolidate. This process, which is involving the Xavánte subsistence economy with wider economic and political systems, as eastern Mato Grosso is drawn into the wider geopolitical system of the Brazilian nation-state, has been going on for centuries and is likely to continue.

Chapter 7

Health Services and Unmet Needs

Early Health Services

Contrary to what one might expect, it was not the Serviço de Proteção aos Índios (SPI), founded in 1918, that made the early attempts to provide Brazilian Indians with health care on a systematic basis. The nearest thing to what we might call an Indian health service was the brainchild of a physician, Noel Nutels, who was introduced to the reality of Indian health as a member of the Roncador-Xingu Expedition organized by the Fundação Brasil Central (FBC) in the 1940s (see chapter 3). At that time Nutels visited the Xavánte, as well as groups of the upper Xingu and Bananal Island, and he was alarmed by the spread of tuberculosis.

In 1952 Nutels drew up a plan "to defend the Brazilian Indian against tuberculosis," in which he stressed the importance of setting up "sanitary cordons" around indigenous territories (Nutels 1952). In this way, the only people allowed into an indigenous area would be submitted to strict health controls. Nutels felt that the health problems of the Indians were closely linked to general conditions of rural poverty and poor health. Recognizing the importance of regional endemic diseases, which could easily be transmitted to indigenous populations, Nutels's plan included the implementation of broad social measures to cover rural areas of the country. According to Nutels, "[while] the cowboy in search of new pastures [continues to] invade the natives' territory with his scrawny herds, and the gold prospector digs there for his dreams, they will be bringing their misery and their diseases to the poor savages" (12).

Although Nutels's plan was apparently well received at the Ministry of Health, it was not until 1956 that it was put into effect as the Serviço de Unidades Sanitárias Aéreas (Service of Airborne Health Units, or SUSA), attached administratively to the Serviço Nacional de Tuberculose (National Tuberculosis Service). Once it was set up, SUSA rapidly expanded its operations beyond the area of the Fundação Brasil Central, reaching Indian villages as far as the south of Mato Grosso, the upper Rio Negro, and other regions of Amazonia. The work of

SUSA's medical teams emphasized the diagnosis, prevention, and treatment of contagious diseases, especially epidemic diseases (Costa 1987; Nutels 1960, 1968; Nutels and Duarte 1961).

In spite of the efforts of a small number of idealists led by Nutels, SUSA was unable to make a great deal of difference in the general health conditions of the country's indigenous populations, even though in specific cases it saved lives by responding to epidemic outbreaks. Soon the major changes that were taking place in Indian policy affected the health services.

Health Services under FUNAI

When the SPI was abolished in 1967 and FUNAI took its place, a health division was created within the agency to plan and provide health services for Indian populations throughout the country. With the guidance of former SUSA technicians (SUSA became extinct along with the SPI) the FUNAI health services were organized around the concept, which originated with SUSA, of "equipes volantes de saúde" (mobile health teams), known as EVSs, installed at strategic points and attached to regional FUNAI offices. Each EVS consisted of a physician, a nurse, a lab technician, and a dentist, with means of transportation to take them on periodic visits to the Indian villages in their jurisdiction. Their backup in the city would be the Casas do Índio, which would lodge Indians who had to come to the city for more specialized medical care. Attached to FUNAI posts in the villages would be small infirmaries with a stock of drugs and staffed by a nurse's aide. That structure, however, never worked well. There were too many factors within the FUNAI health services that together added up to inefficiency and high operational costs.

Although the word *volante* in the name of the FUNAI health teams implies speed and flexibility, the operational costs of the EVSs were high and generally not efficacious. It was not only difficult to guarantee fuel supplies for the teams' jeeps and boats, but they frequently suffered "insoluble" mechanical problems that prevented them from even starting out. Airplanes and helicopters chartered for emergency medical evacuation were very expensive. The teams, which often lacked drugs and the most basic medical equipment, were also plagued by a chronic shortage of personnel.

The primary health care that was supposed to be provided by health workers at the village level was also ineffective. The infrastructure of the village infirmaries and the conditions under which they operated were precarious. The poor living conditions at FUNAI posts, along with the low salaries, discouraged better-qualified people from following FUNAI careers. All of these factors led to a high turnover of health workers at the FUNAI posts. When there was no attendant at the infirmary, long-term medical treatments would be interrupted, sometimes for many months, until he or she could be replaced. This was particularly serious in the case of tuberculosis, since interruptions in medication can lead to the development of resistant strains of the bacteria. The failure of primary health care in the villages put great pressure on the Casas do Índio in the towns, whose infrastructure was overburdened with the influx of patients with health problems that, if not aggravated by delay, could have been solved at the village level.

In the 1980s and 1990s FUNAI went through a period of great instability, with frequent political and structural changes that had a negative effect on its status in the eyes of the Indians, the morale of its functionaries, and how it was regarded by the public. The agency seemed to be drifting, and the successive management changes seemed to produce no clear directives that might make FUNAI's work with the Indians any more effective in the sphere of education, or of health care, or in demarcating reservation lands. Health care for the Indians was especially disorganized and sporadic during this period, with no improvement in the most elementary measures of health, either curative or preventive.

Health Care for the Xavánte

FUNAI health personnel were limited in number and mostly concentrated in urban areas rather than on the Xavánte reservations. Only at the Casa do Índio in Barra do Garças was it possible to find a part-time physician and dentist. The services of these medical personnel were limited to curative intervention; they could not carry out preventive programs in the villages because most of them had other jobs in town, which prevented them from traveling. There was rarely a registered nurse at the post infirmaries or the Casas do Índio. Most of the work was done by *auxiliares de enfermagem* (nurse's aides). Most infirmaries located at FUNAI's Indian posts had been abandoned or were operating under very precarious conditions, with no infrastructure and very limited availability of medical supplies and first-aid kits.

From the 1960s, the Xavánte had been receiving health care in different ways on the various reservations. At Pimentel Barbosa, health services were provided by FUNAI. At Sangradouro and São Marcos, Salesian missionaries provided health care. Nuns, some of whom were registered nurses, were responsible for most preventive and curative care at the missions' infirmaries. The more complicated cases were usually sent to nearby towns, where the Indians were assigned to private clinics. Dentists were hired for certain periods of time to provide basic services at the missions' dental clinics.

By the 1990s it became very clear that FUNAI was not operating with even minimum effectiveness on those Xavánte reservations where it was responsible for health care. Although FUNAI was receiving funding from the World Bank through a project called PRODEAGRO, part of which was intended to implement health services, it was so disorganized that this funding did not result in effective action. At this time several nongovernmental organizations attempted to provide health services to the Xavánte. However, the delivery of these services was seriously handicapped by the lack of central planning and coordination.

Our research team visited Pimentel Barbosa on several occasions in the early 1990s, and we witnessed the absolute lack of FUNAI health services in the villages. The infirmary at Etéñitépa had been abandoned, with a few outdated drugs still on the shelves. In 1990 we made arrangements for FUNAI nurses to accompany our team and vaccinate the villagers. It was obvious that scheduled vaccinations had not been carried out recently, as young children had not received BCG shots.

In 1992, Médecins sans Frontières (MF), an NGO with headquarters in Europe, launched a project aimed at developing programs of preventive medicine for the reservations and training young Xavánte *monitores indígenas de saúde* (Indian health assistants) to carry them out. The MF remained in the area for only a short time (1992–93), for it faced major problems in dealing with Xavánte politics and intervillage factionalism, to the extent that field activities had to be interrupted in some areas. Unfortunately, at the start of its activities in Mato Grosso MF did not seek sufficient advice from Brazilian anthropologists with experience among the Xavánte. Probably they would have been helpful in planning more culturally sensitive ways of entering Xavánte villages and reservations without exacerbating preexisting factionalism. MF also had problems with FUNAI and local clinics when its medical teams started to send a large number of Indians to town for complimentary exams and/or hospitalization. Already unwilling to serve Indians, the local medical infrastructure could not, or did not want to, deal with a sudden increase in demand, which led to a situation in which Indian patients were simply turned down and forced to return to their home villages. In any case, the original plan of MF to stay with the Xavánte for a period of five years was never carried out; it left after only a year and a half.

When Flowers was living at Etéñitépa in 1976–77, FUNAI agents often took patients from the reservation to Barra do Garças or Goiânia for health care. This implied a trip of at least eight to ten hours by road. In the 1980s and 1990s some towns closer to Pimentel Barbosa grew and became more organized. Recently the Etéñitépa Xavánte have come to depend on health services in neighboring towns like Canarana and Água Boa. However, there are very few public health units and there is a deficit of health professionals in the municipalities close to Xavánte reservations. Nova Xavantina, Campinápolis, Canarana, and Água Boa, where the Xavánte reservations of Areões, Parabubure, and Pimentel Barbosa are located, have a population of approximately 63,000. In 1994–95 there were 22 physicians in these four municipalities (3.5 per 10,000 persons), much less than the national average of 13.3. The only public health units are the Pronto Socorro Municipal de Nova Xavantina (Nova Xavantina Emergency Clinic), which had been recently inaugurated at the time of our fieldwork, and some small outpatient clinics located in the municipalities' administrative centers.

Xavánte patients go to those private clinics and hospitals that participate in the federal government system known as the Sistema Único de Saúde, or SUS (Unified Health System). Services provided by these clinics and hospitals are reimbursed by SUS using an authorization form known as Autorização de Internação Hospitalar, or AIH (Authorization for Hospitalization).

Canarana, with a total population of 12,678, is the largest town close to Pimentel Barbosa. In 1994 it had two private hospitals and one public outpatient clinic. The only hospital participating in the SUS had twenty-eight beds. This facility was entitled to ten AIH forms per month, of which apparently only one was reserved by the hospital's administration for Indian patients. This amounts to illegal discrimination, since under the Lei Orgânica da Saúde (no. 8080/90), the law that officially established SUS, there can be no decision regarding the allocation of AIHs within a hospital unit by race or ethnic affiliation of the patients. The restriction on AIH forms for Indian patients stands out as a major problem, since Canarana receives patients not only from many Xavánte villages but from other Indian reservations, including the Xingu Park.

Private hospital administrations often discriminate against Indian patients, using all sorts of excuses, such as claiming that the Indians do not have good manners or proper hygiene habits. Discrimination in these cases is aimed at separating the Indians from the other patients, who usually are of middle-class backgrounds. One very clear example of discrimination was provided by a private hospital in Nova Xavantina. In this case, a special Indian ward was built in the hospital's backyard. There Indian patients stayed under minimal conditions of hygiene and comfort, well below the level of attention provided to non-Indian patients in other sections of the same hospital. In May 1994, the "Indian ward" was closed for renovation and the administration of the hospital refused to accept Xavánte patients in other wards or infirmaries. This particular hospital also made a unique arrangement with the local FUNAI administration: It demanded that FUNAI provide food for the patients, and relatives staying with the patients, in addition to paying for all prescribed medications and the fees of one nurse's aide, whose duties were to care for the patients in the Indian ward. Under these arrangements, Indian patients ended up costing the federal government at least twice as much as other patients, since hospitalization costs were being covered by the SUS as well as FUNAI.

In addition to the barriers that Indians face in their search for health care at regional private hospitals, local authorities are totally uninformed about their living conditions and health needs. For instance, when Rubens Ianelli, a member of our research team, interviewed the mayor of Canarana in 1994, the mayor expressed his concern about the "numerous cases of leprosy and Chagas' disease [in the Xavánte]," unaware of the fact that these infectious diseases are unknown on Xavánte reservations.

Functions of the SUS at the local level are regulated by the Conselhos Municipais de Saúde (Municipal Health Councils), which are intended to bring together representatives of different community sectors. Until very recently, Xavánte leaders did not take part in these councils (fig. 7.1). It should be mentioned, however, that, as only a few Xavánte vote in the municipal, state, and national elections, they do not attract the attention of local politicians, for whom the motto of "medical assistance for all" is a common theme only at election time. Since interest in

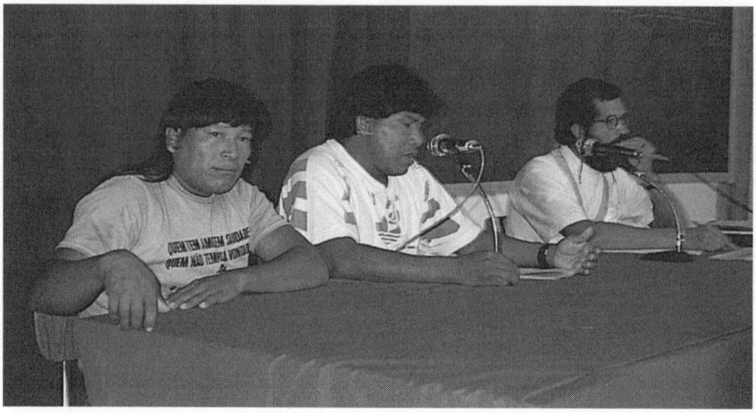

Fig. 7.1 Xavánte discussing the implementation of the new health care system for indigenous peoples. (Photo by R. V. Santos, 1994.)

Indian issues and information about them are almost nonexistent at the municipal level, there is little chance that the Xavánte will become beneficiaries of adequate medical and/or preventive programs through their municipalities.

An incident at Campinápolis in 1994 provides one good example of how the apathy and indifference of local municipal health authorities may prevent the delivery of basic health services to the Xavánte. The town is very close to the Parabubure reservation, which has a population of nearly five thousand. Despite that, the director of the Secretaria Municipal de Saúde (Municipal Health Department) justified the non-inclusion of the Xavánte in the last vaccination campaign with the allegation that the number of doses available was only enough to cover the non-Indian population of the municipality.

In the 1990s, although FUNAI was not as involved in health-related issues as in the past, the Casas do Índio were still operating. There was one in Barra do Garças and another one in Nova Xavantina. They did not fulfill their original aim, which was to provide room and board for patients receiving short ambulatory treatments at local clinics or undergoing other kinds of short-term medical procedures, including laboratory exams and X rays. Both had become places where basic medical and dental care were provided. However, the two Casas do Índio had very elementary infrastructures. In reality, they were operating as small hospital infirmaries, despite the lack of appropriate personnel and infrastructure.

Frustrated by the difficulty of having their health needs attended to locally, the Xavánte are pushed toward larger regional urban centers, where they hope to receive better medical care. A growing number of Xavánte, especially those who are economically better off, prefer to go to cities like Goiânia, Brasília, or São Paulo. In most instances they simply take a bus and pay for their own tickets, bypassing the control of FUNAI, which will only learn about such trips when it comes to paying hospital and hotel bills. The constant traveling of Xavánte families is closely associated with the limited coverage and effectiveness of the services provided at the village and municipal levels. Such shortcomings contribute to the escalating costs (both financial and social) of medical assistance. In many instances families are forced to travel because of complications of relatively simple diseases, such as flu or diarrhea.

Prospects for Indian Health

In the 1980s, the Brazilian health sector went through a deep transformation, which ended with the creation of the SUS along lines laid out at the Eighth National Health Conference, inspired by the Federal Constitution of 1988. Basically, SUS stresses the role of the state in financing, organizing, and implementing a universal and egalitarian health care system. It was set up to integrate all health services, supplementing public with private medical facilities. In structural terms, it was intended to be a decentralized and hierarchical network integrating care at the federal, state, and municipal levels (Buss and Gadelha 1996).

With the decentralization and regionalization of health services, there was no justification for continuing parallel services like those of FUNAI and other federal agencies. In the case of FUNAI there were few arguments that could be used in its favor; it ranked very low in all evaluations and reports on its medical services. Under Brazil's evolving health care system, Indians are to receive "culturally appropriate" health care under SUS through a subsystem for indigenous health.[1] Provisional measure 1911–08, of 29 July 1999, and law no. 9836, of 23 September 1999, transferred from FUNAI to the Ministry of Health the responsibility for Indian health assistance. Under the Ministry of Health the Fundação Nacional de Saúde, or FUNASA (National Health Foundation) was given the responsibility of setting up the subsystem of SUS for Indian health (http://www.funasa.gov.br/ind/ind00.htm [June 2000]). This was a hard blow to FUNAI, which lost a large part of its budget, while most of its health care workers were transferred to FUNASA. As one might expect, the context and implications of these upper-level government decisions led to conflicts among the parties concerned.

In spite of the complex political situation, in 1999 FUNASA began to implement the Subsistema de Atenção à Saúde Indígena (Indigenous Health Care Subsystem), based on the Distritos Sanitários Especiais Indígenas, or DSEI (Special Indigenous Health Districts), linked to the SUS. From August 1999 to the present, thirty-four health districts have been created (FUNASA 2000a, 6). The DSEIs are under FUNASA, and each covers a specific territory, defined in technical and ethnodemographic terms, that does not necessarily coincide with municipal boundaries. According to the DSEI organizational model, each village will have at least one indigenous health worker, attached to a small health post. The health workers will have a number of responsibilities, which will include following the growth and physical development of the children, providing routine prenatal care, administering first aid, and treating common diseases (7).[2]

The Xavánte DSEI, based at Barra do Garças, is divided into four units called *pólos-base.* These are structured as basic health units, each with a medical team that serves as the first level of reference for the Xavánte agents in the villages. According to FUNASA, the prior implementation of local and district health councils will be necessary in order to plan the activities of the Xavánte DSEI, planning made very difficult by the political fragmentation of the Xavánte into six reservations and 100 villages (FUNASA 2000a, 82).

Conclusion

It is hard to foresee what the new structure for indigenous health services will mean for the Xavánte. The plan undoubtedly requires extensive training programs for indigenous health workers and effective articulation of the village level of health care with other levels of the system. We hope that the new proposals for health care that are in the process of being implemented will bring positive results, with greater emphasis on preventive measures. According to FUNASA, the Indians themselves are expected to participate in implementing and managing the DSEI through their leaders, who will sit on the local and district health councils (FUNASA 2000a, 8–9). It is important to stress that the Xavánte are not passive recipients of deficient health care. We have often discussed health-related issues with individual leaders, as well as with the *warã,* and it is clear that the Xavánte, many of whom have traveled and are well aware of discrepancies in health care, are not willing to accept inferior services.

Chapter 8

The Burden of Infectious Disease

A document that was published recently by the Pan American Health Organization (PAHO), a section of the World Health Organization (WHO), paints a gloomy picture of health conditions among the native peoples of the Americas: "Serious and pervasive inequities are known to exist in health status and health service coverage of indigenous population groups in the Americas" (PAHO 1997, 357). Although reliable data on the health and illness of indigenous populations are not uniformly available in countries of the Americas, the report provides several examples that point to the poor health conditions of indigenous peoples. Both in Mexico and Panama, mortality rates are two to three times higher for indigenous than nonindigenous children in the same country; in Belize, Mayan children have the highest rates of growth retardation; and in Honduras, the life expectancies for Indian men and women are close to thirty years less than for their non-Indian counterparts. According to PAHO, a major obstacle to promoting better health among indigenous peoples is lack of data: "Few countries routinely collect and analyze vital statistics or health service statistics by ethnic group, so it has been difficult to develop a good baseline data or to adequately assess the health and living conditions of the indigenous peoples of the Region" (359).

The difficulties that we found in delineating the epidemiological profile of the Xavánte of Etéñitépa exemplify the problems discussed in the PAHO document. As is the case for indigenous peoples in other parts of the Americas, the lack of a reliable system for recording morbidity and mortality data for the Xavánte, as well as for indigenous populations of Brazil in general, handicaps any attempt at detailed epidemiological analysis. Unfortunately we do not have the data to address such fundamental issues as determining the precise morbidity and mortality rates resulting from specific diseases. This information is missing not only for the Xavánte but for most of the country's other indigenous groups (Coimbra 1998; Coimbra and Santos 2000). In the absence of secondary data from which to draw a detailed morbidity and mortality profile of the Etéñitépa population, we have had to collect primary data for a wide range of parameters. The diverse measures that we collected and analyzed, from serological tests for infectious diseases to anthropometric data to evaluate nutritional status, are justified because we had no other sources of information. Of course, we do not consider the information we present here to constitute a complete epidemiological profile of the Xavánte of Etéñitépa. Nevertheless, our set of data certainly touches on important dimensions of the health/disease processes that affect them.

Our purpose in this chapter, therefore, is to present an overview of Xavánte health derived from the field research we carried out in the 1990s. We emphasize the epidemiological and public health aspects of living conditions and disease. Our analysis indicates that infectious and parasitic diseases are the leading causes of sickness and death among the Etéñitépa Xavánte. Our study also points to the inadequacy of the health care system on which the Xavánte depend.

Previous Health Research among the Xavánte of Etéñitépa

There are basically only two studies that give us useful epidemiological information about the Etéñitépa Xavánte before the 1990s; these are the investigations by Freitas-Filho (1955) and Neel et al. (1964).

In 1954, the physician Amaury Sadock de Freitas-Filho, from the federal public health service, carried out a medical and sanitary investigation among the Etéñitépa Xavánte.[1] At that time they were living at a site about one kilometer from their present village, while the Indian post at São Domingos was located sixty kilometers away, on the banks of the Rio das Mortes (see fig. 2.2). It is clear from Freitas-Filho's report that government agencies were concerned about the health conditions, as well as the nutritional state, of the Indians in the Xavánte village and at the SPI post, especially the possibility of the spread of disease (malaria in particular).

Like others who observed the Xavánte soon after contact, the doctor admired the fine physical appearance of the men and their apparent good health, which he ascribed in large measure to their "rational and instinctive diet that contains all the nutrients necessary for life" (1955, 165). He noticed that the women did not seem to be in as good physical condition as the men and believed that their fertility was low.[2]

Freitas-Filho found no signs or symptoms of some diseases that were endemic among rural Brazilians of Mato Grosso at the time: tuberculosis, leishmaniasis, and leprosy. Although he did not make detailed examinations, he found that the Xavánte had excellent teeth, with little evidence of caries, and little skin disease. However, he does note cases of infant diarrhea, which he ascribes to abrupt weaning to an adult diet. Freitas-Filho found evidence of parasitic infection by four different helminths (hookworms, *Trichuris trichiura, Enterobius vermicularis,* and *Ascaris lumbricoides*) but does not give rates of infection or any other details beyond naming the species. The physician was particularly concerned with the possibility of measles reaching the Xavánte at Etéñitépa. At that time, an epidemic of measles was causing high mortality among indigenous groups of the upper Xingu (Nutels 1968) and was also affecting employees of the FBC in the town of Xavantina (Mota 1995).[3]

A good part of Freitas-Filho's report deals with malaria transmission, including an investigation of breeding places for *Anopheles* larvae and the capture of adults. Microscopic examination of ninety-one blood films yielded only one positive case of malaria. However, Freitas-Filho concluded, "We must admit that this is an endemic region—malaria is present, as well as the conditions for its transmission" (1955, 161).

Freitas-Filho recommended that all visitors (and these should be limited to officials in government service) be subjected to strict medical examinations before they were allowed contact with the Indians. The Xavánte themselves should be immunized and their houses periodically sprayed with DDT. He also felt that Xavánte health could be improved through education, especially of the children. He even suggested that a "model hut" might be built by the Indians themselves, like those they already lived in but made more "hygienic" by waterproofing the floor and whitewashing the inner walls.

Freitas-Filho's work was mainly intended to collect information to be used by government agencies providing services to the Indians. It was not, strictly speaking, the result of scientific research. James Neel, Francisco Salzano, and their collaborators were the first to carry out in-depth biomedical research among the Etéñitépa Xavánte (Neel et al. 1964). This research team was mostly interested in questions related to population genetics, but it also carried out some epidemiological research. Neel and his collaborators collected at São Domingos biomedical data covering a wide range of parameters, including some derived from physical examinations. Other data came from blood biochemistry and cell counts, exams of blood films for malaria and filaria parasites, and serological tests for poliomyelitis, measles, influenza, pertussis, *Treponema,* and *Salmonella* infections.

The observations by Neel and his colleagues (1964) are numerous and detailed, and we do not pretend to review all their nuances. But, of the points made in 1962, we would like to stress some that we feel to be particularly informative. In relation to oral health, the authors pointed out that no significant malocclusion was encountered, there was a remarkable absence of obvious caries, and no significant disease of the oral cavity was noted (107). Physical examinations showed relatively slow pulses and low blood arterial pressure readings (108). Approximately one-third of the individuals examined showed hepatomegaly and/or splenomegaly, which the research team related to malaria. The results of blood tests revealed low hemoglobin values in both sexes (131). The leukocyte counts showed marked eosinophilia (131), which was attributed to intestinal parasites, for which no specific examinations were made. As for blood tests for *Treponema* infection (venereal disease research laboratory test, VDRL), none was definitely positive (118). Microscopic examination of blood films revealed three cases of parasitism with *Plasmodium* sp. The analyses also indicated that "a high proportion of persons had antibodies to measles, pertussis, poliomyelitis . . . and *Salmonella*" (132). No cases of obesity were identified, and the researchers found rather low rates of serum cholesterol. Generally speaking, the results showed little or no evidence of chronic noncontagious diseases in the population, while the disease pressure from infection and parasites was certainly considerable.

In the early 1960s, when Neel and his colleagues did their study at São Domingos, the Xavánte were not vaccinated. A high percentage of individuals tested positive for antibodies against measles and pertussis. Although the researchers remarked that they "learned of no major epidemic" (Neel et al. 1964, 91), the finding of the serological tests led them to write that "these findings may provide a clue to the possible recent decrease in the size of the group" and "we cannot exclude the possibility that these diseases have been endemic among the Xavántes and neighboring Indian groups for a considerable period of time" (124). They advanced the possibility that the Xavánte might have been periodically infected through occasional contact with other Indian groups, like the Karajá, who were in permanent contact with non-Indians. It is interesting to observe that Freitas-Filho, working in 1954, was concerned about the possible occurrence of epidemics in the Xavánte, especially because there was an outbreak of measles in eastern Mato Grosso at that time. However, Freitas-Filho suggested in his report that epidemics had not recently affected the Xavánte group living at Etéñitépa. Freitas-Filho's observations, along with those of Neel et al. concerning the results of the antibody studies, may suggest that the Xavánte of Etéñitépa were hit by an outbreak of infectious disease sometime between the late 1950s and the early 1960s.

A particularly interesting aspect that Neel et al. (1964) pointed out was health differential by gender: "by contrast [with men], the women, although in apparent good health and nutrition when young, gave an impression of early aging, an impression more than sustained by the results of the physical examinations. Indeed, one of the most striking impressions of this study was of the different medical worlds of men and women" (110). The investigators were puzzled, moreover, by the very few elders in the population.

In 1963 and in 1964 Neel, Salzano, and their collaborators returned to the Xavánte, on these occasions working in the villages of Simões Lopes and São Marcos; their work resulted in a series of papers.[4] The theory and methodology of these investigations is quite close to those that directed the research at São Domingos, that is, they were primarily studies in population genetics, in the course of which demographic and epidemiological data were collected. The variables investigated in 1963 and 1964 are practically the same as those studied during the research of 1962. It is worth mentioning, however, that the laboratory analyses, especially the immunology, are more thorough and sophisticated (see Neel et al. 1968a, 1968b). The studies made at Simões Lopes and São Marcos are obviously very relevant to the health picture of the Xavánte in the 1960s. However, since they were carried out among Xavánte groups living at a considerable distance from the Etéñitépa population and these groups were already going through different contact experiences with Brazilian society (in the case of São Marcos they were at a Catholic mission), it is even more difficult to make valid comparisons. It is principally for this reason that we will not go into much detail about the research at Simões Lopes and São Marcos, although we may refer to it throughout the chapter.

Health and Disease in the 1990s

When Neel and his colleagues visited São Domingos in the early 1960s almost twenty years had passed since the first Xavánte meeting with the SPI "pacification" team. The conditions of their lives had already been altered by contact with outsiders, from whom they had acquired not only new tools and artifacts but new diseases. The transformations would accelerate in the decades to come, as important factors capable of altering disease pressure changed in intensity or were introduced. For instance, the Xavánte became more sedentary, moving their base village less frequently, which affected sanitary conditions. The Xavánte relied increasingly on agriculture for subsistence, and their nutritional base changed as they adopted new crops, particularly upland rice (see chapter 6). Contact with outsiders intensified. They began to wear clothes and acquired more Western household goods. On the other hand, the Xavánte eventually began to receive vaccines and gain access to some medical services, which, however, do not adequately cover their health needs.

Sanitary Conditions and Major Causes of Ill Health

In the 1990s, sanitary conditions in most Xavánte villages were affected by limited sources of potable water, a situation that becomes aggravated during the dry season. In some villages running water may not even be available at certain times of the year, as streams turn muddy or dry up completely. Wells, even when they exist, are not deep enough, nor are they properly protected from the detritus of all kinds that may fall into them. The sanitary conditions at Etéñitépa are typical of those found in most other indigenous communities in the region, reflecting the scant attention paid to sanitation by the Indian service. In fact, even the installations of the FUNAI post at Etéñitépa, which at one time included a schoolhouse, infirmary, and administration building, have almost never had running water; employees fetched water directly from the stream in buckets. The precarious sanitary infrastructure at Etéñitépa certainly plays a major role in the high rates of infectious and parasitic diseases.[5]

The terrain at Etéñitépa is sloping, with the village center higher than the stream, which meanders through the gallery forest at a distance approximately two hundred meters from the houses. This is the perfect scenario for the continuous contamination of Etéñitépa's most important source of water. It is particularly serious in the wet season, when rainwater washes all sorts of dejecta into the stream from which the Xavánte draw water and where they bathe. The Xavánte bathe frequently, usually in the morning and evening. Families returning from the gardens jump into the water on the way home to refresh themselves and wash off the surface dirt. When in the village, women gather by the stream to bathe their children and wash clothing and household utensils.

As for sewer disposal, suffice it to say that there are no latrines in the village; people defecate behind bushes in the fields some one hundred meters from the houses. The number of domestic animals in the village, chickens and dogs in particular, adds to the increasing level of soil contamination with infective stages of intestinal parasites as well as a number of other gastrointestinal pathogens.

In addition to waterborne agents of diarrheal diseases, the Xavánte are also exposed to other environmental and "behavioral" factors associated with diarrheal infection. Postcontact changes in lifestyle, which include the introduction of clothing, blankets, cloth hammocks, and eating utensils, may have increased the degree of exposure to some pathogens. The household kitchen, now often located in a lean-to at the side of the house, consists of shelving of rough boards on which battered aluminum pots, recycled kerosene cans for water storage, and enamel plates and bowls are kept. These are casually rinsed out after use and only occasionally are taken to the stream for a more thorough wash. Xavánte houses have floors of beaten earth partly covered with palm leaves that cushion the woven basketry sleeping mats, and articles fall on the floor which may later touch food or children's mouths. In the dry season, dust is ubiquitous, blowing into the houses from the surrounding areas where trash is thrown and dogs and chickens wander. Babies do not wear diapers and when defecating are simply held over an unoccupied area of the house floor or yard and then cleaned up with a leaf or anything else handy. These environmental and behavioral factors are not very different from those that have been described in other communities where diarrheal diseases are highly endemic (Bartlett et al. 1992; Bertrand and Walmus 1983; Feachem 1984; Feachem et al. 1983; Khan 1982; Martines et al. 1993).

A major difficulty in characterizing the Xavánte epidemiological profile is the lack of reliable vital statistics, which can be found neither at the local offices of FUNAI nor at those of FUNASA in nearby towns. At the village level one can seldom find any epidemiological or demographic records, since most FUNAI posts are either abandoned or currently operating under very precarious conditions. Under these circumstances, we have to rely on data collected by the regional hospitals to which Xavánte Indians are currently being sent for medical attention. Of course, this information must be handled with great caution, since it is problematic to extrapolate from hospital records to the general population.[6]

Data collected at the Casa do Índio at Nova Xavantina, to which

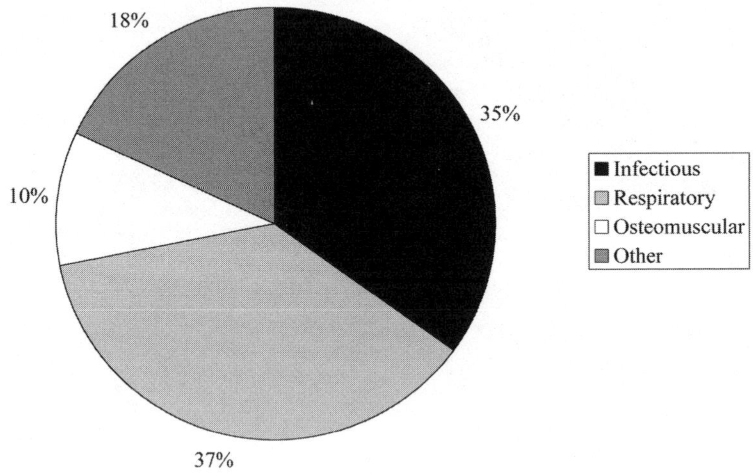

Fig. 8.1. Frequencies (percentages) of diagnosis in 848 consultations at the Casa do Índio at Nova Xavantina, 1994–95. (Data from Ianelli 1997.)

Etéñitépa Xavánte often go in search of medical care, illustrate the importance of infectious and parasitic diseases.[7] In 1994–95, they were the most frequent reasons for consultation, accounting for nearly 70 percent of the total (fig. 8.1).[8]

Other sources of data also show that infectious and parasitic diseases are the most frequent causes of consultation and hospitalization among the Xavánte. In the twelve-month period from June 1993 to May 1994, of the 343 Xavánte consultations at the Pronto Socorro Municipal de Nova Xavantina (Nova Xavantina Emergency Clinic), 30.5 percent related to gastroenteritis and 33.7 percent to pneumonia (Coimbra and Santos 1994a, 110). Of these patients, nearly half were either one year or less (30.5 percent) or one to four years old (19.0 percent).

A report made available by Médicins sans Frontières, based on first-hand data and compilations from various sources, mostly from FUNAI, confirms the general profile presented here. According to this nongovernmental organization, which worked for a short period among the Etéñitépa Xavánte as well as those from neighboring areas, infectious-parasitic diseases accounted for 85 percent of the total morbidity in the Xavánte population between 1989 and 1990 (Ávila-Desser 1993, 10).

Gastroenteritis and respiratory infections lead not only to frequent sickness but high fatality rates in Xavánte children. In 1993, pneumonia

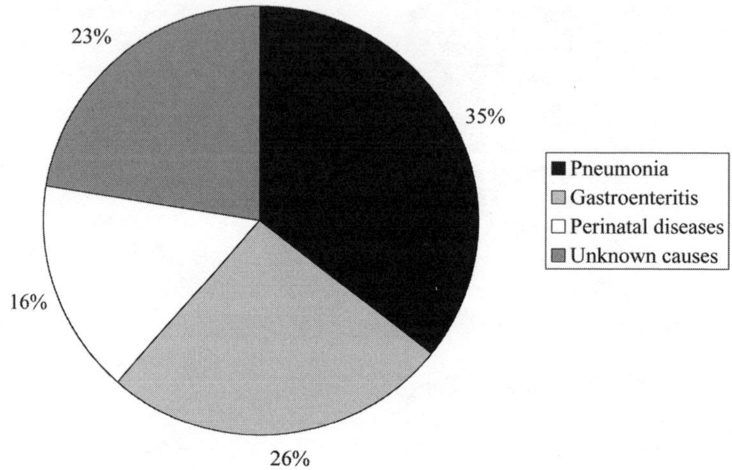

Fig. 8.2. Causes of death in thirty-one Xavánte children age one year or less in 1993. (Data from the State of Mato Grosso, Projeto à Saúde das Comunídades Indígenas Xavánte e Bororo, unpublished report, 1993.)

and gastroenteritis were the most common causes of death in Xavánte infants one year of age or less (fig. 8.2). According to the records of the Hospital Nova Brasília (Nova Brasília Hospital), located in Nova Xavantina, of the ninety-one Xavánte deaths in their wards in 1992, 60.4 percent were of children 5 years of age or less. Pneumonia and gastroenteritis were the major causes of mortality, accounting for 50.5 percent of all deaths (Coimbra and Santos 1994a, 109). In considering these figures, one should be cautious due to the low coverage of the local health services, which may lead to underestimation of the actual situation. The high frequency of causes of deaths classified under "unknown" in children 1 year old or less leads one to question the coverage and quality of health care provided to the Xavánte.

Infectious and Parasitic Diseases

Diarrheal Diseases

Diarrheal diseases have been recognized as leading causes of death in children in developing countries. Key factors in the epidemiology of diarrhea are limited sources of safe water for drinking and personal hygiene and lack of sanitation. Under unfavorable conditions, up to 25 percent of cases of acute diarrhea in children may become persistent, lasting for several weeks (Bern et al. 1992; Martines et al. 1993; PAHO 1980; WHO 1988). Rotavirus, bacteria, and protozoa have all been identified as pathogens causing diarrheal diseases in indigenous populations of Amazonia.[9]

Considering the sanitary conditions that prevail in Xavánte villages, especially the contamination of water sources and the inadequate system of waste disposal, the fecal-oral route plays a critical role in the epidemiology of diarrheal disease. As we already mentioned, at the beginning of the wet season the rains wash the human and animal excreta deposited on the ground into the village stream, suggesting a seasonal pattern of transmission. Unfortunately the epidemiological data available do not permit us to investigate seasonal variations in diarrhea morbidity at Etéñitépa.

In addition to the fecal-oral route, other means of diarrhea transmission must be considered. We have emphasized the importance of game in the Xavánte diet, and it is obvious that, whether in hunting camps or in the village, people frequently come in contact with the carcasses of wild animals. A number of investigators have pointed out the potential of wild animals in Amazonia, where they are commonly hunted for food, to spread bacteria and rotavirus, causing gastroenteritis (Linhares et al. 1986a; Lins 1970; Loureio 1985). This is often referred to as zoonotic transmission. Butchering infected animals, especially removing the intestines, may be an important mechanism for introducing novel strains of these organisms into the human population. This becomes clear when one considers the conditions prevailing in Xavánte villages and camps, where water for washing hands is rarely within easy reach.

Even in the presence of environmental risks for diarrhea, specific behaviors have been associated with a lower incidence and reduction in severity. Emphasis on breast-feeding during the first year of life has been singled out as a key factor in protecting small children against diarrheal diseases (Hoyle et al. 1980; Jeliffe and Jeliffe 1977; Martines et al. 1993; Rowland et al. 1978; Victora et al. 1987, 1992). Breastfeeding is a universal practice among Xavánte mothers at Etéñitépa. Children are breastfed as long as they demand it; it is not uncommon for a woman to breastfeed the child at her knee as well as the younger one in her arms.[10] A Xavánte baby is surrounded by women, sisters, aunts, and grandmothers, most of whom are lactating, who are happy to pick the baby up, comfort, and if necessary feed him or her. Even though there is no abrupt weaning, food supplementation starts quite early, often at three to six months. Babies may be given small morsels of food by their parents or siblings, even if only to suck on as pacifiers. The chances are therefore high that babies will be exposed to foods contaminated by someone's hands. This is a likely route for Xavánte babies to contract diarrhea.

According to some estimates, children under five years of age in developing countries suffer on the average 3.5 episodes of diarrhea per year (Martines et al. 1993; WHO 1988). Similar values have been reported in studies from the Brazilian Amazon. In the state of Amazonas, Giugliano et al. (1986) recorded 4.8 episodes per year for young children. Coimbra et al. (1985b) recorded 3.9 episodes per year in an indigenous group of southwestern Amazonia. Particularly in young children, acute diarrhea, if not treated promptly, may lead to dehydration and early death. Although mortality records for the Xavánte cannot be considered reliable, it is clear that gastroenteritis accounts for a large percentage of death in babies (fig. 8.2). In addition to the mortality directly attributable to such illnesses, chronic and/or recurrent acute diarrhea is also associated with nutritional problems, as we shall see later in this chapter.

During our visits to Etéñitépa, we often observed that diarrhea was a common condition, affecting small children in particular. In search of treatment, worried mothers would often bring babies who were passing watery stools to our attention; many of these women had gone through the despair of losing a child in a similar condition. Unable to provide these children with any effective assistance at the village other than standard oral rehydration therapy (ORT), which is not always easy to administer to small, weak babies, we had to rush the little patients and their mothers in our pickup truck to the nearest clinic in town.

Even though, technically speaking, the management of diarrhea should be relatively easy and cheap through treatment by ORT, the high morbidity and mortality from diarrhea that we find among the Xavánte indicate that much more needs to be done. There are obstacles to the effective use of the packets containing rehydration salts that the health agent delivers to the mother with instructions to dissolve them in water. First of all, it is necessary to have a clean bottle in which to dissolve the salts. Second, the available drinking water is not safe because of the unsanitary conditions in the village that we have described (it is not usual among the Xavánte to boil water, and even if they were to do so it would be hard to find a clean pot). So ORT, which is the main treatment for diarrhea available at the village level, is of limited value since it can hardly be effective if it is prepared with contaminated water.

From the point of view of the health services, ORT is certainly cheaper than intravenous rehydration, mostly because of the cost of hospitalization required for the latter, and it is less traumatic for the child. However, the difficulties involved in applying ORT need to be more carefully evaluated. A child suffering from diarrhea is rarely able to drink a large amount of rehydration fluid at one time. Usually Xavánte mothers feed the fluid by the spoonful, often interrupting the treatment to take care of other domestic tasks, so that it may be a long time before the sick child actually ingests a liter of fluid. According to Leslie (1989), a mother may spend as much as 10 percent of her working time over the year administering ORT to her children with diarrhea. The number of Xavánte children suffering from serious diarrhea and dehydration admitted to regional hospitals and treated with intravenous rehydration (in spite of the availability of ORT salts at the village health post) indicates that the ORT strategy, when used in the absence of investment in sanitation, has serious limitations.

Intestinal Parasitism

Intestinal parasitism is another health problem affecting the Xavánte that is closely related to lack of sanitation. At Etéñitépa, which has been located in the same place for almost three decades without adequate sanitation, a high prevalence of intestinal parasites is to be expected. This is because environmental contamination with the infective larvae of helminths and/or highly resistant eggs builds up in the village surroundings. Of course, contamination is even greater at the defecation sites located behind the village. This leads to high rates of reinfection, since people, especially children, normally go barefoot. Intestinal parasites have far-reaching effects upon human health and nutrition. For instance, they can exacerbate protein-energy malnutrition and lead to anemia, since certain species of helminths may cause serious blood loss.

When we conducted a survey at Etéñitépa in 1990, we found that hookworms (*Ancylostoma duodenale* and/or *Necator americanus,* 33.6 percent) and roundworms (*Ascaris lumbricoides,* 25.0 percent) were common intestinal parasites (table 8.1). Several other parasites were observed, including pinworms (*Enterobius vermicularis*), whipworms (*Trichuris trichiura*), *Strongyloides,* and the dwarf tapeworm (*Hymenolepis nana*). We also observed a well-defined age trend in levels of parasitism by helminths, in particular for species that occur at higher prevalences. Rates were highest in young individuals (less than twenty years), especially in those five to ten years of age, and dropped in the older age cohorts. With regard to polyparasitism by helminths, the majority of individuals examined (58.5 percent) had only one parasite, 26.2 percent had two parasites, 9.3 percent had three parasites, and only 6.2 percent had four parasites. Again, frequencies of polyparasitism were highest in children (R. Santos et al. 1995).

A closer look at the prevalence and distribution of intestinal parasites at Etéñitépa provides further evidence that Xavánte children are particularly vulnerable to fecal-oral infections. For example, *Entamoeba histolytica* and *Giardia lamblia,* which are important protozoan agents for diarrhea, were present, respectively, in 7 and 14 percent of children five years of age or younger at Etéñitépa. The prevalence rate of *Entamoeba coli* infection in this same age group was 21 percent. Although usually not considered pathogenic, its high prevalence is an indication of the widespread environmental fecal pollution of both soil and water that prevails in the village and its immediate surroundings (R. Santos et al. 1995).

We also attempted to evaluate the parasite load by counting the number of eggs per gram of feces. We found that a few individuals, all young children, showed very high counts of roundworm eggs. Epidemiologically speaking, these individuals may play a significant role in maintaining high levels of parasitism in the community (see Haswell-Elkins et al. 1989).

TABLE 8.1. Most Common Intestinal Parasites in the Etéñitépa Xavánte, according to Sex and Age, 1990

Age Groups (years)	Number of Exams	Al	Hk	Tt	Ev	Ss	Hn	Eh	Ec	Gl
0–5	28	21.4	39.3	3.6	—	14.3	3.6	7.1	21.4	14.3
5–10	33	45.5	48.5	—	6.1	21.2	12.1	9.1	33.3	9.1
10–20	26	26.9	46.2	—	19.2	7.7	3.8	3.8	38.5	7.7
20–40	26	15.4	7.7	—	—	3.8	7.7	11.5	38.5	3.8
>40	15	—	13.3	—	—	6.7	—	6.7	80.0	6.7
Total	128	25.0	33.6	0.8	5.5	11.7	6.3	7.8	38.3	8.6

Source: Adapted from R. Santos et al. 1995.

Note: Al = *Ascaris lumbricoides,* Hk = hookworms, Tt = *Trichuris trichiura,* Ev = *Enterobius vermicularis,* Ss = *Strongyloides stercolaris,* Hn = *Hymenolepis nana,* Eh = *Entamoeba histolytica* complex, Ec = *Entamoeba coli,* Gl = *Giardia lamblia.*

In the early 1960s the research team headed by James Neel carried out a survey for intestinal parasites at the village of Simões Lopes that pointed to high rates of infection, particularly for roundworms (70 percent) and hookworms (97 percent) (Neel et al. 1968b). Other than our data from Etéñitépa, the only information on intestinal parasitism recently available for the Xavánte comes from a survey conducted in Parabubure (Ianelli et al. 1995). In this study, roundworms, hookworms, and *Strongyloides stercoralis* were the most common helminths, with prevalence rates remaining relatively high even after mass antihelminthic treatment had been given to the population (28.4, 13.6, and 9.9 percent, respectively). The most common parasitic protozoans found were *E. histolytica* (3.7 percent) and *G. lamblia* (8.6 percent).

Our analysis of intestinal parasitism at Etéñitépa points to rates of infection below those reported for other Xavánte villages. The same may be said in relation to the studies carried out among other indigenous groups in Central Brazil, which, overall, have revealed considerably higher prevalence rates than those reported for Etéñitépa (see Baruzzi et al. 1977 and Kameyama 1985). It is likely that this is related to occasional mass worm treatments given at Etéñitépa by FUNAI or other providers of health care.

Malaria

Malaria is a major cause of sickness and death in Amazonia. It is endemic in the state of Mato Grosso, and malaria caused by both *Plasmodium vivax* and *Plasmodium falciparum* occurs in the municipalities where Xavánte reservations are located (Coimbra and Santos 1994a, 120; Ianelli 1997, 47). Malaria has taken a heavy toll among several indigenous groups of Amazonia, especially after contact with the national society. However, very little is known about the epidemiology of malaria among the Xavánte, either in the past or in the present.

The Xavánte at Areões, approximately 100 km south of Pimentel Barbosa (see fig. 2.2), whose villages are near the Rio das Mortes, are probably those who have suffered most from malaria in recent years. Coimbra (1989b, 10) observed cases of *P. falciparum* malaria during an epidemic at Areões that resulted in the death of six children. In the early 1990s there were several other outbreaks of malaria due to *P. falciparum* and *P. vivax* at Areões. In a population of approximately 600 Indians, FUNASA recorded 206 cases of malaria in 1992 and 78 in 1993 (Ianelli 1997, 48).

For a long time the region around the Rio das Mortes and its tributaries has been known to be malarial, affecting not only Indian villages but farmland along the riverbanks (Freitas-Filho 1955). Neel et al. (1964) observed *Plasmodum* parasites in three out of seventy-six thin blood films examined during their biomedical investigations at São Domingos. They also found that palpable spleens were present in 19 percent of the persons examined. Most of these were women. They concluded, "we are inclined to suspect that malaria is a more significant disease in this community than our limited findings would suggest" (119). In the 1990s Xavánte elders at Etéñitépa recalled the old days when they lived by the Rio das Mortes and were prone to malarial infection that, according to them, gave one "fever, the shakes, and a swollen belly" (Ianelli 1997, 64).

During the 1990s, there were cases of malaria at Etéñitépa, but they were few, even though serological studies in 1990 and 1996 showed that people there were continuously exposed to infection by *P. falciparum, P. vivax,* and *P. malariae* (Coimbra et al. 1995; Ianelli 1997). Several field trips to Etéñitépa for malaria studies were made, but very few individuals were found with positive blood smears (0 to 2 percent). These studies were carried out by a physician, Rubens Ianelli, who was working on his master's thesis in public health at ENSP (Ianelli 1997). Enlarged spleens were also rare; the rates recorded were typical of those found in regions of low endemicity for malaria. However, there is always risk of an outbreak, since the vector mosquito (*Anopheles*) is present in the village. Also, men frequently go on fishing trips to swampy flats and lagoons by the Rio das Mortes, where malaria transmission seems to be more frequent.

Environmental changes in the region of Pimentel Barbosa may alter the present pattern of low malaria transmission. The proposal to construct an inland waterway along the Rio das Mortes and the Araguaia (the Hidrovia Tocantins-Araguaia), which we described in chapter 6, is especially troubling. This waterway would not only have negative environmental impacts but it would probably influence the epidemiology of malaria at the regional level, facilitating transmission and thereby affecting the health of the Xavánte.

A study of the faunal composition and behavior of anopheline mosquitoes at Etéñitépa showed that *Anopheles darlingi* was the most common species in the village and its surroundings (30.9 percent of a total of 350 mosquitoes captured and identified) (Ianelli et al. 1998). *A. darlingi* is the most important malaria vector in Amazonia due to its anthropophylic propensities. The great majority of the mosquitoes were captured outside the houses, most of them near the banks of the village stream. This means that the behavior of anophelines at Etéñitépa was markedly exophilic, that is, they tended not to be inside the houses.

Few other studies on the anopheline fauna of Indian reservations of Central Brazil have been carried out. Freitas-Filho (1955) reported that *A. darlingi* was the most common species he found in breeding places by the Rio das Mortes (65 percent of the total anophelines captured). A more recent investigation by Lourenço-de-Oliveira (1989) in the Xingu National Park, which is located north of the Xavánte area, also identified *A. darlingi* as the most common species (90 percent of the total) and clearly pointed to its predominantly exophilic behavior (87 percent were collected by the rivers). Earlier studies carried out in other areas of Central Brazil also pointed to the exophilic propensities of *A. darlingi* in the region (Rachou 1958).

The recurrent observation of *A. darlingi* as exophilic may stem, at least in part, from the peculiarities of indigenous house architecture in Central Brazil. The closed and dark interior of these dwellings may hamper the adaptation of local *Anopheles* populations showing propensity to bite and seek shelter within the houses (i.e., with endophilic/endophagic preference; see fig. 8.3). The constant presence of cooking fires may also contribute to the apparent insect-repellent character of the traditional Indian house. However, one cannot rule out the possibility that the predominantly exophilic behavior of anophelines on Indian reservations may be associated with changes in mosquito behavior due to house spraying with residual insecticides such as DDT. Employees of FUNAI confirmed that house spraying was practiced in Xavánte villages as early as the late 1940s, particularly in those close to the Rio das Mortes.

Certain Xavánte practices facilitate outdoor exposure to mosquitoes. Xavánte women and children often start their daily activities by going to the village stream at daybreak for bathing and to draw water. On their way to the gardens, Xavánte couples also take time for an early swim in the creek. Xavánte children and adults go to the stream at the end of the day and often stay until the sun sets. By the water, the Xavánte wear little or no clothing. Considering that anopheline mosquitoes are most active at sunrise and sundown, one may conclude that the Xavánte pattern of water use exposes individuals of both sexes and all

Fig. 8.3. Xavánte houses are covered with overlapping palm leaves, with a single small entrance facing the center of the village and no windows. (Photo by C. E. A. Coimbra Jr., 1990.)

ages to numerous mosquito bites, some of which may well be infected. A very large percentage (93.5 percent) of the *A. darlingi* captured at Etéñitépa were by the banks of the village stream (Ianelli et al. 1998).

The tendency of *A. darlingi* at Etéñitépa to bite outside the houses, along with the people's timing of outdoor activities, raises doubts as to the effectiveness of applying the more commonly recommended malaria control strategies. These include biannual in-house spraying with residual insecticides, screening windows, and using impregnated bed nets. All of these measures are intended to hamper human-mosquito contact inside dwellings, as they take for granted that anophelines will frequent the interior of houses and bite people there. On the one hand, the Xavánte seem to have relatively "mosquito proof" houses, not requiring bed nets or screens. Their houses actually do not have windows to be screened nor beds to be netted since they sleep on mats (see fig. 8.3). On the other hand, their pattern of water use, which brings them to the banks of the village stream at the time when the mosquitoes are most active, exposes them to a large number of bites. Conventional means of mosquito control alone are not likely to be effective in Amazonian Indian communities. There is a need to develop alternative control measures that take into better account living conditions and subsistence activities as these relate to the observed behavior of mosquito vectors.

Chagas' Disease

Chagas' disease (or South American trypanosomiasis) is not a health issue for the Xavánte. This zoonosis, which is caused by the protozoan *Trypanosoma cruzi,* may be transmitted to humans by blood-sucking insects, mostly in the genera *Triatoma* and *Panstrogylus* (known in Portuguese as *barbeiros* and in English as kissing bugs or assassin bugs). A number of different mammals, including armadillos, opossums, and rodents, have been implicated as reservoirs for the protozoa.

Although potential insect vectors and reservoir animals infected with *T. cruzi* are found in the *cerrado* region of Central Brazil (Mello 1981), the Xavánte do not have the disease. We carried out a serological survey at Etéñitépa and found that all sampled sera tested negative for *T. cruzi* antibodies (Coimbra et al. 1992).

The absence of Chagas' disease among the Xavánte, as well as other Amazonian indigenous peoples, may be because their living conditions still include a number of "protective" factors that help to prevent the bugs that may carry *T. cruzi* from adapting to their dwellings. These include small, often mobile settlements and traditional housing as well as the absence of domestic animals. Chagas' disease is endemic in indigenous populations in the Andes. In the highlands, animal domestication, especially the habit of raising guinea pigs inside houses, is important in the transmission of the disease. When infected, domestic guinea pigs help to maintain the domestic cycle of transmission (Coimbra 1988; Coimbra and Santos 1994b).

Chagas' disease has been a historically important endemic disease in Central Brazil, especially in the regions of Goiás, Tocantins, and Southern Mato Grosso. Recently, however, the National Program for the Control of Chagas' Disease has been very successful in controlling its most important vector, *Triatoma infestans,* which transmits *T. cruzi* in the most affected areas of Central Brazil (PCDCh/MT 1999). The program is also likely to have benefited the Xavánte, since it lessens the pressure of disease transmission in their territories as well.

Skin Diseases

Scabies and pyoderma (impetigo and boils) are the most common skin diseases afflicting the Xavánte, affecting children in particular. Although usually not life threatening, they cause great discomfort. Usually when scabies occurs in small babies it spreads all over the body and produces intense itching and irritability. It is not uncommon to find bacterial infections associated with the sores of scabies as well as those of impetigo. Louse infestation also causes intense itching, and the insect has been shown to be directly implicated in the transmission of bacteria involved in pyoderma (Taplin and Meinking 1988).

A dermatological survey conducted in Etéñitépa in 1990 reported that 66.6 percent of the population had scabies, 37.4 percent had lice, and 18.0 percent suffered from some kind of pyoderma infection (Alvarez et al. 1991). These authors also observed a large number of children with perlèche (24.0 percent), which is a cracking of the labial comissures often associated with riboflavin deficiency superimposed on fungal or bacterial infection; perlèche is commonly seen in undernourished children.

Impetigo in Xavánte children is most often seen on the scalp, ears, and neck, manifested by numerous itching blisters that soon after breaking form characteristic hardened yellowish scabs. The disease causes great discomfort to the children, whose mothers crop their hair very short and submit them, according to the advice of the health agent, when one is present, to all kinds of ineffectual treatments, from painting the affected area with thimerosal to applying antibiotic creams from the drug supplies available at the post.[11]

The overall poor skin health at Etéñitépa can be attributed largely to the limited availability of water of acceptable quality, to which we have already referred. There are other contributing factors that may open the way to infection, especially by staphylococci, which are the most important bacterial organisms associated with pyoderma in children. These bacteria are extremely resistant to environmental conditions; they can survive outside the human body for a long time (even years), remaining viable on fomites, like clothes and towels, or even on dust particles (Baldy and Lopes 1976). For this reason, conditions that have been observed at Etéñitépa, as in other Amazonian indigenous communities, like continued wearing of unwashed clothing, use of the same blanket by different persons, close living conditions, and numerous sores and scratches on the skin from insect bites (especially in children), are key factors in spreading staphiloccocal skin infections (see Coimbra et al. 1985a; Lawrence et al. 1979).

It is important to note the absence of leprosy among the Xavánte, despite the fact that the disease is considered moderately to highly endemic in the municipalities of Mato Grosso where Pimentel Barbosa reservation is located (Munhoz et al. 1997; Penna 1994).

American cutaneous leishmaniasis, another endemic infectious disease in Central Brazil (CENEPI 1992; Marzochi 1992), is not a problem among the Etéñitépa Xavánte. Alvarez et al. (1991) found no cases of the disease in a dermatological survey. Leishmaniasis is a zoonotic protozoan disease caused by species of *Leishmania.* Wild rodents are the most important reservoirs of *Leishmania,* and the protozoa are transmitted to humans by the bite of infected sandflies of the *Lutzomyia* genus.

The only Xavánte reservation where records for leishmaniasis were found is Parabubure, with approximately forty cases diagnosed in the late 1970s (Cardoso 1979). This apparently exceptional situation may have been associated with environmental disruption of zoonotic foci of *Leishmania* sp. due to a recent move of the village to a new site and/or the intensification of agricultural activities in conjunction with the FUNAI rice project of the time (see chapter 6). The only other outbreak of cutaneous leishmaniasis in an indigenous population in Central Brazil that has been reported in the literature was in an area relatively close to Xavánte reservations, among the Waurá in the Xingu Indigenous Park. It broke out shortly after they had moved their village site, which led the authors of the report to advance the hypothesis that the Indians had been infected with some local strain of *Leishmania* to which they had no cross-immunity (Aston and Thorley 1970; Carneri et al. 1963). We might suggest that the epidemiology of cutaneous leishmaniasis among the Xavánte is similar to that described for other Amerindian populations of the Brazilian Amazon, that is, endemic infection with rare cases of disease, most of which are benign. These circumscribed outbreaks are often associated with some sort of environmental disruption of the forest due to population movement into new sites and/or forest clearing (see Coimbra et al. 1996a).

Endemic pemphigus foliaceus (EPF) or fogo selvagem, an autoimmune exfoliative dermatosis characterized by blisters that form beneath the skin, is a major health concern to the Xavánte. This disease of yet unclear etiology is endemic in Central Brazil, with most cases occurring in adults (Castro et al. 1983; Diaz et al. 1989). For the Xavánte, fogo selvagem is a potentially fatal disease and a major cause of stigma and suffering due to the appearance and discomfort of the lesions, which are distributed all over the body. As early as in the 1960s, when the research team coordinated by James Neel and Francisco Salzano visited the Xavánte village at Simões Lopes, they not only described a case of fogo selvagem but they advanced the hypothesis that it was an autoimmune skin disease (Weistein et al. 1967).

Several independent studies have concluded that fogo selvagem is precipiated by the continuous exposure of genetically susceptible individuals to a specific environmental factor (virus?) or factors (Auad 1972; Castro et al. 1983). Blackflies (family Simuliidae), locally known as *pium,* have been pointed out by various authors as the most likely vectors; farmers exposed to their bites were 4.7 times more likely to develop the disease than those who were not exposed (Lombardi et al. 1992; see also Eaton et al. 1998). Blackflies are an extremely common insect pest at Etéñitépa, especially in the rainy season and along the banks of rivers and streams.

The distribution of fogo selvagem at Etéñitépa is far from random. All known cases are clustered within three lineages, all descendants of the same ancestral couple; four are their grandchildren, five are great-grandchildren, and two are great-great-grandchildren (Friedman et al. 1995). Available evidence strongly suggests the inheritance of HLA (histo-compatibility antigens) genes predisposing to fogo selvagem as an explanation for the familial clustering of cases in Etéñitépa (Cerna et al. 1993a, 1993b; Friedman et al. 1995).[12]

Records of the Nova Brasília Hospital in Nova Xavantina for 1992 indicated forty-eight cases of fogo selvagem in Xavánte patients; the records available at the FUNAI office in Nova Xavantina gave a slightly lower figure (forty-one cases). According to the FUNAI data, frequencies were highest in the population from Pimentel Barbosa, most patients being from the village of Etéñitépa, where prevalence reaches 3 to 4 percent (Friedman et al. 1992, 1995).

Corticoids are the most widely used drugs for fogo selvagem treatment in the general population, where they have contributed significantly to a reduction in the number of deaths due to the disease (Anhalt et al. 1989). However, the ways corticoids are often used in treating Xavánte cases may cause additional harm. This is because patients are sent back to their villages under high doses of medication and follow-up procedures are not carried out. Treatments are interrupted abruptly when patients run out of medication in the village, and it may be months before the patient has the opportunity to visit a medical clinic again. The way fogo selvagem has been handled in Xavánte areas is clearly inadequate. It also results in higher costs to FUNAI and FUNASA, since chronic patients end up being sent to hospitals in towns as far away as São Paulo, Campo Grande, Cuiabá or Brasília. These trips must certainly exacerbate the psychological and economic impact of fogo selvagem on their lives and the community.

Hepatitis B

Hepatitis B infection and disease are highly endemic in South American indigenous populations. Positive rates for hepatitis B virus (HBV) serological markers (HBsAg, anti-HBs and/or anti-HBc)[13] are very high in Amazonia, and it is not uncommon for these markers to be found in all age groups and in both sexes (Bensabath et al. 1987; Hadler et al. 1987). The epidemiology of this ancient and ubiquitous viral disease is closely related to a complex of cultural practices that enhance the likelihood of HBV transmission (among others, bloodletting, scarification, tattooing, and orally processed food).

We conducted a survey among the Xavánte of Etéñitépa in which we found high rates of positivity to HBV serological markers (Coimbra et al. 1996b).[14] Considering three serological markers (HBsAg, anti-HBs, and anti-HBc), 68.3 percent of the males and 49.3 percent of the females were positive, indicating either current or past contact with the virus. However, only a small number of individuals (a man and a woman in their twenties) were HBV carriers, that is, they had the virus at the time of examination. The remaining had had contact with the virus in the past but were no longer infective. Our results also found high rates of infection in children and young adults and no sex differences in positive rates for HBV serological markers.

Hepatitis B is a major public health problem both in rural and urban Amazonia. Rates of infection in indigenous groups seem to exceed those of non-Indian populations (A. Santos et al. 1995; Soyano et al. 1976). Although the epidemiology of hepatitis B in Indian populations is still poorly understood, it is clear that there are effective mechanisms of transmission. Anthropological accounts of native Amazonian groups are rich in descriptions of cultural practices that are regarded as playing an important role in the epidemiology of hepatitis B.[15]

The hepatitus B virus is found not only in blood but in other body fluids such as saliva, semen, menstrual discharge, colostrum, and milk

(Gust 1980; Mazzur 1973; Zuckerman 1983). The virus is very resistant to extreme environmental conditions, remaining infective for a considerable amount of time after removal from a host. In addition, HBV transmission requires only a very small amount of infective material. Therefore, hepatitis B is highly contagious and its endemicity may be maintained in a population by means of both horizontal (person to person) and vertical (parent to offspring) transmission.

Practices that involve breaking the skin are probably a major route of horizontal transmission of HBV among Amazonian indigenous peoples, who hold bloodletting and scarification to be prophylactic or therapeutic. Scarification has also been implicated in hepatitis B transmission in Africa (Tsega et al. 1986; Williams and Williams 1972). The Xavánte consider bleeding to be a major therapeutic resource. According to Maybury-Lewis, "they bleed each other as the commonest way of treating ailments; when they are tired; in order that they may not feel tiredness" (1967, 70). Xavánte children are also bled when their parents want to strengthen them. Bleeding involves incisions in the flesh made with small wooden sticks or animal claws. When an adult decides to bleed his or her children, they are all done on the same occasion, including "any other small boys that happen to be about" (70). We observed in the 1990s that bleeding is still a very common practice among the Etéñitépa Xavánte (fig. 8.4).

As we mentioned in chapter 2, among the Xavánte ear piercing is part of the male initiation ritual. This is done in one day on a group of boys, age twelve to eighteen, as they leave seclusion in the *hö,* the bachelors' hut (Maybury-Lewis 1967, 116). This is done by an elder, who uses the same instrument, a sharp piece of bone, on all the boys. According to Giaccaria and Heide (1984, 166), "the ear-piercer dampens the bone with saliva by passing it through his mouth . . . then he begins to pierce the ear by twisting the bone."

In the surroundings of Indian villages, people often scratch their skin on sharp grasses, twigs, and thorns along forest and garden paths. Since HBV can survive for a long time outside the body, it may stay viable on the surface of the plants on which they scratch themselves. In an experimental assay, Soares and Bensabath (1988) found indications of the presence of HBV on plants collected along paths heavily used by rural populations in the eastern Amazon. In Scandinavia, the importance of this form of transmission was demonstrated among fell runners and track finders (Vellar 1964). In the Amazon, it may be facilitated by

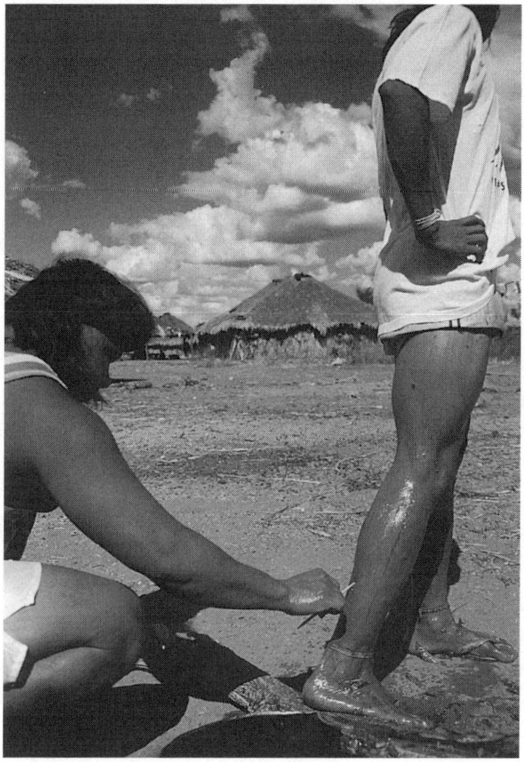

Fig. 8.4. Xavánte often scarify their legs using a sharp stick, at the same time bathing them with herbs soaked in water. They do this especially before log racing, in the belief that bleeding endows them with extra strength and swiftness. (Photo by C. E. A. Coimbra Jr., 1995.)

the fact that people wear little clothing, thus leaving much bare skin exposed to vegetation while walking or working.

Although the role of insects and ticks in transmitting HBV is still debatable, the constant scratching of sores from bites can facilitate transmission from one person to the other. This factor can be of particular importance in Amazonia due to the plethora of blood-feeding arthropods to which individuals are constantly exposed. According to Black et al. (1986, 341) the lower rate of HBV infection observed among the Chilean Mapuche, when compared to Amazonian groups, may in part be attributed to the "relative freedom from biting insects and full clothing cover."

As was described earlier, the skin of the Xavánte is often affected by rashes, abrasions, sores, and eventual bleeding. Pyoderma ranks among the most common skin diseases in the Xavánte, especially in children, and is of particular importance in the epidemiology of hepatitis B. Evidence of the presence of the virus has been detected in samples of the fluid from the lesions of impetigo taken from HBV carriers (Petersen et al. 1976).

Close living conditions are associated with increased risk of the spread of a number of infectious diseases, including hepatitis B. For instance, among the Maori and Pacific islanders, the number of people in the household was pointed out as a major risk factor for HBV transmission to children (Milne et al. 1987). Xavánte houses shelter large extended families, with individuals sharing many implements, and children especially come into frequent close contact both with adults and other children.

Oral-oral HBV transmission has been demonstrated in a number of studies (Alter et al. 1977; Petersen et al. 1976). Maybury-Lewis reported that it was common practice among the Xavánte to supplement a baby's diet with premasticated food. Xavánte babies "may even be fed slops from the mouth of one of the women . . . [and] grandmothers are particularly fond of feeding their tiny grandchildren in this way" (1967, 67). Saliva may transmit HBV, and any bleeding that takes place during chewing may enhance the infective potential of the item being shared.

Sexual behavior may also be related to the spread of HBV. For instance, sexual activity has been recognized as a major mode of HBV transmission (Alter et al. 1989; Petersen et al. 1992). Transmission from men to women and from women to men seems to be equally efficient (Petersen et al. 1992). Having multiple sexual partners is a major risk factor (Alter et al. 1989; Petersen et al. 1992; Szmuness et al. 1975). Since our investigation found high rates of HBV infection in children, and so many different social practices may result in HBV transmission, it is possible that sexual transmission plays a less important role among the Xavánte than in societies where first exposure to the virus is likely to occur later in life.

A common theme in the literature on the epidemiology of hepatitis B is the identification and quantification of practices favoring exposure to HBV. It is well known that in Western societies needle sharing, tattooing, and a high frequency of sexual encounters with different partners are associated with higher risks for HBV transmission. These and other risk-related practices are often associated with specific social segments or "subcultures." It is expected that the likelihood of becoming infected will be associated with the frequency and types of risk behavior in which individuals engage.

We have seen that many cultural practices and ecological factors may be involved in HBV transmission in the Xavánte population. In Amazonian societies, the issue of "risk behavior" with regard to the transmission of hepatitis B is closely tied to social organization and culture. Unlike Western societies, where there is high internal sociocultural heterogeneity, with the formation of a number of subcultures, indigenous peoples of Amazonia tend to present much less within-group sociocultural differentiation. Therefore, compared to Western societies, risk behaviors are not as easily connected to specific social segments. One result is that when Amazonian Indians engage in practices that facilitate the spread of the HBV the universality of these practices favors the transmission of the virus within the community as a whole.

Respiratory Diseases

Acute respiratory infections (often referred to as ARIs) are leading causes of sickness and death among the Xavánte. As we wrote in a previous section, approximately a third of the Xavánte patients assisted at the Pronto Socorro Municipal de Nova Xavantina in 1993–94 were treated for diseases of the respiratory system. Data collected from the Casa de Saúde do Índio at Nova Xavantina for 1994–95 reinforce this picture (fig. 8.1). In 1993, pneumonia accounted for 35 percent of deaths in Xavánte infants (fig. 8.2). Even without taking into account the probability that the statistics just cited are seriously underreported, the conclusion that acute respiratory infections have a tremendous impact on Xavánte health is inescapable.

Respiratory diseases may be caused by viruses or bacteria, and they are most effectively transmitted under crowded living conditions (Denny and Clyde 1983; Stansfield and Shepard 1993). No matter what the specific etiology, respiratory infections are easily spread within the household by droplets from a cough or a sneeze. The likelihood that they may become fatal is highest among children weakened by malnutrition and associated infections such as diarrhea, which, unfortunately, are common among Xavánte children. Malnutrition is, in itself, a major risk factor for pneumonia in children; some authors estimate the risk to be twelve to twenty times higher in undernourished than in well-nourished children (Berman 1991; Tupasi et al. 1988). As Stanfield and Shepard (1993, 68) point out, "While nutritional diseases augment the chances of ARI episodes, episodes of ARI contribute to nutritional deficiency, thus further increasing the risk of subsequent infection and death."

Of the acute respiratory infections that affect the Xavánte, pneumonia is the most dangerous because of its life-threatening potential. Even nonspecific "influenzalike" respiratory diseases may rapidly become serious, particularly in children, due to increased production of mucus and bronchospasms.

The dry season months appear to be particularly favorable to the rapid spread of acute respiratory diseases among the Xavánte. It is typical to see children with runny noses, ear infections, moderate to high fevers, and hacking coughs when one approaches the entrance of a Xavánte house at this time of the year. Unfortunately, we were unable to break down the data obtained from consultation and hospitalization records by month in order to see seasonal variations. However, considering such contributing conditions to respiratory diseases as living in close quarters, dry air (relative humidity in Central Brazil may drop below 10 percent in the dry season), increased dust particles in suspension due to lack of rain (in some years there is no rainfall in July and August at all; see fig. 8.5), and continual exposure to smoke from indoor fires lit for warmth during the cold nights, one would expect an increase in the incidence of respiratory conditions during this season.

Indoor air pollution has become a leading concern in the epidemiology of respiratory diseases. Wood smoke and soot are important pollutants in societies where biomass fuels are used for cooking and heating. In Xavánte houses, the cooking fire is usually in the center of the house floor, and since there is no chimney smoke drifts around the house until it finally filters to the outside through the blackened palm fronds of the roof. Even though some Xavánte families have built separate cook houses, they are likely to make fires indoors for warmth, as well as cooking, during the cold, dry nights of the winter months. A number of studies have pointed to the link between pneumonia and chronic exposure to wood smoke (Campbell et al. 1989; Koning et al. 1985).

Tuberculosis

Physical examinations conducted by Neel and his coinvestigators in the 1960s found no clinical signs or symptoms of tuberculosis at São Domingos (Neel et al. 1964). In the villages of Simões Lopes and São

Fig. 8.5. Dust blowing across the village in the dry season. Upper respiratory infections in children often become more serious at this time because of dust, low humidity, and cold nights. (Photo by C. E. A. Coimbra Jr., 1990.)

Marcos they found only one positive response to a tuberculin test (Neel et al. 1968a, 487).

Tuberculosis did spread throughout most Xavánte villages in the 1960s and mid-1970s, when it reached epidemic proportions. Unfortunately, there is very limited epidemiological data about tuberculosis among the Xavánte at that time. When Flowers lived at Etéñitépa in 1976–77, some twelve people were under tuberculosis treatment. Today, tuberculosis remains in Xavánte reservations as an important endemic disease, affecting the elders in particular. Most patients have gone through various series of treatments at different periods. It is not uncommon that treatments are interrupted and the disease manifests itself again in a couple of months or years.

The FUNAI administration in Nova Xavantina recorded nine cases of pulmonary tuberculosis in the reservations under its jurisdiction between 1988 and 1990 (Areões, Parabubure, and Pimentel Barbosa). This information is very likely to be underreported. Unfortunately, we were not able to obtain reliable data on the incidence of tuberculosis on

Xavánte reservations. It is known, however, that in some villages tuberculosis is particularly serious. In Água Branca, a village located on the Pimentel Barbosa reservation, twelve cases of pulmonary tuberculosis were confirmed by bacteriological sputum examination between January and June of 1996 (Amarante et al. 1996). This indicates that previous figures reported were probably far too low.

It is clear that the picture of tuberculosis among the Xavánte is extremely grave, considering that we see only the fragmentary reports that come to light. The dismantling of the FUNAI health services has led to the disruption of tuberculosis treatment, either because the necessary drugs are not provided or because there is no health worker in the village to administer them. During our visits to Pimentel Barbosa in 1990–91 we were alarmed to find that a large number of children under the age of five had no BCG scar, which showed how rare routine medical visits to the villages there had become, even though the Brazilian Ministry of Health requires BCG vaccination for all newborns.

Syphilis

Sexually transmitted diseases (STDs) are a cause for concern among many Amazonian indigenous populations. Physicians who visited Etéñitépa in the 1990s and carried out physical examinations did not report finding STDs (Alvarez et al. 1991). Neel et al. (1964, 118), who tested sixty-three specimens for syphilis, concluded that the Etéñitépa Xavánte were free of treponemal infection.[16]

Our genealogical studies show that the Etéñitépa Xavánte have maintained a high level of village endogamy over recent generations. Also, there is no observation or record of prostitution either within the community or with outsiders. Even during the period of postcontact social disruption, family structure remained largely intact. At present, with young men visiting neighboring towns as well as distant urban centers, there is increasing risk that they may bring STD infection to the village.

Arbovirus Infections

Arbovirus infections are a group of diseases caused by numerous viruses that multiply in blood-sucking arthropods (mostly commonly mosquitoes). The bites of these insects transmit the viruses to vertebrate hosts (birds and mammals).

In 1990 we carried out a survey for arbovirus antibodies at Etéñitépa.[17] Flaviviruses produced by far the highest number of positive reactions (22 percent of the sample). This may be explained by the high number of positive reactions to the yellow fever virus (15.8 percent), which is a flavivirus. Since no cases of yellow fever among the Xavánte have been reported, we may suspect that the number of positive reactions is due to vaccinations against yellow fever. One would expect this figure to have been even higher if the public health services had provided thorough coverage of indigenous populations in the region.

Among the alphaviruses, only the Mayaro was observed (in 3.2 percent of the sample). Finally, only one positive reaction to the bunyaviruses was observed, specifically to the Maguari virus. The presence of antibodies to the Mayaro and Maguari viruses was reported in the 1960s by Neel et al. (1968a) among the Xavánte of Simões Lopes and São Marcos, with a much higher prevalence (17 to 20 percent) than we found among the Etéñitépa Xavánte. In addition to these viruses, Neel and his colleagues detected several other important arboviruses, including Ilhéus, to which all our samples tested negative. It is difficult to explain why the Etéñitépa Xavánte, since they are highly exposed to mosquito bites, had much lower prevalences of arbovirus infection than those observed in the previous studies among the Xavánte.

It should be noted that ecological modifications that lead to the proliferation of mosquitoes may alter transmission patterns of arboviruses. This is a concern, considering the environmental disturbances that the planned Araguaia-Tocantins waterway might cause (see chapter 6). Environmental changes could lead to outbreaks of important disease-producing sylvatic arboviruses, such as Oropouche and Mayaro, in the Rio das Mortes region, as has occurred in other areas of Amazonia (Dégallier et al. 1992; Vasconcelos et al. 1989, 1991).

Nutritional Status

The nutritional status of populations and of individuals is closely related to their biological and socioeconomic surroundings. In poor regions of the world, inadequate dietary intake of energy, micro- and macronutrients, and exposure to infectious-parasitic diseases are major immediate (or "proximal") determinants of malnutrition. But beyond these there is a whole range of socioeconomic factors, which, although they may not directly determine nutritional status, have a great impact on it, since in various ways they increase the risk of becoming sick, and/or not consuming sufficient food. These factors, usually referred to in the literature as "distal determinants," include, among other things, precarious sanitation, which increases exposure to diseases like diarrhea and respiratory infections; inadequate health care; insufficient agricultural production; and low purchasing power (Pinstrup-Andersen et al. 1993; Waterlow 1992; WHO 1995).

Insufficient consumption of calories, proteins, and other nutrients, especially if prolonged, directly affects bodily functions, resulting in a state of malnutrition. In children, food deprivation can also directly compromise physical growth, leading to slowing or a complete halt in linear growth (in stature) and/or slow weight gain or loss of weight. Sickness, especially when caused by infection, interacts with malnutrition in various ways. Sick children usually lose their appetites, lowering food intake. Reduction in nutrient absorption also commonly occurs with diarrhea and intestinal parasitism. Respiratory infections in children may also affect physical growth (Pinstrup-Andersen et al. 1993; Waterlow 1992; WHO 1995).

If, on the one hand, episodes of sickness can lead to malnutrition, on the other hand malnourished children are more susceptible to infection (Chen and Scrimshaw 1983; Martorell 1980; Martorell et al. 1980; Nabarro et al. 1988; Scrimshaw 1977; Scrimshaw et al. 1968). The synergism between infection and malnutrition is seen as possibly the most damaging combination for child health in the less economically favored regions of the world (Briend 1990). Malnourished children tend to have longer and more severe episodes of diarrhea, with increased risk of dehydration and hospitalization. One reason is because undernutrition has a negative effect on the immunocompetence of the child's organism (Chandra 1979; Chandra and Scrimshaw 1980). The association between malnutrition and increased mortality is also well established, especially in children with severe growth deficits (Chen et al. 1980; Gómez et al. 1956; Kasongo Project Team 1983; Pelletier et al. 1995; Van Lerberghe 1988).

The epidemiological data previously presented in this chapter show that we can find at Etéñitépa both the proximal and the distal determinants associated with malnutrition. Diarrhea, intestinal parasites, and respiratory infections are common among Etéñitépa children. Sanitation is inadequate. As we will demonstrate, rates of anemia and protein-energy malnutrition, possibly two of the most common nutritional problems in the world, are high at Etéñitépa.

Anemia

Anemia is a major public health problem in many areas of the world[18] and in poor populations living in less developed regions in particular. It has been estimated that the total prevalence of anemia in the world was probably close to 30 percent in 1980. Prevalence of anemia was 8 percent in the more developed regions of the world, while in developing countries it reached 36 percent. Young children and pregnant women were the most affected groups. In developing countries, it was estimated that anemia affected 51 percent of children four years of age and younger and 59 percent of pregnant women (DeMaeyer and Adiels-Tegman 1985).

Although a number of factors may play a role in the causation of anemia, from the public health point of view nutritional deficiencies and infectious and parasitic diseases are among the most common causes (Baker 1978; WHO 1975). Iron deficiency is by far the commonest cause of nutritional anemia all over the world. Diseases of the gastrointestinal tract are also associated with anemia. Blood-sucking intestinal parasites, such as hookworms, may accelerate the development of anemia because of direct blood loss associated with bites on the intestinal mucosa (Crompton and Stephenson 1990; Roche and Layrisse 1966; Variyam and Banwell 1982).

Anemia in pregnant women increases maternal morbidity; it may also lead to low fetal birth weight and increased risk of premature delivery. The ability of the individual to resist infections may be reduced by anemia and iron deficiency. In young children, there is evidence of growth retardation associated with anemia. Anemia may also reduce work tolerance and productivity, especially in energy-demanding tasks (Baker 1978; WHO 1975).

In 1995 we carried out a study to determine the prevalence of anemia among adults and children at Eténitépa.[19] The results are presented in table 8.2. Approximately 40 percent of the individuals tested were anemic. Frequencies of anemia were found to be quite high in children and adolescents, often above 50 percent. In adults, anemia was much more frequent in women than in men. There were few cases of adult men with anemia. On the other hand, anemia is very frequent in women during the period when they are bearing children (fifteen to thirty years), when nearly 45 percent were anemic.[20]

Considering the population as a whole, the prevalence of anemia at

Etéñitépa, at 39.8 percent, is high. However, it is important to mention that the majority of cases of anemia are mild or moderate. Of the 231 individuals tested, only 20 had hemoglobin concentrations below 10 mg and 8 below 9 mg. The majority of those with very low hemoglobin concentrations (below 9 mg) were children 10 years of age or younger. Anemia is widespread in Xavánte adult women. However, only two women of reproductive age (both 15 to 29.9 years) had hemoglobin values below 10 mg.

Anemia is frequent among women at Etéñitépa, and this is certainly associated with their reproductive patterns. For the period between 1972 and 1990 our demographic study indicates that women at Etéñitépa bore an average of eight children during their lifetimes (see chapter 5). This implies that during most of their reproductive years Xavánte women were pregnant and/or lactating. Early and Peters (1990, 51) made an interesting analysis, using demographic data from the Yanomámi, which showed the proportion of their reproductive years that women spent in pregnancy and lactation. Combining the nine months of each pregnancy with an average lactation period of nearly two years, these authors concluded that the average Yanomámi woman is pregnant or nursing for 92 percent of her reproductive years. Considering that the fertility of Xavánte women is similar to that of Yanomámi women, the percentage of time that they spend in pregnancy and lactation must be

TABLE 8.2. Frequencies of Anemia in the Etéñitépa Xavánte, according to Sex and Age, 1995 (in percentages)

Age Groups (years)	Males	Females	Total
0–5	43.5 (*n* = 23)	58.3 (*n* = 24)	51.1 (*n* = 47)
5–10	56.0 (*n* = 25)	65.0 (*n* = 20)	60.0 (*n* = 45)
10–15	50.0 (*n* = 10)	45.5 (*n* = 22)	46.9 (*n* = 32)
15–30	8.0 (*n* = 25)	45.2 (*n* = 31)	28.6 (*n* = 56)
30–40	— (*n* = 8)	21.4 (*n* = 14)	13.6 (*n* = 22)
40+	33.3 (*n* = 12)	17.6 (*n* = 17)	24.1 (*n* = 29)
Total	34.0 (*n* = 103)	44.5 (*n* = 128)	39.8 (*n* = 231)

close to that found by Early and Peters (1990). In other words, during most of their adult life Xavánte women are in physiological states characterized by increased needs for micronutrients, such as iron and folate, whose deficiencies are closely related to nutritional anemia.

Protein-Energy Malnutrition

When the human organism is not able to meet its energy and protein requirements, growth may falter. Protein-energy malnutrition (PEM) results from infectious diseases, insufficient intake of energy and protein, or, most commonly, a combination of the two. With iron-deficiency anemia, PEM is the most common nutritional deficiency affecting children in developing countries (Pinstrup-Andersen et al. 1993; Waterlow 1992; WHO 1995).

The analysis of physical growth by anthropometry is the most common procedure for evaluating the nutritional status of children. At present the World Health Organization recommends analyzing the physical status of children through three principal indices: height for age, weight for age, and weight for height (WHO 1995). According to the WHO definition, *malnutrition* "refers to the syndrome that results from the interaction between poor diet and disease and leads to most of the anthropometric deficits observed among children in the world's less developed areas" (163).[21]

A major difficulty in attempting to use anthropometry to characterize the nutritional status of Etéñitépa children is the lack of reliable information about their ages. Without this information the height-for-age index, which is one of the most informative measures of nutritional status, cannot be used. This is a common limitation in nutritional studies of indigenous children in Amazonia (Dufour 1992; Santos 1993). During our fieldwork at Etéñitépa in 1990 we were not able to use anthropometry to investigate nutritional status because we did not know the children's ages. Although they measured children at São Domingos in 1962, James Neel and his collaborators were also unable to describe physical growth in detail because age information was missing (Neel et al. 1964).

The nutritionist Silvia Gugelmin, as part of the research for her master's thesis in public health at ENSP, collected anthropometric data at Etéñitépa in 1994–95. She carried out, according to WHO (1995) recommendations, a nutritional evaluation of approximately fifty children aged

4 years or younger. She was able to do this because our team kept track of the birth dates of children born between 1990 and 1995. Also, a growing number of parents at Etéñitépa are noting the birth dates of their children.

Part of the results of the nutritional evaluation that Gugelmin carried out are presented in table 8.3. Two sets of cross-sectional data were collected, in 1994 and 1995, using basically the same group of children. We can see that there are differences in the frequencies of stunting, wasting, and low weight for age when we compare the data for 1994 with those for 1995. These differences may be due to seasonal influences on morbidity (diarrhea and other infections) or subsistence (food availability) and/or random variation because of the small sample size.[22]

Overall, the data presented in table 8.3 show high rates of low height for age or "stunting" (22.0 to 27.7 percent) in Etéñitépa children. From the nutritional evaluation it is clear that approximately one-quarter of these Xavánte children suffer from "chronic malnutrition." The frequency of low weight for height, or "wasting," however, is only zero to 4.2 percent. Wasted children are often those who have experienced a recent and severe process that has led to significant weight loss, often due to acute starvation and/or severe disease.

In Etéñitépa children PEM is closely associated with age. Frequencies of stunting and wasting are particularly high in children between twelve and twenty-four months, the second year of life. Innumerable studies from many parts of the world have shown that, after a period of relatively good health and nutritional status in first few months of life,

TABLE 8.3. Frequencies of Malnutrition in Etéñitépa Xavánte Children 0–48 Months Old, Sexes Combined, 1994–95 (in percentages)

Age (months)	Stunting (low height for age)		Wasting (low weight for height)	
	May 1994	February 1995	May 1994	February 1995
0–12	7.7 (*n* = 13)	0 (*n* = 03)	0 (*n* = 13)	0 (*n* = 03)
12–24	45.5 (*n* = 11)	45.8 (*n* = 24)	20.0 (*n* = 10)	0 (*n* = 24)
24–48	19.2 (*n* = 26)	10.0 (*n* =20)	0 (*n* = 25)	0 (*n* = 20)
Total	22.0 (*n* = 50)	27.7 (*n* = 47)	4.2 (*n* = 48)	0 (*n* = 47)

Source: Data from Gugelmin 1995, tables 3.19–3.21.

both are likely to deteriorate after six months and especially in the second year (see Chen and Scrimshaw 1983; Martorell and Habicht 1986; Pinstrup-Andersen et al. 1993; and Waterlow 1992). Growth retardation is often closely associated with an increase in exposure to infection, especially diarrhea. Gordon et al. (1963) coined the term *weanling diarrhea* to refer to the "diarrheas and the dysenteries that come with the transition of babies from a breast-fed existence to a mixed diet." What tends to happen with Xavánte children, as is common in other populations, is that their nutritional status improves during the third year. The frequency of malnutrition among Etéñitépa children aged twenty-four to forty-eight months is significantly lower than that found in children aged twelve to twenty-four months (table 8.3).

The results of the nutritional evaluation point to a situation of long-term, cumulative inadequacies in health and nutrition at Etéñitépa (high frequencies of stunting but relatively low frequencies of wasting). Anthropometric studies done by the nutritionist Maurício Leite in the village of São José on the Xavánte reservation of Sangradouro (1998) show the same problem of infant and child malnutrition. Frequencies of stunting (35.8 percent) and wasting (2.0 percent) found at Sangradouro in 1997 point to a situation even more serious than that at Etéñitépa.

It is informative to compare the results of the nutritional evaluation of children at Etéñitépa with data for Brazil. Table 8.4 presents the prevalence of PEM in Brazilian children five years of age and younger. These data are from national-level research on health and nutrition carried out in 1989 in the Pesquisa Nacional Sobre Saúde e Nutrição, or PNSN (National Survey of Health and Nutrition) (Monteiro et al. 1992).

TABLE 8.4. Frequencies of Malnutrition in Brazilian Children 0–5 Years of Age, according to Geographic Region, 1989 (in percentages)

	Stunting (low height for age)		Wasting (low weight for height)	
Regions	Urban	Rural	Urban	Rural
North	23.0	—	3.1	—
Northeast	27.3	30.7	2.4	2.8
Southeast	8.1	12.7	1.9	1.3
South	8.7	11.7	1.4	0.9
Center-West	8.2	10.2	2.0	3.2
Brazil	12.3	22.4	1.6	3.2

Source: Adapted from Monteiro et al. 1992.

The frequencies of stunting (22.0 to 27.7 percent) among Xavánte children are closer to those found in children from the North and Northeast, economically the least favored regions of Brazil, than they are to those found in the Center West, Southeast, and South. The North and Northeast regions have the highest infant mortality, the lowest life expectancy at birth, the poorest sanitation indicators, the lowest vaccination coverage, and the lowest levels of maternal education in the nation (Monteiro et al. 1992). If we compare the number of children classified as chronically undernourised at Etéñitépa with information from the Center West, where all the Xavánte reservations are located, we can see that the rate of stunting among Xavánte children is two to three times higher than that generally found in rural parts of the Center West region (10.2 percent).

Dental Health

The subsistence and food production strategies that human groups have practiced at different periods in history are reflected in conditions of oral health. Many studies in dental anthropology, mostly based on the analysis of skeletal remains, show a relationship between oral health and changes in diet and subsistence strategy (Cohen 1989; Cohen and Armelagos 1984; Hillson 1996; Scott and Turner 1988). The increased prevalence of caries after the introduction of agriculture is usually attributed to the adoption of a diet rich in carbohydrates. Studies of present-day indigenous populations also point to deterioration in oral health when people begin to consume foods typical of Western diets (Donnelly et al. 1977; Grim et al. 1994; Tsubouchi et al. 1995).

Neel and his collaborators collected data on oral health at São Domingos in 1962, where they found a low frequency of caries. They reported that, out of a total of thirty-seven men and women who they examined, twenty-eight were caries free, which led them to conclude that, "the observed frequencies [of caries in São Domingos] are comparable to those reported for Australian aborigines who have not been exposed to European foods" (1964, 107–8). During the 1990s master's students in public health from ENSP carried out two epidemiological surveys on oral health at Etéñitépa (Pose 1993; Arantes 1998; Arantes et al. 2001). Comparison of the results of these recent surveys with data from 1962 shows a clear tendency toward deterioration in dental health, with an increase in caries and tooth loss.

The decayed, missing, and filled tooth (DMFT) index is widely used in epidemiological research on oral health (WHO 1987). This index indicates the number of decayed (component D), missing (component M), and/or filled (component F) teeth (T). Thus, an adult who had, of his thirty-two teeth, three with caries, two filled, and one missing, would be given a DMFT value of 6.

The mean DMFT values for permanent teeth at Etéñitépa in 1997 are given in table 8.5. The twelve to fourteen age group has a DMFT of 3.7. The DMFT values practically double in each subsequent age group, reaching the highest value (19.6) in the oldest group (fifty plus years), demonstrating a clear tendency for oral health to deteriorate with age. The "missing" component (M) is relatively low until the fifteen to nineteen age group, but increases markedly in people who have reached their thirties. Among the Xavánte tooth loss is primarily due to caries, trauma, and the common practice of extraction, which is usually the only treatment option available. The "filled" component (F) is rare in all age groups, which can be explained by the lack of access the Etéñitépa Xavánte have to dental services. The only individuals with filled teeth were young men in their twenties who had dental work done during trips away from the village.

Since Neel et al. (1964) have given us a description of the dental status of the Xavánte in 1962, we can see how the DMFT index has evolved (table 8.6). We see a constant tendency over time for increased values at all ages. In some age groups there was a more than tenfold increase from the 1960s to the 1990s.

The experience of the Etéñitépa Xavánte is typical of changes in

TABLE 8.5. Mean Number of Healthy Permanent Teeth (H), Decayed (D), Missing (M), Filled (F), and Mean DFMT Values by Age, Sexes Combined, at Etéñitépa, 1997

Age (years)	Sample Size	H	D	M	F	DFMT
2–5	45	0.44	0.04	—	—	0.04
6–11	52	10.67	0.84	—	—	0.84
12–14	20	23.75	3.30	0.30	0.10	3.70
15–19	25	24.80	4.20	0.40	—	4.60
20–29	39	22.92	5.28	2.28	0.97	8.53
30–39	20	16.95	6.75	7.95	0.35	15.05
40–49	17	17.70	6.82	7.41	0.06	14.29
50+	10	12.40	5.90	13.70	—	19.60

Source: Adapted from Arantes et al. 2001.

TABLE 8.6. Mean DFMT Values from Surveys of the Etéñitépa Xavánte at Different Periods, by Age, Sexes Combined

Age (years)	1962	1991	1997
6–12	0.21	0.37	1.08
13–19	0.30	1.16	4.54
20–34	0.71	8.13	9.72
35–44	2.40	9.10	14.25
Over 45	3.60	13.79	17.75

Source: Adapted from Arantes et al. 2001. Data for 1962 collected by Neel et al. (1964), reanalyzed by Pose (1993). Data for 1991 from Pose 1993.

oral health associated with lifestyle changes. The consumption of certain industrialized foods, such as sugar and starches that may ferment, is considered the main factor leading to changes in the frequency of caries in indigenous peoples, since a close relationship between sugar and the prevalence of tooth decay has been confirmed (Gustafsson et al. 1954; Newbrum 1982; Woodward and Walker 1994). In the past twenty years the Xavánte consumption of industrialized foods has greatly increased (see table 6.5). Changes in the Xavánte diet became more pronounced in the 1970s at the time of the Xavánte Rice Project. By the 1990s rice had become the basis of the Xavánte diet. According to Sreenbny (1983) there is evidence that maize, which was in the past an important element in Xavánte diet, is less cariogenic than rice.

Conclusion

The Xavánte are exposed, like people in other societies, to many health risks on a daily basis, some of which are found within their own homes (like indoor air filled with smoke) or in their village (like waste-contaminated water from the stream where they drink and bathe). Again, like people in other societies, the Xavánte actively affect the process of disease transmission through their behavior and cultural practices. While some of these practices apparently protect them against certain diseases, others clearly facilitate disease transmission.

We have tried to show that the Xavánte carry a heavy burden of infectious and parasitic disease and that the consequences affect many aspects of Xavánte life. Approximately ten out of one hundred children born at Etéñitépa do not reach their first birthday, and five more die before reaching their fifth birthday (see chapter 5). Most of these deaths are due to environmentally related (and thus preventable) diseases such as diarrhea and respiratory infections. Not only death rates but the quality of life are affected by the widespread occurrence of disease. Signs of malnutrition, including anemia and growth faltering, are evident at Etéñitépa, where they result from infectious disease in combination with insufficient intake of nutrients. Not surprisingly, many health indicators for the Xavánte are well below the averages reported for Brazilian non-Indian populations.

In this chapter we have not explicitly addressed the question of whether Xavánte health over time has become more, or less, affected by infectious and parasitic diseases. We have purposely avoided this discussion for various reasons. In the first place, it would be difficult to determine the direction and intensity of change because we do not have dependable diachronic data for the Xavánte. But there are other reasons as well. The epidemiology of infectious and parasitic diseases affecting indigenous populations in contact with the national society is so multifaceted and complex that general statements about improvement or worsening over time are necessarily oversimplifications. On the one hand, the introduction of medical technologies, such as vaccines, in recent decades has certainly diminished, or even eliminated, the threat to the Xavánte of previously epidemic diseases like measles, polio, and whooping cough. Fortunately, the danger of biological extinction through epidemics, which seemed a real possibility during the early years of contact from the 1940s through the 1960s, seems to have been averted.

However, if some menaces to Xavánte health and survival have lessened, ecological and socioeconomic changes have created conditions that favor transmission of certain infectious diseases. For decades, the Xavánte have been engaging in close interactions with non-Indians. At present, people from Etéñitépa visit nearby towns many times in a week. This constant flow increases the likelihood of people contracting and bringing to the village contagious diseases, from flu to sexually transmitted diseases. The village of Etéñitépa has remained in the same place for over thirty years, which, without sanitation, means more opportunities for exposure to a number of pathogens, especially intestinal parasites. Also, sedentism places increased pressure on hunting and gathering in the vicinity of the village, affecting the quantity and diversity of food supplies. Even these changes are not in one direction only, since for food the Xavánte at present do not depend solely on the resources of their reservation.

It is very important that the health services for indigenous peoples that are being implemented in Brazil (see chapter 7) take into consideration the complexities of the indigenous health situation. As the Xavánte case shows, this situation is not static but constantly changing as economic, social, and environmental pressures create new conditions that may not only increase the burden of "old" diseases but could lead to the emergence of "new" ones.

Chapter 9

The Emergence of New Diseases

The physical condition of Xavánte men made a powerful impression on those who saw them in the years soon after contact. A journalist who visited the Xavánte at São Domingos in the 1940s wrote: "Although they are not really giants, as was rumored at first, the Xavánte physical presence is admirable. Bronzed, of medium height, among them we can find some individuals who are tall, others who are short. However, what seems to be most characteristic of the tribe are their fine, at times even Olympian, physical proportions. They are strong, vigorous, athletic" (Souza 1953b, 99).

The Xavánte physique impressed not only journalists and visitors but also the physicians and biologists who did research among them. When reporting on health conditions among the Xavánte in 1954, Amaury Sadock de Freitas-Filho referred to their good health and "fine physical appearance" (1955, 155). The doctor even suggested that the Xavánte physique represented "the physical standard that should be the Brazilian ideal" (165). At the beginning of the 1960s, when James Neel, Francisco Salzano, and their collaborators made a detailed biomedical study of the Xavánte at São Domingos, almost twenty years after first contact, they were hardly less emphatic. They wrote of the physical state of the men: "The general impression of the men was exuberant health and vitality. They were erect in carriage, deep-chested, and very well muscled, with a notable absence of adiposity" (Neel et al. 1964, 110).

The most recent reports paint a very different picture from the one Souza, Freitas-Filho and Neel described in the 1950s and 1960s; they suggest that the Xavánte, especially those in certain communities, are experiencing an accelerated nutritional and epidemiological transition. The endocrinologist João Paulo Botelho Vieira-Filho, from the Escola Paulista de Medicina, recently wrote, based on his many years of experience in providing medical care in the Xavánte reservations of Sangradouro and São Marcos:

> The Xavánte, who were slim before contact . . . have become overweight or obese (some weighing more than 100 kg) because of dietary changes that came with the government rice-growing project . . . [and] high consumption of rapidly absorbed carbohydrates. Their consumption of vegetable fiber is drastically reduced. Every time I visit the western Xavánte I see more cases of diabetes. Twenty years ago there was no diabetes among them. (2000, 2)

In the previous chapter we described the infectious and parasitic diseases that affect the Xavánte at present, their nutritional and oral health status, and sanitary and other environmental conditions related to their health. Our intent in this chapter is to continue our discussion of Xavánte health, focusing on some chronic noncontagious diseases that have emerged as Xavánte society in recent decades has passed through broad social, economic, and environmental transformations. Recent data on health and disease show that, overlying a pattern still dominated by infectious and parasitic diseases, obesity, cardiovascular conditions, and diabetes mellitus are now becoming serious health problems for the Xavánte.

The Theory of Epidemiological Transition

In 1971, Abdel Omran wrote an influential article on the "theory of epidemiological transition." *Epidemiological transition* refers to how the demographic parameters and health and disease patterns of a population shift over decades or centuries, changing from a configuration dominated by infectious and parasitic diseases and their effects to one largely characterized by noncommunicable disorders. According to him, "the theory of epidemiological transition focuses on the complex change in patterns of health and disease and on the interactions between these patterns and their demographic, economic and sociologic determinants and consequences" (1971, 510).

Omran divided the epidemiological history of human societies into three phases, which follow one another in linear progression. The first is the "age of pestilence and famine," characterized by very high and fluctuating mortality, frequent epidemics, low life expectancy at birth (twenty to forty years) and little or no population growth. Major causes of death are malnutrition, infections, and complications of reproduction. Phase 2 is the "age of receding pandemics," when epidemics become less frequent, mortality is reduced, life expectancy at birth increases to thirty to fifty years, and demographic growth is sustained. During this period infectious diseases and malnutrition still predominate, but rates tend to decline. The third phase, the "age of degenerative and man-made diseases," is characterized by a great decline in mortality, an increase in life expectancy at birth (over fifty years), and the rise of chronic noncontagious diseases (such as cardiovascular disease, cancers, and diabetes) as the primary disorders that afflict human populations (Omran 1971).

Omran also noted that, even though human societies will find themselves in one stage or another, one would expect to observe "peculiar variations in the patterns, the pace, the determinants and the consequences of population change" (1971, 532). On the basis of this observation he proposed three basic models of transition: (1) "the classical or Western model," experienced mainly in Europe; (2) "the accelerated model," represented by the Japanese experience; and (3) "the contemporary or delayed model," characteristic of present-day developing countries.

The main differences among the three models are time of onset of the transition and the relative importance of the mechanisms that drive the process. In some countries of Western Europe the epidemiological transition was already under way by the middle of the eighteenth century. In countries that experienced the classical or Western transition, the fall in mortality was primarily due to socioeconomic, political, and cultural determinants, which included transformations in standards of living and health habits as well as in hygiene and nutrition (Omran 1971, 520). The timing and causes of the contemporary or delayed transition in the so-called developing countries is characteristically different. In these countries mortality only began to fall in the twentieth century and its decline has been more due to the adoption of imported public health measures and key biomedical interventions (e.g., vaccines and antibiotics) and less to improvements in economic and social factors. The accelerated transition experienced by Japan and some Eastern European countries combines aspects of the other two models. In Omran's own summary, "although it would be naive to attempt precise identification of the complex determinants in each case, it does seem apparent that the transition in the new developed countries was predominantly socially determined, whereas the transition in the 'Third World' is being significantly influenced by medical technologies" (521).

The epidemiological transition involves much more than a shift from a pattern characterized by infectious and parasitic diseases to one in which chronic diseases predominate. Omran called attention to the fact that the drop in mortality (the recession in pandemics) favors the survival of more children and young women. Over the long term not only is the epidemiological pattern of the population transformed but so is its age structure. As fertility rates decline, the age structure of the population changes from a distribution heavy in young age cohorts to one in which adults and elders are in the majority. Older people are more exposed to chronic noncontagious diseases than are young people, which tends to increase the prevalence of these health problems in the community.

Omran's theoretical formulation of the epidemiological transition presents it as a universal process destined to take place in the most diverse human societies. Even though the pattern, pace, determinants, and consequences may vary from one region to another, the result is a final displacement of infectious and parasitic diseases by chronic diseases. The transition must eventually reach the countries of the third world, even though in the second half of the twentieth century they may still remain in a condition of "yet-to-be completed transition" (Omran 1971, 535).

In the late 1980s and the early 1990s, a number of investigators, including many from Latin America, began to raise important objections to Omram's formulation of the epidemiological transition (Barreto et al. 1993; Bobadilla et al. 1993; Frenk et al. 1989, 1991). A general criticism concerned the limitations of linear and unidirectional models to explain disease and mortality trends in human societies over historical time. According to Frenk et al., "the health and disease patterns of a society evolve in diverse ways as a response to broader demographic, socioeconomic, technological, political, cultural, and biological changes . . . [thus] health conditions are continuously being transformed, as different diseases disappear, appear or re-emerge" (1994, 27). These same authors voiced specific criticisms of Omran's formulations, basing them on epidemiological data from what they called "middle-income countries," including Mexico and Brazil. An important characteristic of the epidemiological experience of these countries was the coexistence of infectious and parasitic diseases with chronic noncontagious diseases, with both groups contributing significantly to death rates.

Frenk and his collaborators also called attention to the phenomenon of "epidemiological polarization." According to them "the coexistence of pre- and post-transitional diseases leads, in certain countries, to an epidemiological polarization . . . the poorer sector of the population would not only present higher rates of disease, but these would be of different kinds [compared to the well-to-do social segments], mostly either infections or nutritional disorders" (1989, 31). Instead of a homogenous transition taking place, different disease patterns might prevail in countries of the same region, and even within some countries, pointing to the limitations of a model based on presuppositions of linear and universal progression.

Frenk and his collaborators then proposed a new model that they called the "protracted-polarized model of epidemiological transition." Some distinctive aspects of this model are, first, that, despite significant reductions in mortality from infectious diseases, these are not fully brought under control. This situation, together with an increase in chronic noncontagious diseases, produces an "overlap of eras." Second, the reemergence of infectious and parasitic diseases that had been controlled or eradicated may occur, resulting in "counter-transitions." Third, the unequal distribution of wealth and inequitable health care coverage result in a widening gap in health status among social classes and geographic regions, thus leading to "epidemiological polarization" (1989; see also Bobadilla et al. 1993, 53). Frenk and his collaborators use the Mexican epidemiological transition to characterize the protracted-polarized model, stating that "in the decade of the 50s, the ten main causes of death already included many chronic and degenerative diseases, such as heart conditions, tumors and cerebrovascular problems; this was the time when infectious diseases caused more than 30% of all deaths" (Frenk et al. 1989, 32).

Epidemiological Transition in Brazil

In Brazil, demographic data clearly point to a significant drop in mortality, beginning in the mid–twentieth century (Barreto and Carmo 1998; Carvalho 1997–98; Duchiade 1995; Mello-Jorge and Gotlieb 2000; Prata 1992). Between the 1940s and the 1990s, life expectancy at birth in Brazil increased from approximately forty-two to approximately sixty-six years (Mello-Jorge and Gotlieb 2000, 27). As a consequence of declining mortality and fertility, the Brazilian population is aging (Veras 1988). Between 1940 and 1996 the proportion of elders (sixty-five years and older) in the total population grew from 2.4 to 5.4 percent (Mello-Jorge and Gotlieb 2000, 20). In spite of serious socioeconomic and political setbacks, in recent decades there have been major improvements in sanitation and general infrastructure, along with expanded access to basic educational and health services. Between 1960 and 1991, the percentage of houses with piped water increased from 21 to 71 percent, and the percentage connected to sewer lines grew from 13 to 35 percent. Between 1970 and 1991, illiteracy dropped from 57 to 20 percent of the population (30–34). During the same period, immunization programs expanded and deaths due to infectious diseases declined markedly, especially those that were, like diarrheas, related to poor sanitation and those that, like measles, could be prevented by immunization.

The drop in fertility is much more recent than the mortality decline, as it began only in the 1960s. The decrease has been very rapid—between 1970 and 1996, the total fertility rate (TFR) in Brazil fell from 5.8 to 2.5 (Carvalho 1997–98, 6). This reduction has occurred both in urban and rural areas. The most important proximal determinant of this decline is the widespread use of modern contraceptive technology, especially oral contraception and female sterilization. Urbanization, along with improved education and increased participation of women in the job market, are important distal determinants of the fertility decline (Carvalho 1997–98; Duchiade 1995; Patarra 1995).

The Brazilian epidemiological experience seems to conform quite well to the predictions of the "protracted-polarized" model. Although cardiovascular disease, cancers, and external causes (injuries) have become the leading causes of mortality in Brazil, infectious and parasitic diseases remain prominent causes in most of the country, with the exception of the economically better off Southeast and South regions (table 9.1). Therefore, the coexistence of infectious diseases with chronic noncontagious disorders is clearly evident in the present epidemiological scene.[1]

Baretto and Carmo (1995) have summarized the complexity of the Brazilian epidemiological picture, pointing to the peculiar features of the epidemiological transition taking place in the country, especially its inherent contradictions, which make it difficult to trace tendencies and foresee its future course.

> The striking change in some health statistics for the population of Brazil in recent decades, especially the significant increase in life

TABLE 9.1. Five Major Causes of Death (in percentages) in Brazil in 1993–95 in Order of Importance

Causes of death (rank)	North		Northeast		Center-West		Southeast		South	
	Men	Women	Men	Women	Men	Women	Men	Women	Men	Women
(1)	Injuries (19.5%)	Circulatory (20.8%)	Circulatory (18.0%)	Circulatory (21.6%)	Circulatory (24.5%)	Circulatory (30.0%)	Circulatory (27.4%)	Circulatory (35.2%)	Circulatory (29.5%)	Circulatory (37.1%)
(2)	Circulatory (16.4%)	Neoplasias (9.4%)	Injuries (14.4%)	Neoplasias (7.4%)	Injuries (23.4%)	Neoplasias (12.1%)	Injuries (17.9%)	Neoplasias (13.4%)	Injuries (16.1%)	Neoplasias (15.7%)
(3)	Perinatal (7.4%)	Perinatal (8.0%)	Infectious (6.3%)	Respiratory (6.1%)	Neoplasias (8.8%)	Respiratory (8.8%)	Neoplasias (10.9%)	Respiratory (11.0%)	Neoplasias (14.7%)	Respiratory (10.6%)
(4)	Infectious (6.8%)	Respiratory (7.4%)	Respiratory (5.6%)	Infectious (5.9%)	Respiratory (7.3%)	Injuries (8.7%)	Respiratory (10.2%)	Endocrine (6.9%)	Respiratory (11.0%)	Injuries (5.7%)
(5)	Neoplasias (6.7%)	Infectious (7.1%)	Neoplasias (5.5%)	Endocrine (4.5%)	Infectious (6.0%)	Infectious (6.9%)	Endocrine (6.2%)	Injuries (5.3%)	Digestive (5.0%)	Endocrine (4.9%)

Source: Adapted from Mello Jorge and Gotlieb 2000, 93.

Note: "Injuries" include all deaths not related to natural causes (accidents of all kinds, violence, suicide, and so on)

> expectancy due to a drop in infant mortality and in deaths caused by infectious disease, may lead to the conclusion that there has been a great improvement in general health. On the other hand, the unequal effect of these improvements on different social sectors of the population, the appearance of new problems, the persistence or recrudescence of major endemic parasitic diseases [countertransition], as well as a generalized crisis in the health care system, reveal the limitations of these early indications, calling for careful analysis to understand this complex and paradoxical situation and its implications for health policy. (17)

Are the Xavánte Undergoing an Epidemiological Transition?

In the following sections of this chapter we will analyze a body of data that points to the emergence of obesity, hypertension, and diabetes mellitus among the Xavánte. The explication of these data will allow us to discuss some specifics of the epidemiological transition that they are undergoing.

Weight Gain and Obesity

Overweight and obesity are major health concerns throughout the world, both in industrialized countries and in regions of the developing world undergoing rapid urbanization (Popkin 1993, 1994; WHO 1995, 312–44; WHO 1998). Some authors refer to obesity as a worldwide epidemic (see Popkin and Doak 1998). High-fat and low-fiber diets, combined with low levels of physical activity, play an important role in the increase of overweight and obesity. Obesity may have a number of different impacts on health; it is associated with diabetes, cardiovascular disease, hypertension, gallbladder disease, and dyslipidemia. (NIH 1985; WHO 1995, 323–27; WHO 1998).[2]

Until recently investigations had not suggested that obesity might be a collective health issue for indigenous peoples in Brazil. But, while there are no reliable epidemiological data to characterize the incidence and distribution of obesity in these populations, in recent years isolated studies have reported the occurrence of obesity in a few populations such as the Xavánte and the Boróro (Leite 1998; Vieira-Filho 1981, 1996, 2000; Vieira-Filho et al. 1983, 1984), Suruí from Rondônia (Santos and Coimbra 1996, 1998), and Gavião-Parakatejé (Capelli and Koifman

2001; Tavares et al. 1999). Santos and Coimbra (1996) attempted to place the rapid weight gain experienced by Suruí adults in the context of the socioeconomic and environmental transformations that the group had recently experienced. They found that the segment of the population that gained the most weight was the one that had become involved with introduced economic activities (e.g., coffee farming and logging) and withdrawal from traditional subsistence activities (see also Coimbra 1989a). It is very likely that the still small but fast-growing epidemiological literature on obesity among native peoples of Brazil has documented no more than the tip of the iceberg.

In 1962, James Neel and his collaborators carried out an anthropometric survey at São Domingos, collecting data on weight and stature (Neel et al. 1964, 64–68). The sample is small (only twenty-five adults, ages twenty to fifty years). Nevertheless, a comparison of the measurements that they made in 1962[3] with those that we collected at Etéñitépa in 1990 is revealing (table 9.2). The mean weight of adult men has increased from 68.3 to 72.9 kg and that of adult women from 53.3 to 60.0 kg, showing a trend toward weight gain over the three decades that separate the two surveys. There was an increase in the body mass index (BMI)[4] from 23.8 to 25.8 kg/m^2 in men and from 22.0 to 25.2 kg/m^2 in

TABLE 9.2. Comparison of Blood Pressure Levels and Anthropometric Parameters of Etéñitépa Xavánte 20–50 Years of Age in 1962 and 1990

	Men		Women	
Parameters	*N*	Mean	*N*	Mean
Height (cm)				
1962	13	169.3	12	155.3
1990	25	168.1	26	154.0
Weight (kg)				
1962	13	68.3	12	53.3
1990	25	72.9	26	60.0
BMI (kg/m^2)				
1962	13	23.8	12	22.0
1990	25	25.8	26	25.2
Systolic pressure (mmHg)				
1962	13	114.0	12	106.7
1990	26	120.9	27	120.6
Diastolic pressure (mmHg)				
1962	13	63.5	12	66.2
1990	26	76.5	27	75.0

Source: Adapted from Coimbra et al. 2001.

women. Interestingly, the results show little change in mean stature. In fact, mean stature for both sexes was slightly lower in 1990 than it was in 1962.

The World Health Organization (WHO 1998) recently divided BMI values into four categories: low weight (BMI <18.5); normal weight (18.5 to 24.9), overweight (25.0 to 29.9) and obesity (≥ 30.0 kg/m^2). When we look at the data collected in 1962, we observe that 88 percent of the adults measured by Neel and his associates fall into the normal weight category (table 9.3). Only three young men are classified as overweight.[5] A different picture emerges from the 1990 data collected at Etéñitépa, as the percentage of individuals classified as normal weight is only 46.2 percent. The majority is classified as overweight (48.4 percent), and there were a few obese (4.4 percent). More women than men were above normal weight.

The finding that so many people were above normal weight in 1990 came as a surprise to us because judging from the appearance of the Xavánte at Etéñitépa it did not seem to us that so many people were overweight. This apparent contradiction was resolved when we looked at the data in more detail. As we can see from table 9.2, in 1990 mean BMIs for men and women were very close (25.8 and 25.2) to the lower cutoff point that defines “overweight” (25–29.9), perhaps suggesting that a shift from normal weight to overweight at Etéñitépa was recent at the time the data were collected.

The comparative analysis of the anthropometric data for Etéñitépa Xavánte adults collected in 1962 and in 1990 is quite revealing. The anthropometric profile of the Xavánte at Etéñitépa changed substantially over that time, with a trend toward weight increase in adults of both sexes. Is the same process taking place on other Xavánte reservations? If so, is it possible to compare differences in magnitude and can we uncover the determinants of the process?

The nutritionist Silvia Gugelmin recently carried out a short-term investigation on another Xavánte reservation, where she collected comparative anthropometric and ecological data (Gugelmin 2001; see also Gugelmin and Santos 2001). The study was done at São José, a village on the Sangradouro reservation (see fig. 2.1). The village is located within walking distance of a Salesian mission compound that includes a church, a school, and an infirmary. The Xavánte at Sangradouro have a very different contact history from those at Etéñitépa, with strong influences from Catholic missionaries (see chapter 3). In the 1950s a group of

TABLE 9.3. Comparison of BMI Values in Etéñitépa Xavánte 20–50 Years of Age in 1962 and 1990

	1962			1990		
BMI Classification	Men	Women	Both Sexes	Men	Women	Both Sexes
Low weight (BMI < 18.5 kg/m^2)	—	—	—	—	1 (1.1%)	1 (1.1%)
Normal (18.5–24.9 kg/m^2)	10 (76.9%)	12 (100%)	22 (88.0%)	22 (53.7%)	20 (40.0%)	42 (46.2%)
Overweight (25.0–29.9 kg/m^2)	3 (23.1%)	—	3 (12.0%)	17 (41.5%)	27 (54.0%)	44 (48.4%)
Obese (BMI ≥ 30.0 kg/m^2)	—	—	—	2 (4.9%)	2 (4.0%)	4 (4.4%)
Total	13 (100%)	12 (100%)	25 (100%)	41 (100%)	50 (100%)	91 (100%)

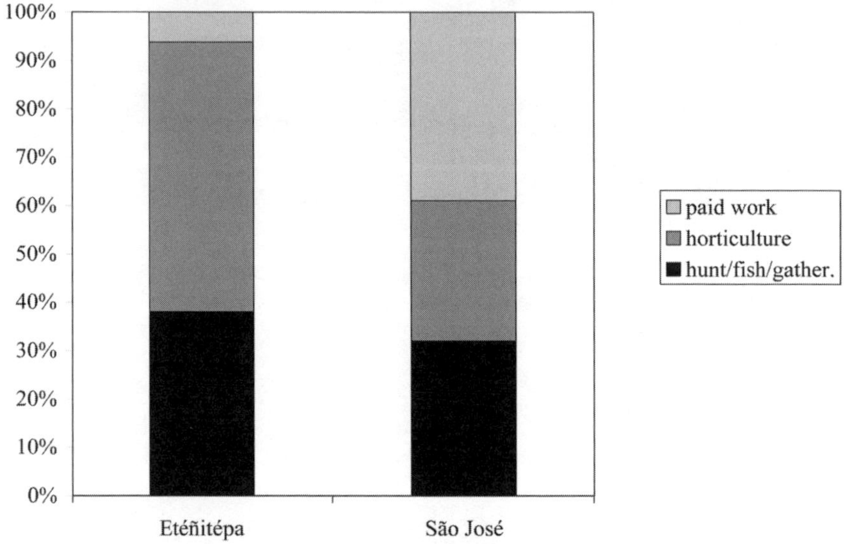

Fig. 9.1. Percentage of subsistance labor time allocated to different activities by Etéñitépa and São José Xavánte. (Data from Gugelmin 2001.)

Xavánte who were weakened by disease, and whose land was increasingly being invaded by ranchers and threatened by settlers, sought refuge at a Salesian mission (Silva 1992; Souza 1999; Souza and Santos 2001). Salesian missions for the Xavánte, which were eventually established at both Sangradouro and São Marcos, have had numberous influences on the Xavánte settled there, from changing their demographic pattern by discouraging the practice of polygyny to introducing them to a monetary economy (Menezes 1984; Souza and Santos 2001).

Unlike the Etéñitépa group, the Sangradouro Xavánte were unable to remain on the land they occupied before contact. There are also important differences in the size of the two reservations and the consequent relation of population to natural resources. Sangradouro covers 100,280 ha and in 1997 had a total population of around 920. The size of Pimental Barbosa is three times greater (328,966 ha), with a population that is only slightly larger, 1,060 in 1995 (see table 2.1). Even though we should not make population density a direct indicator of availability of resources, it is closely related to environmental degradation, intensity of land use, and pressure on fish, wild plants, and especially game.

Time allocation data reported by Gugelmin (2001) show that, indeed, there are marked differences between the subsistence patterns of Etéñitépa and São José (fig. 9.1). The Etéñitépa Xavánte have much greater involvement in agricultural activities and in hunting, fishing, and gathering. At São José approximately 40 percent of the observations of subsistence activities are related to paid work, a much higher figure than what was observed at Etéñitépa. Paid labor at São José is in large part work at the Salesian mission and the mission school as well as employment as head of the Indian post, driver, tractor driver, and nurse's aide.

Gugelmin (2001) also compared the anthropometrics of the two communities. The results revealed substantial differences between them (table 9.4). While there were no major differences in stature between Etéñitépa and São José adults of either sex, the mean weights of the men and of the women at Sangradouro were, respectively, 5.1 and 12.6 kg higher than those at Etéñitépa.[6] Not surprisingly, BMI averages are much higher in São José.

In Figures 9.2 and 9.3 we compare mean height and weight for Xavánte men and women collected in the 1990s with measurements from earlier decades.[7] These comparisons make it clear that the mean statures of adult men and women have changed very little over time, though with a slight tendency to diminish (fig. 9.2). However, the mean weights for men and women measured in the 1960s are consistently lower than those collected in the 1990s (fig. 9.3). It is also evident that the mean weights at São José are a great deal higher than the others.

TABLE 9.4. Comparison of Anthropometric Parameters of Adult Xavánte ≥ 20 Years of Age from Etéñitépa and São José

	Men		Women	
Parameters	*N*	Mean	*N*	Mean
Height (cm)				
Etéñitépa	40	167.3	42	154.6
São José	65	166.8	63	155.3
Weight (kg)				
Etéñitépa	40	70.4	42	59.0
São José	65	75.5	63	71.6
BMI (kg/m^2)				
Etéñitépa	40	25.2	42	24.6
São José	65	27.1	63	29.7

Source: Adapted from Gugelmin 2001.

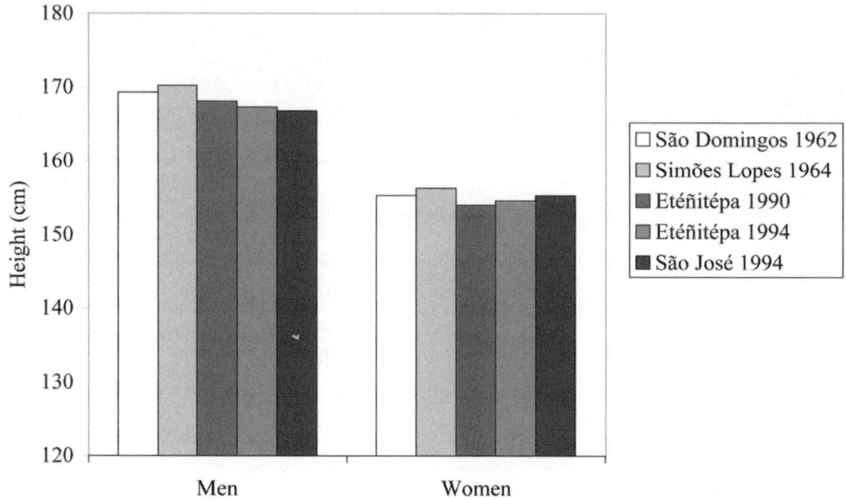

Fig. 9.2. Mean height (in cm) for adult Xavánte men and women in different periods

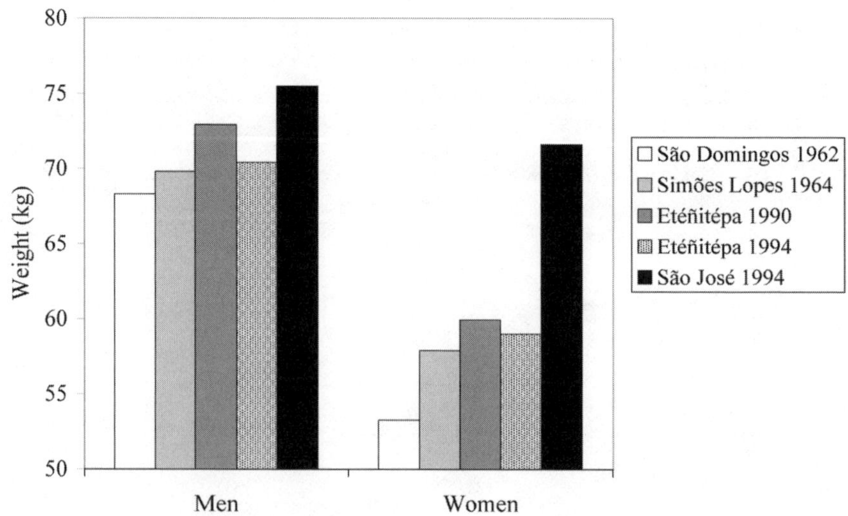

Fig. 9.3. Mean weight (in kg) for adult Xavánte men and women in different periods

How can we explain these differences in Xavánte body morphology? The earlier Xavánte lifestyle was characterized by high levels of physical activity and a low-fat, high-fiber diet. As Neel et al. noted in the 1960s, they were "very well muscled, with a notable absence of adiposity" (1964, 110). The diet of the Etéñitépa Xavánte has certainly changed in recent years. Although in the 1990s it was still largely based on food items obtained within reservation boundaries from horticulture, hunting, fishing, and gathering, it included more industrialized products, including canned food, pasta, soft drinks, refined sugar, wheat flour, and cooking oil, than in the 1970s (see table 6.5). Rice has become a staple in their diet (fig. 9.4). Cash income, mostly derived from social security pensions paid to retirees, has made it possible for them to purchase industrialized, energy-dense foods on a regular basis.

In explaining the trend toward weight gain at Etéñitépa, altered patterns of physical activity are perhaps of even greater importance than nutritional shifts. The village of Etéñitépa has remained at the same location since 1972, and people do not move around on foot within the reservation as they did in the past. Groups composed of up to several dozen people would leave the village, especially in the dry season, and go on treks lasting several weeks or even months. Present-day hunting and fishing practices of the Etéñitépa Xavánte illustrate how certain subsistence patterns may continue while physical exertion decreases. With game near the village becoming scarce, it is common for a group of hunters to take the village truck to the more remote parts of the reservation, like the Rio Corixão region (40 to 45 km from the village) for a one- or two-day hunting expedition (see fig. 9.5). They may even go out in the morning and return in the evening. The truck is also often used to transport groups of men to the Rio das Mortes for a day of fishing. In this way the physical effort of reaching the hunting or fishing grounds is practically eliminated. This means that an important aspect of their subsistence system, which was the intense physical activity that mobility demanded from adults and children, is no longer what it was a few decades ago. However, the Etéñitépa Xavánte are still very active physically. Their daily activities include long walks to the gardens and on expeditions to collect wild tubers and other plant foods (see fig. 9.6). Ceremonial activities, especially those demanded of the young men, who are expected to engage in frequent log races (see fig. 9.7), foot races, and dancing, can be exhausting. What Maybury-Lewis wrote of initiation rituals is as true today as it was in 1957.

Fig. 9.4. In recent years upland rice has been grown extensively by the Xavánte and has become their dietary staple. It is very common to see both children and adults making a meal of plain rice boiled in water with a little salt. (Photo by N. M. Flowers, 1977.)

Fig. 9.5. Today the Etéñitépa Xavánte often go to hunting grounds by truck. They can reach the more distant parts of the reservation and still return the same day with fresh meat. Now that the Xavánte have become more sedentary they seldom go on extended hunting and gathering trips. (Photo by N. M. Flowers, 1977.)

Fig. 9.6. Women returning from the gardens. The woman in the foreground is carrying her youngest child in the "baby basket" on her back, while a toddler is perched on top. (Photo by N. M. Flowers, 1977.)

> the initiation ceremonies themselves consist largely of ceremonial exercises designed to develop physical resistance and fleetness. They are in fact the culmination of a series of athletic rites which are interspersed over the entire period of seclusion in the bachelors' hut. (1967, 114)

Reports by several investigators indicate that the changes in subsistence, diet, patterns of physical activity of the Xavánte at Sangradouro, and the ongoing effects on their nutritional status are more pronounced

Fig. 9.7. Log racing is a ceremonial sport, requiring speed and stamina, that has been practiced for centuries by many Jê groups. For the race two lengths of palm trunk are cut, each weighing up to 80 kg. Two teams race, with one member of each team carrying the log on his shoulder for a short distance and then rolling it, while running at top speed, onto the shoulder of another team member. (Photo by N. M. Flowers, 1977.)

than at Etéñitépa (Gugelmin 2001; Leite 1998; Vieira-Filho 1981, 1996, 2000).[8] Early in the 1980s the endocrinologist João Paulo Vieira-Filho called attention to the social and environmental changes that were affecting the Xavánte at Sangradouro and São Marcos, especially territorial restriction; a decline in hunting, fishing, and indigenous agriculture; and the dietary changes that resulted from the intensification of rice cultivation through the Xavánte Rice Project (see chapter 6).

> The Xavánte and Bororo Indians have experienced structural changes in their diets. With the invasion of their lands by ranchers and squatters, and the settlement of neighboring areas, the game disappeared and, consequently, so did protein food from their diet. . . . I have seen boys lunching on nothing but a large dish of

> boiled rice or, at other times, on rice with pumpkin and corn, foods that are rich in starch and poor in protein. (1981, 38–39)

The anthropologist Claudia Menezes, who carried out research at São Marcos, also attached to a Salesian mission, called attention to the difficulty the Xavánte there faced in continuing their practices of hunting, fishing, and gathering. She relates the problems to population growth, new economic practices (the Xavánte Rice Project), and environmental degradation (1984, 400). At the beginning of the 1980s rice already held a prominent place in the diet of the Xavánte at São Marcos.

> The Xavánte classify rice (*adsaró*) as a "weak" food, favored by young children who, when there is no rice, now refuse to eat traditional gathered foods. . . . In 1979, when there was a serious food shortage, households were forced to intensify craft production in order to buy rice for their children to eat. (401)

Changing Blood Pressure

Indigenous peoples from many parts of the world have attracted the attention of epidemiologists and anthropologists interested in the impacts of socioeconomic and environmental changes on cardiovascular disease. Indigenous populations have aroused particular interest due to their low blood pressure levels compared to those of Western (or "Westernized") urban populations (Fleming-Moran and Coimbra 1990; Mancilha-Carvalho et al. 1989; Page 1974; Vaughan 1978). Moreover, indigenous peoples lack the classic association between age and increased blood pressure and have a very low proportion of individuals with hypertension and other cardiovascular diseases (Burkitt 1973; James and Baker 1990). The most common explanation focuses on the absence of, or lower exposure to, risk factors for developing such diseases.

The incidence of cardiovascular disease in indigenous peoples tends to change rapidly when these groups intensify their contact with Western societies. They quickly adopt new eating habits, including the consumption of salt, saturated fat, and alcohol; they often begin smoking cigarettes and have a tendency to reduce their physical activity. Numerous studies indicate a rapid emergence in these populations of arterial hypertension associated with such lifestyle changes (McGarvey and Schendel

1986; Page 1974; Sinnett et al. 1992; Vaughan 1978). In addition, changes tend to occur in their broader psychosocial environment that have negative impacts on their social support networks; these factors have also been associated with hypertension and other cardiovascular diseases (Kaplan et al. 1994; Marmot 1985).

Little is known about the epidemiology of hypertension and other cardiovascular diseases among indigenous peoples in Brazil. We should, however, stress that the majority of available studies on blood pressure levels have focused on groups that are still relatively isolated (or were at the time data were collected) and whose lifestyles differ markedly from Western standards, especially since they still consume no salt and limited amounts of saturated fat (Fleming-Moran and Coimbra 1990; Crews and Mancilha-Carvalho 1993; Lima 1950). To date, few studies on the epidemiology of hypertension in indigenous peoples of Brazil have been carried out that take into account the impacts of the changes they are experiencing (Cardoso et al. 2001; Fleming-Moran et al. 1991; Mancilha-Carvalho et al. 1985).

The Xavánte case allows for a comparative perspective over time, since James Neel and his collaborators collected blood pressure readings at São Domingos in 1962 and we collected the same data in 1990 at Etéñitépa (Coimbra et al. 2001; Neel et al. 1964). The results of this comparison clearly show that Xavánte diastolic and systolic arterial blood pressure readings have altered over this time frame. As we can see from table 9.2, in the 1960s systolic and diastolic pressures in adults from twenty to fifty years of age were in the 94 to 126 and 48 to 80 mmHg ranges, respectively. It is particularly noteworthy that in 1962 no hypertensive individuals were observed. Our 1990 readings show that there has been an increase in mean systolic and diastolic blood pressure in both sexes.[9] The increase in mean systolic pressure was particularly pronounced in women, and the increase in diastolic pressure was slightly higher in men. In 1990 some 70 percent of the study population had excellent or normal blood pressure levels, with 25 percent in the high-normal range. An important finding, however, was that six individuals (8.3 percent of the men and 11.7 percent of the women) were classified as hypertensive (three between twenty-eight and thirty-seven years of age and three over forty-four years). No one was classified as stage 3 hypertensive (table 9.5).[10]

Over the course of fifty years of contact the Etéñitépa Xavánte have acquired new habits that may predispose them to arterial hypertension and other cardiovascular diseases. A lower level of physical activity is reflected in the increase in the mean weight and body mass index between the 1960s and the present. Salt is consumed daily, even being used on foodstuffs the Xavánte obtain from hunting, gathering, and fishing. A significant portion of the men now smoke cigarettes. Unlike many other indigenous groups in Brazil, including some Xavánte ones, alcoholism is not present in Etéñitépa as a community issue, although a few men have begun drinking heavily on occasion, especially when they go to town.

Diabetes among the Xavánte

Diabetes mellitus is a metabolic disorder characterized by the presence of abnormally high levels of glucose in the blood, a condition known as hyperglycemia, which is secondary to inefficient insulin action and/or secretion.[11] Non-insulin-dependent diabetes mellitus (NIDDM), also known as type 2 diabetes or maturity onset diabetes, is by far the most common form of diabetes among indigenous peoples throughout the world (see Kunitz 1994a).[12]

Some investigators have suggested that before the middle of the twentieth century diabetes was rare among native peoples of the Americas (West 1974). Later it became one of the most serious health problems for these societies. In Canada and the United States, where the largest number of studies have been done on the epidemiology of diabetes mellitus in indigenous peoples of the Americas, prevalence is, in general, much higher than in the nonnative population (Ghodes 1986;

TABLE 9.5. Classification of Blood Pressure Levels among Etéñitépa Xavánte, Age $\geq$ 18 Years, 1990

	Men		Women	
Classification	*N*	%	*N*	%
Excellent	12	31.6	18	46.2
Normal	18	47.4	9	23.1
High-normal	6	15.8	9	23.1
Hypertension (stage 1)	2	5.3	2	5.1
Hypertension (stage 2)	—	—	1	2.6
Hypertension (stage 3)	—	—	—	—
Total	38	100.0	39	100.0

Source: Adapted from Coimbra et al. 2001.

Narayan 1996; Szathmáry 1994; Young 1993). The Pima of the southwestern United States, have been shown to have one of the world's highest frequencies of diabetes. Their age and sex adjusted incidence is nineteen times that of the population of Rochester, Minnesota (Knowler et al. 1978).

Referring specifically to native North Americans, Szathmáry (1994, 470) observed that "In general, the sequence of possible changes [that lead to the emergence of NIDDM] could be increasing sedentism, diminishment of physical exercise, increase in total caloric intake and/or possible specific nutrients, shift in the distribution of body fat, and development of obesity." Obesity and its duration are both closely associated with the occurrence of NIDDM. In Pima Indians, Everhart et al. (1992) observed that the incidence of the disease is twice as great in individuals who have been obese over ten years compared to those who have been obese for less than five years. Pima Indians with a BMI in the 25 to 30 kg/m^2 have a 21.6-fold higher incidence of NIDDM compared to those with a BMI below 20 kg/m^2 (Knowler et al. 1991).

Accounts of diabetes mellitus in indigenous peoples in Brazil are anecdotal and limited to a few case studies. The first medical report of NIDDM among an indigenous group dates to the 1970s, when endocrinologist João Paulo Vieira-Filho described the disease among the Palikúr and Karipúna in the northeastern Amazonian state of Amapá (Vieira-Filho 1977). He expressed surprise about the number of obese individuals in both groups: "obesity had become common, and among the Palikúr I saw more Indians with a heavy layer of abdominal fat than I had on any other reservation" (177). According to Vieira-Filho, the emergence of diabetes in these populations was linked to changes in their diet, in which manioc flour and refined sugar (two food items unknown to them before contact) had become basic foods. These dietary changes were connected to commercialization of the Karipúna and Palikúr economies, as they had become much involved with the local market, primarily through the production and sale of manioc flour.

During the 1970s and 1980s, the image of obese Indians suffering from diabetes was foreign to most Brazilian anthropologists and physicians. Highly influential research conducted in the upper Xingu region depicted a complex of culturally rich societies with physically fit members practicing a subsistence economy based on horticulture, hunting and fishing, and gathering wild plant foods. The Escola Paulista de Medicina medical team led by Roberto Baruzzi, which for two decades has been carrying out research and providing medical assistance to indigenous peoples in that region, has found no diabetes (Baruzzi and Franco 1981; Franco 1981, 1992). A number of other studies on glucose metabolism have been carried out in indigenous groups that adhered to traditional diets and maintained relatively high levels of physical activity. None of these found any indications of diabetes (Spielman et al. 1982; Bloch et al. 1993; Vieira-Filho 1975).

Vieira-Filho has called attention to the health implications of dietary changes for indigenous peoples and the emergence of diabetes mellitus (and/or of altered levels of blood glucose) in various populations of eastern Amazonia and Central Brazil, including the Xavánte (Tavares et al. 1999; Vieira-Filho et al. 1983, 1984).[13] Vieira-Filho's observations of the occurrence of diabetes mellitus at Sangradouro and São Marcos are particularly important, as he has made regular visits and given medical assistance to the Xavánte on those reservations since the 1970s. They give us a qualitative view of how the situation of the Xavánte has changed over recent decades.

> Twenty years ago I observed that [the Xavánte] were slim and very physically active, and there has not one case of diabetes mellitus among them. . . . In recent years we have seen cases with declared symptoms, and the Xavánte have become obese. . . . With the introduction of the Rice Project their diet changed. . . . The Xavánte grew to like rice, which became the basis of their diet. . . . Rice growing limited the area in their fields available for other crops, like corn, pumpkins, manioc and yucca. . . . Soft drinks were consumed in families that had access to cash. The adults increased in weight, with many individuals developing a heavy layer of fat on the abdomen. (1996, 61)[14]

We do not have any consistent and detailed epidemiological information that would allow us to map the occurrence of diabetes in the different Xavánte areas. The only information relative to diabetes that we were able to find was in a recent report of the FUNASA health team, which provides the Xavánte with health care at Barra de Garças (FUNASA 2000b).[15] This document lists seventy-two cases of diabetes known to the local health service, which are followed up on a monthly basis (table 9.6). The cases reported are concentrated on three reserves (São Marcos, Parabubure, and Sangradouro); these together make up 94.4 percent of

TABLE 9.6. Cases of Diabetes Mellitus in Xavánte Reservations, according to Age Group and Sex, 2000

	<20 Years		20–39 Years		40–59 Years		+60 years		Unknown Age		Subtotal		Total
Reservation	M	F	M	F	M	F	M	F	M	F	M	F	(all ages)
São Marcos	—	1	1	12	2	3	2	6	—	—	5	22	27 (37.5%)
Parabubure	—	—	3	7	2	1	1	2	4	4	10	14	24 (33.3%)
Sangradouro/ Volta Grande	—	—	2	7	2	2	1	3	—	—	5	12	17 (23.6%)
Other[a]	—	—	—	—	1	2	—	1	—	—	1	3	4 (5.6%)
Subtotal	—	1	6	26	7	8	4	12	4	4	21 (29.2%)	51 (70.8%)	72 (100%)
Total (sexes combined)	1 (1.4%)		32 (44.4%)		15 (20.8%)		16 (22.2%)		8 (11.1%)		72 (100%)		

Source: Data from FUNASA 2000b.

[a] "Other" refers to one case from Marechal Rondon (Apertado village), one case from Areões (Buriti village), and two from Pimentel Barbosa (Água Branca and Caçula villages).

the total. There is a striking difference between these areas and the others (Areões, Marãiwasede, Marechal Rondon, and Pimentel Barbosa), which together only had four cases. As for age and sex distribution, 44.4 percent of the patients are between twenty and thirty-nine, and 70.8 percent are women. Approximately one-third of the cases are women between the ages of twenty and thirty-nine. The FUNASA document shows no cases of diabetes at Etéñitépa.[16]

Conclusion

Despite the growing effort to understand epidemiological trends in Brazil, little can be said about indigenous populations. Either Brazilian vital statistics do not allow for the proper identification of the various ethnic groups in a given region or the relevant information is not available. Not surprisingly, recent analyses of health and disease trends in Brazil do not include Brazilian Indians (Mello-Jorge and Gotlieb 2000; Minayo 1995; Monteiro 1995).

In this chapter we have presented data, based on field research, showing that certain chronic noncontagious diseases are already part of the Xavánte epidemiological picture. Information gained from our research among the Xavánte of Etéñitépa clearly points to increases in overweight and obesity, as well as hypertension, that were not present in the recent past. In other Xavánte communities, where ecological and dietary changes have been greater, high rates of obesity and some diabetes are already present. Swift changes in lifestyle, including diet, levels of physical activity, and smoking threaten to make chronic diseases increasingly important causes of morbidity and mortality. Nevertheless, we cannot say that a disease replacement is taking place, since infectious and parasitic diseases remain primary. Some aspects of the Xavánte epidemiological experience we have observed, which we will discuss in chapter 10, do not seem to fit any of the current models of epidemiological transition.

Chapter 10

The Xavánte in Transition

In our Xavánte case study we have used analyses of bioanthropological, ecological, demographic, and epidemiological data; through these data we have tried to draw a picture, however partial, of the historical experience of the Xavánte with Western expansion in central regions of lowland South America.

Although the interaction of indigenous peoples with Europeans in Central Brazil, as indicated by maps, reports, and other historical sources, has a time depth of centuries (the Xavánte are mentioned in colonial documents of the eighteenth century), we have paid most attention to the Xavánte confrontation with expansion fronts during the last fifty years, that is, from the 1940s, when the Brazilian government implemented a range of measures designed to integrate into the national geopolitical scheme vast regions of Central Brazil that were perceived to be a "demographic vacuum." This may seem only a short period—and in fact it is, allowing us to retrieve no more than fragments of the Xavánte historical trajectory—but it covers a length of time that is unusual in bioanthropological, ecological, and epidemiological research among indigenous peoples in Brazil.

In writing this book, one of our intents has been to locate the Xavánte case in the reiterative and traumatic history of the experience of native peoples with Western expansion and later with the consolidation of the Brazilian nation-state in the latter half of the twentieth century. But our analyses have a further aim. We also argue that the course of Xavánte contact with Western society has been mediated by specific historical, social, economic, and cultural contexts that developed not only from the characteristics inherent to expansion fronts but also in large measure out of Xavánte society itself, with its own culture, social and political organization, and economy. The Xavánte were not passive recipients of outside pressures; rather these pressures reacted with elements of Xavánte society, resulting in unique patterns of interaction and change.

Therefore, it is central to our argument to emphasize that while the demography, ecology, and health/disease processes of the Xavánte over the centuries have been influenced by local conditions they have also always been linked to broader influences. On this point our theoretical focus is similar to that recommended by Goodman and Leatherman (1998, 20), who sought

> to understand how particular local histories shape everyday realities of anthropological subjects, and moreover, how separate communities are connected through larger historical political-economic processes that affect human biologies. Understanding humans "under the skin" can enrich our understanding of the link between global change and the everyday struggles of human groups.

A number of the anthropologists and historians who have recently worked in Amazonia have recognized and documented the time depth and geographical breadth of Western expansionism and its effect on indigenous peoples (Castro 1996; Cunha 1992b; Hemming 1987; Roosevelt 1994). Epidemiological and bioanthropological research in the region is lagging woefully behind in this respect. By taking the Xavánte as a case study, one of our objectives has been to show how a historical view can add an important dimension to them.

Local Contexts and Particularistic Approaches

In *Disease and Social Diversity: The European Impact on the Health of Non-Europeans,* Stephen Kunitz (1994b) makes a comparative analysis of what he calls the "changing epidemiological regimes" of indigenous peoples in different parts of the world, drawing his examples from North America, Australia, new Zealand, and Polynesia. Kunitz emphasizes the ways in which processes of health and disease overlap with colonial policies, political institutions, and indigenous cultures and patterns of social organization in order to demonstrate that "diseases rarely act as independent forces but instead are shaped by the different contexts in which they occur" (5).

Stressing the importance of recognizing diversity and giving attention to local contexts, Kunitz argues that although it is possible to find parallels in the situations of indigenous peoples in different parts of the world—when placed in a national context they invariably have lower life expectancies, higher morbidity and mortality, and higher rates of violent death compared to nonindigenous peoples—this inequality is produced and reproduced in multiple and diverse forms. Kunitz does not believe that a few explantory schemes can be made to render, in any satisfactory way, the complexities faced by human groups living in different historical, social, economic, and political situations. This is even more true of indigenous peoples, which have such diverse histories of interaction with Western expansion and colonialism, for

> at our present stage of knowledge and in the wake of the recent collapse of many old certainties, it is more useful to understand in detail the myriad ways in which different causes of morbidity and mortality in populations are affected by social processes, rather than to strive to build grand theories. (1994b, 4–5)

In dealing with the Xavánte experience, we have tried to adhere to an analytic focus that recognizes the influence of local history, or, as Kunitz writes, to adopt a "particularistic approach to the study of diseases in populations" (1994b, 4). This approach is especially pertinent to our discussion of the changing health and disease patterns of the Xavánte. When considering the various forms that the "epidemiological transition" may assume, we find that they are made up of elements that include political history, demography, human ecology, and disease ecologies—in short, the themes of our book.

The epidemiological configuration of the Xavánte is undoubtedly changing. Infectious and parasitic diseases still remain paramount, causing individual and social suffering by leading to impairment and early death. In children, diarrhea and respiratory infections are the most important reasons for sickness, hospitalization, and death. During the 1970s and the 1980s around 15 percent of Xavánte children died in the first ten years of life, a very high figure if we compare it to statistics from even the most marginalized sectors of nonindigenous Brazilian society.

At the same time we see the rapid and alarming emergence of chronic noncontagious diseases such as obesity, hypertension, and diabetes. This group of diseases is becoming increasingly visible only a few decades after the time when severe epidemics of parasitic and epidemic disease drastically reduced the Xavánte population, threatening it with extinction. The intensity as well as the speed of this process is striking. On some Xavánte reservations, like Sangradouro, the prevalence of obesity is much greater than it is in any other segment of the Brazilian population, whether urban or rural, of high or low socioeconomic status.

Therefore, it does appear that the Xavánte are experiencing an epidemiological transition. However, it seems to us that there are important differences between the Xavánte case and the models of epidemiological transition that we find in the literature. This applies as much to the classical formulations of epidemiological transition theory proposed by Omran (1971) as to the more recently advanced schemes, using data from Latin America, that attempt to account for the epidemiological experience of third world countries (see Frenk et al. 1989, 1991).

Of the various models that have been proposed, the so-called protracted polarized model developed by Julio Frenk and his collaborators is perhaps the one nearest to the Xavánte epidemiological experience. In contrast to other models, which emphasize unilineal progression, the protracted polarized model allows for superimpositions, dissonant tendencies, uncertain predictions, and internal variation. It is manifested through (1) the simultaneous occurrence of parasitic and infectious diseases with chronic noncontagious diseases; (2) the persistence, or even the increased incidence, of infectious and parasitic diseases; and (3) epidemiological patterns that vary by geographic region or socioeconomic class (polarization). According to some authors, this is the model that best describes the transition now taking place in Brazil (Barreto and Carmo 1995; Bobadilla and Possas 1993; Sabroza et al. 1995).

The coexistence of infectious and parasitic diseases with chronic nontransmissible diseases is already a reality for the Xavánte. However, even while the Xavánte are beginning to suffer from chronic diseases there is little indication that infectious diseases are lessening their impact. In any future that we can foresee, unless sanitation programs are implemented, their environment will become increasingly contaminated by various infectious disease agents that will have negative effects on their health. A number of the areas where Xavánte live have conditions that favor the outbreak of vector-borne diseases like malaria. Therefore, chronic diseases are not "substituting" infectious and parasitic diseases; rather they are adding to the Xavánte's already heavy burden of disease. Unfortunately, the current situation leads us to vizualize a Xavánte patient, in the not too distant future, malnourished as a child, becoming an obese, diabetic, and hypertensive adult, who may also be diagnosed with tuberculosis.

When we look at the various Xavánte groups, each with a different history of contact and interaction with national society, we can also see epidemiological polarization. There is a clear gap in health status between the Xavánte at Etéñitépa and those living at Sangradouro and São Marcos, where there are many more cases of obesity and diabetes.

These diverse situations are closely linked to particular conditions of contact and to the succeeding socioeconomic and environmental transformations that created conditions favoring the appearance of "new" diseases. To a greater or lesser degree, all Xavánte groups have experienced a nutritional shift from self-produced, low-calorie, high-fiber foods, to industrialized, high-calorie, low-fiber food items. More cases of obesity and diabetes are found in those communities that have undergone greater changes in diet and physical activity.

This increasing epidemiological complexity has serious repercussions. On Xavánte reservations, efforts to control infectious and parasitic diseases rely mainly on technical measures (vaccination, antibiotics, oral rehydration, and mosquito spraying). Structural measures intended to improve sanitation, housing, and nutrition are seldom appropriate or effectively applied. In the case of chronic diseases, the relationship with technology is especially problematic. As Kunitz (1994b, 39) ironically points out, "circulatory diseases, neoplasms, and accidents . . . cannot be prevented with vaccines or cured with anything analogous to antibiotics." The health services on which the Xavánte depend, apparently unable to cope effectively with infectious diseases, fall even farther behind when confronting the added burden of chronic disease.

Although in some ways the epidemiological experience of the Xavánte comes close to the conditions of a polarized protracted transition, in others it differs. For example, the rapid transition that they are experiencing occurs in a rural environment, not in the context of urbanization and industrialization that is held to be inherent to the transition in Latin America and other parts of the world. Also, we do not find among the Xavánte some of the demographic changes that have been considered essential to the epidemiological transition. Thus,

> the progressive increase in survival beyond infancy raises the level of exposure to risk factors associated with chronic diseases. . . . The drop in fertility affects the age structure and has repercussions on the morbidity pattern, because the growing proportion of older individuals increases the importance of chronic and degenerative disorders. (Frenk et al. 1991, 487)

One of the most distinctive aspects of the Xavánte epidemiological transition, and possibly one of the most important finds of our study, is that the emergence of chronic, nontransmissible diseases is occurring without a drop in fertility and consequent aging of the population. We have evidence that fertility at Etéñitépa is high, and it shows no sign of dropping. If we wanted to give a name to the Xavánte transition we might call it "nonaging protracted polarized."

Is it plausible that other indigenous groups in Brazil may be experiencing a transition resembling that of the Xavánte? In other groups for which it has been demonstrated that chronic, noncontagious diseases are emerging as important causes of sickness and death, as among the Suruí (Coimbra and Santos 1994a; Santos and Coimbra 1996) and Gavião-Parkatejé (Capelli and Koifman 2001; Tavares et al. 1999), infectious diseases remain, and there is no clear indication that fertility is dropping or that the population is aging. However, we cannot even conjecture on general trends becaue of the paucity of information available for most indigenous groups in Brazil.

The specificity of the Xavánte epidemiological experience amply confirms the argument that Stephen Kunitz has advanced regarding the primary role of local contexts and the necessity of valuing diversity. It would be hard to find better words to express this than Kunitz' own:

> diseases in populations are so diverse that attempts at generalizing may often — but not always — be vacuous and go badly astray. There is still a lot of mileage to be gotten out of thinking about diseases in their local or national contexts, even understanding that some processes are universal. Generalizations may be parismonious. They may also be impoverished. (1994b, 6)

Visibility, Invisibility, and the Future

While only a few decades ago doubts were raised about the survival of indigenous peoples in Brazil, we can now see a clear tendency toward population increase. In Amazonia this is partly because of the recovery that followed the traumatic situation of the 1960s and 1970s, when the expansion of economic and demographic fronts threatened, through the spread of infectious and parasitic diseases, the biological and cultural survival of many groups. At present the indigenous population in Brazil is made up of over two hundred groups that total 280,000 persons. A striking aspect of this diversity is that more than half of these societies have 500 or fewer members while only about half a dozen have more than 10,000 (Ricardo 1996, xii).

The powerful presence of Indians in Brazilian culture, even though they make up less than 0.5 percent of the country's population, is a recurrent theme in Brazilian anthropology. Alcida Ramos (1998, 3) asks why "being so few, [Indians in Brazil] have such a prominent place in the national consciousness." She replies: "they have the power to burrow deeply into the country's imagination" due to the place they occupy in the construction of the national identity. Other anthropologists have already called attention to the ambiguity inherent in ascribing so much symbolic value to Indians in Brazil while at the same time representing them as obstacles to "progress and development." In spite of progressive legislation that protects Indians, their rights are systematically disrespected (Castro and Andrade 1988, quoted in Ramos 1998, 285). This inconsistency has important effects on the daily lives of indigenous peoples. "While the law guarantees differentiated measures of federal assistance (in education, health and community development) respecting the culture and customs of indigenous societies, very little of this has been implemented by the Brazilian government" (Oliveira 1999, 207).

It is undeniable that indigenous people have become visible players on the sociopolitical scene in Brazil. Now it is the Indians themselves who are spokesmen for their own interests, moving with greater and greater ease in national and international circles. The political activity of the Xavánte exemplifies this trend.

However, indigenous societies in Brazil continue to be "invisible" in certain crucial areas that affect their well-being. There is little understanding of their specific health conditions and needs, understanding that is essential if any health care system intended to meet these needs is to be effective. The situation becomes even more complex as indigenous peoples in Brazil undergo changes that affect their health in unprecedented ways.

The transitions that indigenous peoples in Brazil are experiencing today have their roots in a long history of interaction between local systems and larger social, economic, and political institutions and processes. Some of the domains of change that stand out are the ways in which they relate to the land, engage in new economic and labor relations, and even experience urbanization. The Xavánte, like other indigenous societies, seem increasingly determined to take an active role in shaping these transformations, which will have far-reaching consequences not only for their demography, ecology, and health but for many other aspects of their lives.

Notes

Chapter 1

1. See Moran 1996, 12.

2. The term *terra indígena* (indigenous land) has officially replaced the previous *reserva indígena* (indigenous reservation), which is, however, still widely used.

3. The designation "primitive" was widely used in anthropology through the 1960s (Diamond 1993; Hsu 1964). In their own words, the way James Neel and his collaborators used *primitive* was to refer to "society which is preliterate, employs very simple agriculture and 'manufacturing' techniques, and is primarily organized around concepts of kinskip" (Neel et al. 1977, 109; see also Neel 1994, 120).

4. The interest in populations believed to be "relatively untouched" by Western history, and that therefore could help to elucidate aspects of preurban human adaptation to the physical environment, was in accordance with a line of thinking that guided not only the research in Amazonia but a significant proportion of the investigations in human biology carried out in the 1960s and 1970s. This was the logic behind the so-called Human Adaptability Component of the International Biological Program (IBP) in which Neel, Salzano, and their colleagues participated after the Xavánte project (Neel 1968; Neel 1994, 130; Neel and Salzano 1967b). The IBP was launched at the beginning of the 1960s, with the objective of carrying out bio-ecological studies of communities of plants and of human and nonhuman animals, in order to obtain a "comprehensive global understanding of the processes and forces responsible for the properties of our complex planetary shell" (Collins and Weiner 1977, 1). The perspective was that science was facing a crucial moment in time that "represented probably a last chance of making a concerted study of the still remaining communities of hunters and gatherers and simple agriculturalists" (3–4). The Human Adaptability Component was an undertaking on an international scale, and when it was in its operational phase it brought together nearly three hundred projects, from forty different countries, covering a wide spectrum of themes in human physiological, developmental, and genetic adaptability.

5. In this overview of epidemiological and human biological research in indigenous groups living in Brazilian Amazonia there are a large number of specific studies that we have not cited. Most of these are cross-sectional descriptions of epidemiological situations (diseases of varied etiologies, nutritional evaluations, and so on) or specific human biological case studies, most of which were

not part of broader research programs. Coimbra (1995) and Salzano and Callegari-Jacques (1988) have reviewed some of these studies.

6. Studies of indigenous peoples in the context of the national society are not new in Brazil, but they expanded in the 1980s and 1990s. A generation of Brazilian anthropologists, in the 1950s and 1960s, studied the effects of contact and "ethnic friction," showing that many indigenous groups in Brazil, in spite of decades of interaction with the national society, kept a strong sense of identity (e.g., Matta and Laraia 1979; Melatti 1967; Oliveira 1960, 1972). During the 1960s and 1970s several studies looked at the effects of economic articulation with the outside world (e.g., Aspelin 1975; Leacock 1964; Murphy 1960; Tavener 1973). In the mid-1970s Daniel Gross and his students (Gross et al. 1979) carried out a comparative ecological study of acculturative pressures on four Central Brazilian societies that were in contact with the national society for different periods of time.

7. This trend is also made clear in the issue of *L'Homme: Revue Françiase d'Anthropologie* edited by Philippe Descola and Anne Christine Taylor and entitled "La remontée de l'Amazone: Anthropologie et histoire des sociétés Amazoniennes" (Descola and Taylor 1993).

Chapter 2

1. In this book we will be referring to the Xavánte as located in the Amazon region. Generally speaking, the Amazon region is associated with the tropical rain forest. However, Amazonia has a much more complex set of vegetation types, including areas of *cerrado.* It can be argued, on the basis of hydrography, that Xavánte lands are part of Amazonia. Some Xavánte reservations are on the headwaters of the Xingu River, a tributary of the Amazon. The reservations of Marechal Rondon and Parabubure are on the Curisevo and Culuene Rivers, which feed the Xingu. However, a major river that borders Xavánte reservations, including Pimentel Barbosa, is the Rio das Mortes, which feeds the Araguaia, a tributary of the Tocantins. The Tocantins empties not into the Amazon but into Marajó Bay on the Atlantic Ocean. The region where the Xavánte live has also been defined as part of Amazonia for economic reasons. Xavánte reservations are considered to be in "Amazônia Legal" (Legal Amazonia), a political and economic construct that has been utilized since the 1960s for economic development planning. Amazônia Legal includes the states of Amapá, Roraima, Pará, Amazonas, Rondônia, Acre, Mato Grosso, and portions of Tocantins north of 13° S latitude and Maranhão west of 44° W longitude (Fearnside 1986, 2, 236).

2. The Xakriabá, who are closely related to the Xavánte and Xerénte, were in the captaincy of Goiás in the eighteenth century. They were almost wiped out in colonial wars (see chapter 3), but a group of Xakriabá settled in 1775 at Santana do Rio das Velhas in Minas Gerais (Chaim 1974: 116). There they mixed with the peasants of the region, and by the early nineteenth century, when Saint-Hilaire (1847–48, 283–93) and Eschwege (1830, 95–96) visited the Xakriabá village, they found only a few who still spoke their native language. At

present there are nearly five thousand Xakriabá living on a reservation in northwestern Minas Gerais (ISA 1996, 694).

3. São Domingos was the place where the Xavánte had their first peaceful contacts with agents of the Serviço de Proteção aos Índios (the Indian Protection Service, or SPI) in 1946. When the anthropologist David Maybury-Lewis carried out his fieldwork among the Xavánte in the late 1950s they were living at São Domingos (see chapter 3).

4. A few individuals live long enough to see the full cycle take place. When this happens, there will be *wapté* and elders belonging to the same age set.

5. In the 1960s, the *hö* was fully operational at Etéñitépa. This is not the case in several other Xavánte communities, however, where young boys no longer spend a period of their lives in the *hö*. Sometimes parents from other villages send their sons to Etéñitépa to experience the traditional Xavánte upbringing. This may happen when the boy has relatives at Etéñitépa.

6. According to Graham (1995, 67, and personal communication), the Poridza 'õno and the Öwawẽ are not clans in the usual sense but exogamous moieties, and the Topdató is a lineage of the Öwawẽ moiety. However, to avoid confusion, since we make a number of references in our text to Xavánte "clans" and "lineages" as Maybury-Lewis (1967) used the terms, we have retained his terminology throughout the book.

7. See chapter 5 for a discussion of how political disputes in the 1950s and in the 1960s resulted in the elimination of some lineages.

8. Laura Graham (1995, 66–70) has a description of Suptó's marriage and the process he went through to become part of Agostinho's household.

Chapter 3

1. See Sereburã et al. 1998, 88–89.

2. See Flowers 1983a, 133–37; Giaccaria and Heide 1984, 36–37; Ravagnani 1978, 104; Silva 1992, 365; and Graham 1995, 29.

3. This argument is largely derived from Flowers 1994b.

4. According to the *Oxford University Dictionary, Illustrated* (1970), a fluid dram is equal to one-eighth of a fluid ounce, so Anhangüera's haul was one thousand ounces.

5. According to Giraldin (1977, 69), offensive war, which meant that native villages could be attacked, could only be waged on the king's order. The principal argument employed to justify offensive war was hostility on the part of Indians. Defensive war, on the other hand, prohibited attacks on native villages. Attack was only permitted on warriors who appeared to be taking an aggressive stance.

6. *Bandeiras* were expeditions organized by *bandeirantes* (backwoodsmen), mostly from São Paulo, who made a profession of exploring the hinterlands, seeking gold, and capturing Indians (see Morse 1965). Many of the defeated groups were settled in *aldeias,* settlements of pacified Indians administered until 1758 by Catholic missionaries of various orders and later by appointed directors. The *aldeia* system was generally supported by the Crown throughout the colonial

period as a means of subordinating the Indians, "civilizing" them, and making them "productive subjects," but it was resented by the colonists, who coveted the land set aside by the Crown for maintenance of pacified Indians and who claimed for themselves the use of Indian labor (see Hemming 1978).

7. It was not until after the Xakriabá defeat of 1762 that the Xavánte, who had been living peacefully north of the mining towns, began to attack the settlements (RIHGB 1918, 83). The attacks may well have been instigated by Xakriabá whom Xavánte absorbed into their villages.

8. Under the Directorate, the religious orders no longer had temporal or civil control over pacified Indians; they were to limit their activities to religious conversion and instruction. Indians were to be settled in colonies, or "secular missions," also called *aldeias,* administered by "directors," officials appointed by the governors (see Almeida 1997; and Hemming 1987, 40–61).

9. The title of the map depicted in figure 3.3 reads: "Equinographic Plan of the Aldeia of S. José de Mossâmedes, Habitation of the Acroa Indians, Who, with incomparable Zeal in the Catholic Faith, to Increase the Vassals of his Most Faithful Majesty, were brought to Civilization by the Efforts of His Excellency General José de Almeida e Vasconcellos de Soveral e Carvalho in the year of 1774. This Settlement was built between the 15th of November of that year when the land was marked out and the 28th of April of 1778 when the said gentleman ordered this plan to be drawn of the Aldeia, which is 5 Leagues from Vila Boa toward the Southwest." The buildings are identified by number:

1 Church with high chapel and sachristy
2 Towers of the church
3 Storehouse or granary
4 Houses of two stories that look like towers
5 Two-story house at the front of the settlement, the residence of His Excellency
6 Guest quarters for the persons who accompany His Excellency
7 His Excellency's dining room
8 Kitchen
9 Quarters of slaves and pages
10 Stables
11 Dwelling houses of the Indians
12 Manager's quarters with kitchen
13 Chapel of His Excellency
14 Quarters of the farm administrator
15 Mill
16 Manioc flour mill
17 Blacksmith
18 Quarters of the foreman of wagons
19 Houses of the officers who work in the *aldeia*
20 Houses of the couples belonging to the same
21 Quarters of the foreman of farm animals
22 Quarters of the carpenter Caetano
23 Corral with a shed for calves
24 Kitchen garden
25 Large banana plantation with two thousand trees
26 Quarters of the nurse, schoolmaster, and schoolmistress for the children
27 Bath house
28 Workshop with forty spindles
29 Chaplain's quarters
30 Kitchen for the same
31 Carpenter's shop

32 Storage
33 Storeroom for things belonging to the church
34 Weaving room with looms
35 Gates resembling the arches

10. Carretão was not the only *aldeia* where Xavánte were settled. In the late eighteenth century some were at Mossâmedes (Chaim 1983, 118–20). In the nineteenth century, Italian missionaries founded *aldeias* along the Araguaia at Estiva, Salinas, and São Joaquim do Jamimbu, where a few Xavánte were settled (Hemming 1987, 386, 393–94; Silva 1992, 363–64). It should be pointed out that some Xavánte groups never accepted sedentarization.

11. Pohl's eyewitness account of the Xavánte he met in August of 1819, as his expedition sailed down the Tocantins, shows that the Xavánte of this period, in spite of their reputation for hostility, on occasion might trade, and even make a little sport, with the whites. Pohl hoped for a meeting when his party began to see fires in the distance. He was told that these fires were set by the Indians, "who frequently leave their villages in the dry season and go along the river in groups of forty or fifty men with women and children and practice a kind of circular hunt" ([1837] 1951, 2:121). Pohl (126–30) and his party finally met some Xavánte hunters near the spot where the Rio do Sono enters the Tocantins. The Indians appeared to be unarmed, for they knew that if they were captured with weapons they could legally be enslaved (128). As a sign of friendship, one that is still customary among the Xavánte, they took off their cotton neck cords and tied them around the necks of the members of Pohl's party. Soon they were challenging one of Pohl's rowers, a Xavánte who had been captured by the whites as a child, to foot races along the riverbank, but Pohl called his man back, as he feared he might be drawn into the forest by the "wild" Xavánte and would disappear with them. Later, as Pohl and his party traveled back upriver, a group of about two hundred Xavánte appeared near the same place (173), and a few came to the boats with maize, peanuts, and palm nuts, which they exchanged for salt, manioc flour, and beads.

12. In the early 1950s an elderly Xerénte told Maybury-Lewis (1988, 100), "Some men have visited them. It is said that they still speak our language and remember the old ways. . . . Perhaps if we too had fought the whites and refused to accept them . . . perhaps we would now be strong like the Shavante. We would have our own lands and the white men would fear us and respect us. But we made peace, and look at us now."

13. In 1858, a group of Indians was reported near a town not far from Vila Boa, alarming the townspeople, although they showed no hostility. According to the governor of Goiás, these Indians probably belonged to "the numerous horde of Xavante inhabiting, as I have been informed, populous villages which are said to exist on the left bank of the Araguaia" (see Ravagnani 1978, 132). The governor added that the Indians intended to supply themselves with food at the expense of local farmers. This group, which remained "on the eastern side of the river [das Mortes,] may have become extinct since there are no further references to them in documents of the period" (Silva 1992, 365).

14. A legend told in the 1970s to Salesian missionaries by an elderly Xavánte

describes the experience of a man overcome with a sickness that could well have been smallpox (Giaccaria and Heide 1975, 36–43).

15. See Carone 1976; Fausto 1994; Levine 1970; and Skidmore 1967 on Vargas and the "Estado Novo," the political regime of the Vargas dictatorship.

16. The SPI was a Brazilian federal government agency, founded in 1910, with an army officer, Cândido Mariano da Silva Rondon, as director. The declared purpose of the agency was to contact isolated Indian groups and protect them against the destructive effects of frontier expansion. On the history of the SPI, see Lima 1995 and Stauffer 1955.

17. See Garfield 1996, 126–31, for a description of the pacification of the Xavánte by Meireles.

18. See Silva 1992, 369–72; and Maybury-Lewis 1967, 13–30, for accounts of first contact among the different Xavánte groups.

Chapter 4

1. In several other chapters we will be referring to the study by Neel et al. (1964). In chapter 5 we compare demographic data collected in the 1990s at Etéñitépa with information obtained at São Domingos in 1962. In chapters 8 and 9 we discuss health-related issues.

2. Results of these studies appeared in Gershowitz et al. 1967; Neel and Salzano 1967a; Neel et al. 1968a, 1968b; Niswander 1967; Niswander et al. 1967; Salzano et al. 1967; Shreffler and Steinberg 1967; Tashian et al. 1967; and Weinstein et al. 1967.

Chapter 5

1. In order to evaluate how consistent the responses obtained in the two surveys were, we compared the demographic data obtained from a group of twelve women who were interviewed in both 1977 and 1990. We selected women who had all their children before 1977. The youngest woman in this sample was born in 1942 and the oldest in 1924. Since the second interview was thirteen years after the first, it might be expected that in 1990 some women would fail to recall some children they had mentioned in 1977, especially those that were born and died in the more distant past. When the women's responses were tabulated and compared, however, they did not show the anticipated direction. Overall, the women in 1977 reported eighty-three birth events (seventy-five live births and eight stillbirths); in 1990, they reported eighty-five birth events (eighty live births and five stillbirths). Thus the average number of birth events was remarkably close (6.9 from the 1977 data and 7.1 from the 1990 data), even more so if one considers that these women were remembering birth events that sometimes took place several decades before the interviews. However, we observed that there was some ambiguity in reporting a birth as a stillbirth or a live birth, as seems to be a problem in demographic investigations in general (see Palmore and Gardner 1996, 64). A few of the birth events reported as stillbirths in 1977 were reported as live births in 1990. The result of

this inconsistency probably had little effect upon the computation of fertility rates (which are based on live births only). If we consider only the live births, the average number of children born per woman according to the 1977 data was 6.3 and according to the 1990 data 6.7.

2. Among the Yanomámi, for instance, attempts to evaluate the reliability of the reported reproductive histories have included urine tests for pregnancy and physical examinations (Neel and Weiss 1975) as well as statistical mathematical models (MacCluer et al. 1971).

3. Population sizes and sources of the census data are as follows: 210 for 1958 (Maybury-Lewis 1967, 172), 195 for 1962 (Neel et al. 1964, 90), 195 for 1969 (Giaccaria and Heide 1984, 281), 249 for 1977 (data collected by Flowers), and 332 for 1980 (Queiroz 1980, 29). When Neel and his collaborators did fieldwork at São Domingos in 1962, the population had recently split, with one group staying at São Domingos (110 persons) and the other moving to a place called Ötõ (estimated at 80 to 90 persons). These two groups joined again a few years later. For the 1962 figure, we are considering these two groups combined (see Neel et al. 1964, 90).

4. In the 1990s there was a fourth village at Pimentel Barbosa, known as Água Branca. The Xavánte who settled there originally constituted the northernmost Xavánte group and came from a place called Marãiwasede, from where they were expelled. They were allowed to settle on Pimentel Barbosa reservation because some are related to Etéñitépa Xavánte. They do not have a reservation of their own, but they are seeking to reclaim their lands (see chapter 3). Although they live on the same reservation, there is little interaction between the Xavánte from Etéñitépa, Caçula, and Tanguro and the Água Branca group. The Água Branca Xavánte are not included in our demographic study. In 1996 a new village (Pe'azarupré) was founded (see chapter 2).

5. This computation was carried out as follows: (212/317) × 1,000, which is equal to 660. This figure was divided by 13, which is the number of years between the censuses, resulting in a crude birth rate of 50.8 births per 1,000 inhabitants.

6. This computation was carried out as follows: (42/317) × 1,000, which is equal to 132.5. This figure was divided by 13, which is the number of years between the censuses, resulting in a crude death rate of 10.2 deaths per 1,000 inhabitants.

7. Data in table 5.4 for São Domingos in 1962 come from Neel et al. (1964, 97), who, however, do not provide the actual number of live births. To obtain these figures we multiplied the number of women by the mean number of live births, which are provided in table 23 of Neel et al. 1964.

8. Data in table 5.5 for São Domingos in 1962 come from Neel et al. (1964, 98), who, however, do not provide the actual number of surviving offspring. To obtain these figures, we multiplied the number of women by the average numbers of surviving offspring, which are provided in table 24 of Neel et al. 1964. The discrepancy between the results of surviving children born to São Domingos women fifteen to nineteen years of age presented in this table compared to table 24 in Neel et al. 1964 derives from different ways of obtaining the number of

surviving offspring. Neel et al. seem to have utilized fifteen women instead of thirteen, that is, a difference of two women. Apparently, these authors included all women (fifteen), and not only those who had at least one live child (thirteen women) in their computation. So, fifteen women times 0.7 (average number of surviving offspring) equals 10.5, which is the number of surviving offspring. This figure in relation to the number of live births (18.0) results in a percentage of survival of 58 percent, which is very close to the value obtained by Neel and his coworkers (41.7 percent of decrease in relation to the average number of live births).

9. Data presented in the abridged life tables are for boys and girls combined. When we computed abridged life tables for each sex in the three periods, we did not observe a consistent pattern of higher mortality in boys or girls.

10. Among the Boróro, another group of Central Brazil, very high infant mortality and failure to have children has resulted in continuous population decline; women often declare their wish not to have any more children, justifying this with the phrase, "I do not want to see the child die" (Crocker 1985, 47).

11. An example of fertility-mortality interaction is given by Brainard in her demographic research on East African pastoral populations (1991, 109–25). She observed that the mean completed parity of a group of nomadic Turkana women (age fifty and older) was 7.23, thus higher than the value detected for a sample of settled women (5.65). According to Brainard, these were unexpected findings, as much of the literature on the demography of pastoral populations points to higher fertility rates in settled communities. Her argument is that the higher fertility of nomadic women is largely explained by a higher level of child mortality, which shortened birth intervals (140–45).

12. Between 1974 and 1976 the Indian agent living at Etéñitépa recorded the dates of births and deaths. Flowers did the same during her stay at the village in 1976–77. These data show an association between the occurrence of early child death and length of birth interval. During these thirty-three months of accurate record keeping, forty-two of the sixty-five women between the ages of fifteen and forty-four gave birth at least once (not counting stillbirths). Fourteen women gave birth twice and one woman a third time. The mean birth interval for women whose babies lived at least a year was 20.9 months, and for women whose babies died it was 16.6 months.

13. Data in table 5.9 for 1958–1962 come from Maybury-Lewis's genealogies and household censuses (1967, 317–42).

14. Data in table 5.10 for 1958–1962 come from Maybury-Lewis's genealogies and household censuses (ibid.).

Chapter 6

1. This research effort was closely related to the establishment of the nation's new capital, Brasilia, in the heart of *cerrado* land in the late 1950s and subsequent economic use of the *cerrado* for agribusiness (see Ferri 1963, 1971, 1977a, 1977b; Marchetti and Machado 1979; and Pinto 1994).

2. Latin names of mammals, birds, and reptiles used in this section follow

Costa et al. 1981; Emmons 1997; and Mares et al. 1989. For more information on *cerrado* fauna, in particular in the Roncador–Upper Xingu area, see Pine et al. 1970; Sick 1955, 1965; and Vieira 1951.

3. Maybury-Lewis (1967, 38) mentions that various species of monkeys were eaten by the Xavánte at the time he did fieldwork.

4. Although Xavánte hunters obtained nearly twice as much meat in 1993 as in the two other years, this was largely because they were hunting over a larger area and spent more hours hunting. "Compared with 1991, the harvest in 1992, as measured by number of kills per km^2 decreased by more than 50% for 5 of the 11 major game species. . . . The same comparison shows that the harvest per km^2 in 1993 [again] decreased by 50% for 5 of the 11 major game species" (Leeuwenberg and Robinson 2000, 388). Productivity in terms of number of game animals killed per one hundred hunter-days increased from 33.22 kg in 1991 to 37.51 in 1992, but it then declined to 26.36 in 1993 (385). Overall, this indicates that more intensive hunting did not result in a corresponding increase in game harvested. It should be mentioned that Xavánte hunters in the 1990s were making use of truck transportation, which allowed them to explore hunting areas at some distance from their base village.

5. Latin names of fishes used in this section follow Géry 1984 and Lowe-McConnell 1975.

6. Rice, manioc, papayas, watermelons, mangoes, bananas, and some other fruits became part of the Xavánte food repertoire after permanent contact in the late 1940s. It is uncertain whether beans and pumpkins were introduced after contact or whether the Xavánte were already cultivating them.

7. There is a large body of literature on development policies in the Amazon. See Barbira-Scazzocchio 1980; Becker 1982; Bunker 1985; Hecht and Cockburn 1989; Hees et al. 1987; Mahar 1979; Moran 1981, 1983; Santos 1980; Schmink and Wood 1984, 1987, 1992; and Velho 1981, among others. See chapter 2 for more information about the concept of Amazônia Legal (Legal Amazonia).

8. The Xavánte Project was only one of a number of "community development" projects implemented by FUNAI on Indian reservations during the 1970s (see Coimbra 1989a; Garfield 1996, 491; Picchi 1991; and Picchi 2000, 86–89).

9. No investigation aimed at an evaluation of the "Xavánte Project" has been carried out. It seems, however, that general diagnoses of government policies implemented in the Amazon from the 1960s on apply (Mahar 1979; Moran 1983; Schmink and Wood 1992, 344–55). Moran stresses that three biases have been consistently present in development programs in the Amazon, resulting in what he refers to as "growth without development": a bias toward macromodels in plan formulation to the neglect of microplanning; a bias toward the quantitative aspects of planning to the neglect of crucial but unquantifiable aspects such as human resource development; and a bias toward detailed planning to the neglect of implementation. In his assessment, the most serious constraint has been "the lack of management capacity at all levels of institutional functioning and the structural inability to use expertise at lower levels in adjusting the production process to specific environmental and social conditions" (1983, 10).

10. See Gross 1984 and Johnson 1975 for overviews on time allocation methodology. For the specific methodology used at Etéñitépa in 1976–77 and 1994, see Flowers 1983b, 390.

11. See Arnt and Schwartzman 1992; Benavides 1996; Brown 1993; Conklin and Graham 1995; Fisher 1994; Jackson 1995; Schmink and Wood 1992, 95–135; and Turner 1993.

12. Dias's (1994) statements about the *cerrado* refer to all the regions of the country covered by *cerrado* vegetation, not only to Central Brazil. Other regions of Brazil, particularly in the northeast and southeast, are also covered with *cerrado*.

13. Opposition to the waterway is bringing together a number of indigenous peoples and environmentalists, who are appealing to the interests of small farmers and other people living along the rivers who would not profit from the waterway. In early 1999, after the Xavánte met at São Felix do Araguaia with representatives of six other native groups whose lives and lands would be affected by the Hidrovia, they issued a declaration describing themselves as "Sons and Daughters of the Rivers," which was designed to broaden their appeal by showing that not only the interests of indigenous people were at stake (International Rivers Network 1999a). After pointing to the negative impact that the Hidrovia would have on the ecology of the rivers, the statement goes on: "We have concluded that the Hidrovia project will not contribute to advancing the quality of life of the populations of the Araguaia and the Tocantins regions. For this reason we want the resources destined for this project to be applied in works that will guarantee more jobs, in the improvement of already existing roads (BR–158 and BR–142), in the conclusion of the North-South and Ferronorte railroads, in improving education and health care, in reforestation of degraded areas, in agricultural projects, and in ecotourism. . . . We would like the progress that is so often talked about to really be for all. But for this to happen it is necessary that small farmers, indigenous peoples, and the riverbank dwellers participate in it. That is why we will fight so that projects designed for the region can bring the progress that we all want."

14. In August 1999 an environmental impact statement was finally issued and public hearings were announced (to be held in places where public opinion was likely to be in favor of the project), but the scientists who carried out the research on which the statement was based cried cover-up when negative information was found to have been suppressed in the final document (Baptista 1999; Valente 1999). In October 1999, lawyers from ISA representing the Xavánte obtained a court order suspending the licensing process for the second time. The judge stated, "If the public hearings were permitted to take place, the public would know only those facts 'chosen' to be shown to them, rather than the complete studies carried out by qualified professionals" (International Rivers Network 1999b).

Two years later the project was still halted. Questions raised about the honesty and technical credibility of the environmental impact statement submitted to IBAMA by the company, which seem to justify the opposition of environmentalists and indigenous groups, have made the Araguaia-Tocantins Waterway project increasingly controversial, and even placed in doubt its eventual implementation (Lerrer 2001). A recent editorial in the newspaper *Estado de São Paulo* (Novais 2001), citing critical defects found in the environmental impact statement, called for abandonment of the project: "Given this evidence the only responsible move is to halt immediately all public spending on this project and exclude it from future federal development plans."

Chapter 7

1. For brief descriptions and critical views of the new health care system for indigenous peoples in Brazil, see Langdon 2000 and Athias and Machado 2001.

2. In order to implement the DSEIs, FUNASA has signed contracts with municipalities and/or various NGOs. These agreements imply a transfer of funds to the contracting institution, which will assume the responsibility for planning and providing health care for the indigenous areas under its jurisdiction.

Chapter 8

1. An abridged version of Freitas-Filho's report was presented in Freitas-Filho and Oliveira 1955.

2. Freitas-Filho presented some inconsistent demographic data in his report. He wrote that the total population at Etéñitépa was 618, including 218 children under age fifteen, 317 adult men, and only 83 women (1955, 148). He makes no comment about this strange sex ratio of adult Xavánte. As we saw in chapter 5, our demographic data show that the fertility of Xavánte women was not low at this time.

3. Freitas-Filho's concern about measles reaching the Etéñitépa Xavánte was justified. Nutels (1955, 129) wrote that a group of thirty Xavánte who had been living near Xavantina rejoined the Etéñitépa group, and this suggests that groups who had closer contact with settlements could bring infection to more isolated groups.

4. See Gershowitz et al. 1967; Neel and Salzano 1967a; Neel et al. 1968a, 1968b; Niswander 1967; Niswander et al. 1967; Salzano et al. 1967; Shreffler and Steinberg 1967; Tashian et al. 1967; and Weinstein et al. 1967.

5. In 1994–95 a donation from a German NGO paid for digging a well near the village houses. Because of inappropriate technology, which results in high maintenance (fuel and repair) costs under village conditions, the well has never provided water on a reliable basis.

6. As we saw in chapter 7, the responsibility for indigenous health in Brazil was transferred from FUNAI to FUNASA in 1999. When FUNASA restructured the Indian health system an information system was implemented intended to provide data on morbidity, mortality, and other vital statistics (http://www.funasa.gov.br/ind/ind00.htm, accessed Nov. 2000). At the end of 2000 the process was not yet completed, and therefore its effects were limited. Among other difficulties, it has not been easy to collect systematic data at the village level because of the lack of health personnel trained to record data.

7. These data refer not only to Xavánte from Etéñitépa but to all Xavánte living in the Nova Xavantina area, including those from other reservations. The Casas do Índio are hostels run by FUNAI where Indians from the villages may lodge when they come to town for health care. There is usually a physician and a nurse in part-time attendance or patients may go to a hospital for more specialized care while their families stay at the Casa do Índio.

8. In figure 8.1, which shows frequencies of diagnosis in 848 consultations at the Casa do Índio in Nova Xavantina, 1994–95, "other" includes endocrine-metabolic (3.9 percent), genitourinary (1.8 percent), skin diseases (3.9 percent), and other (8.8 percent). See Ianelli 1997, 42.

9. See Coimbra et al. 1985b; Coimbra and Santos 1991; Eveland et al. 1971; Linhares 1992; Linhares et al. 1981, 1986b; Loureiro and Lins 1976; Neel et al. 1968b; Salzano and Callegari-Jacques 1988, 94–95; Rees and Shelley 1977; and Santos et al. 1991.

10. Xavánte mothers keep their small babies in large covered baskets that serve as cradles and are carried on the mother's back by means of a handle across her forehead. So, even though they are frequently held, babies do not have the permanent access to the breast that they do in societies where they are carried against the mother's body.

11. Neel et al.(1964, 109) mentioned that scalp infections affected a high proportion of young children in 1962.

12. Six of ten Xavánte patients shared the *HLA-DRB1*0404* allele, a gene that was present in only five of the seventy-four healthy Xavante control subjects. Since the other allele present among the four remaining patients was *DRB1*1402,* and the two alleles share a sequence of eight amino acids in the third hypervariable region of the first domain of the molecule, it is possible that this specific epitope is the one involved in the susceptibility to the disease (Cerna et al. 1993a, 1993b; Friedman et al. 1995).

13. The disease is caused by the small hepatitis B virus (HBV). The core of the virus contains two antigens, the core antigen per se (HBcAg), and the "e" antigen (HBeAg). A third antigen is located on the outer surface, known as surface antigen or HBsAg (originally termed the Australian antigen). These antigens elicit the production of specific antibodies that can be detected in the serum of individuals infected with HBV. These antibodies are called anti-HBcAg, anti-HBeAg, and anti-HBsAg. The latter antibody confers protection to subsequent HBV infection (see Coimbra et al. 1996b, 1735).

14. In our study, we investigated not only the Xavánte but three other groups from the Brazilian Amazon, the Gavião, Suruí, and Zoró (Coimbra et al. 1996b). With regard to the age distribution of HBV serological markers, we observed two important aspects: (1) children ten years of age and younger were highly exposed to the hepatitis B virus, as nearly 40 percent showed evidence of past infection; and (2) the highest rates of HBV carriers were observed in the ten to twenty-nine age cohort. These findings may suggest that, as none of the HBV carriers showed signs or symptoms suggestive of acute hepatitis nor a history of jaundice, hepatitis infection in these populations is probably acquired in early childhood and does not lead necessarily to a state of disease. In addition, once

infected, individuals seem quite capable of clearing up the virus. The only longitudinal investigation on HBV infection carried out among Amazonian Indians reported that the great majority of the Kayapó who were HBV carriers in the first examination proved to have eliminated the virus in the second testing (Black et al. 1986).

15. Cultural practices, which include sexual customs, bleeding for therapeutic purposes, body piercing, and close living conditions, have been implicated as playing an important role in the transmission of the hepatitis B virus in other parts of the world (see Blumberg and Hesser 1975; Brabin and Brabin 1985; Dickie et al. 1982; Langendorfer et al. 1984; and Milne et al. 1987).

16. One hundred and forty serum samples collected in 1990 at Etéñitépa, including individuals of both sexes (57 men and 83 women), ages one to sixty-six, were tested for antitreponemal activity. Samples were analyzed by the laboratory of clinical pathology at Evandro Chagas Hospital, Fundação Oswaldo Cruz, in Rio de Janeiro, in accordance with VDRL (Veneral Disease Research Laboratory) procedures, using the slide qualitative flocculation test. Only five samples were weakly reactive to VDRL, and these were subjected to FTA-Abs (fluorescent antibody absorbed test), with all tests negative.

17. In our 1990 study we tested 127 serum samples, which included individuals of both sexes (50 men and 77 women), ages one to sixty-six years. The samples were tested for hemagglutination-inhibition antibodies against arbovirus. These samples were analyzed by the laboratory of virology at the Instituto Evandro Chagas in Belém, following Clarke and Casals 1958. The following antigens were used: Eastern Equine Encephalitis (EEE), Western Equine Encephalitis (WEE), Mayaro (MAY), Mucombo (MUC), Yellow Fever (YF), Dengue-1 (DEN-1), Dengue-2 (DEN-2), Dengue-4 (DEN-4), Ilhéus (ILH), St. Louis Encephalitis (SLE), Rocio (ROC), Cacipacoré (CPC), Carapuru (CAR), Guaroa (GRO), Tacaiúma (TCM), Oropouche (ORO), Maguari (MAG), and Catu (CATU).

18. *Anemia* is defined as a condition in which the concentration of hemoglobin, the oxygen and carbon dioxide carrying substance in the blood, is below normal (WHO 1972).

19. In order to determine the prevalence rates of anemia, we utilized the cutoff points of hemoglobin concentration suggested by the World Health Organization (WHO 1972), which are the following: children 0.5 to 5.9 years, 11 g of hemoglobin/100 ml of blood; children 6 to 13.9 years, 12 g; males above age 14, 13 g; females above age 14, 12 g (29). The HemoCue equipment was utilized for measuring hemoglobin (Hudson-Thomas et al. 1994; Johns and Lewis 1989).

20. In their study of the Xavánte in 1962, James Neel and his collaborators tested the blood samples for hemoglobin concentrations (Neel et al. 1964). We reanalyzed their data utilizing the same cutoff points that we used in analyzing our 1995 data. Frequencies of anemia were lower in 1995 (39.8 percent) than in 1962 (55.8 percent). Exposure to malaria may have been higher in the 1960s since they were living by the Rio das Mortes, where malaria transmission seems to be more intense. Also, as we have seen in chapter 5, our demographic data indicate that the 1960s were especially difficult years for the Xavánte. The

percentage of survival of children to the age of ten years was significantly lower from 1957 to 1971 (43 percent) than from 1972 to 1990 (83 percent). Epidemics and social disruption affected the Xavánte, which may have exposed them to increased risk of anemia through infection and nutritional deficiencies.

21. There are important differences in the biological, nutritional, and epidemiological interpretations of the various anthropometric indices. Height for age is regarded as an anthropometric indicator of long-term, cumulative inadequacies in health and nutrition. Children low in height for their age are said to be "stunted," meaning that they have gained insufficient height relative to their age; often they have been undernourished for a relatively long time. Children with low weight for height are said to be "wasted," implying recent or continuing weight loss. The evaluation of anthropometric status requires the use of growth standards or references. The World Health Organization has recommended the use of height and weight reference data gathered in populations studied by the U.S. National Center for Health Statistics (NCHS). For the Xavánte, we followed WHO recommendations. A child was classified as presenting low height for age, or low weight for height if he or she was −2 Z-scores below the reference mean or median value of NCHS curves (WHO 1995, 161–262). According to the WHO (225), "while it is accepted that there are some variations in the growth patterns of children from different racial or ethnic groups . . . these are relatively minor compared with the large worldwide variation that relates to health, nutrition and socioeconomic status. For this reason, a common reference has the advantage of uniform application, allowing international comparisons without loss of its usefulness for local application." It is important to mention that there is an ongoing debate in the literature as to whether those recommendations are applicable to children throughout the world (Eveleth and Tanner 1990; Santos 1993; van Loon et al. 1986).

22. In 1976–77 Nancy Flowers (1983a, 1983b) conducted a study of weight gain on a sample of children aged two years and younger at Etéñitépa. Her results suggest that the seasonal subsistence activities of adults affect the weight gain of children. The nutritional status of children, especially those between twelve and twenty-three months, improved in the period after the harvest. Flowers's data permitted a detailed analysis of seasonality because the growth of children in the planting season (September and November) could be compared to growth after the harvest (March and July).

Chapter 9

1. Epidemiological polarization in Brazil is closely related to the extreme social and economic inequity that prevails in the country. The persistence of infectious and parasitic diseases may be partly due to the expansion from rural into urban areas of certain endemic diseases (e.g., malaria and visceral leishmaniasis in Amazonia and northeastern towns) and/or their recrudescence in areas where control had previously been achieved (Araújo 1992; Barreto and Carmo 1998; Merchan-Hamann 1997; Sabroza et al. 1995; Waldman et al. 1995). Sabroza et al. (1995, 216) carried out a detailed analysis of factors related to

these "countertransitions" in Brazil. They found that the increased incidence of many transmissible diseases can be attributed in large part to the fact that in Brazil, as in many other middle-income countries, the reduction in mortality was largely the result of specific preventive measures and improved access to health care, with relatively little change in the overall environmental, social, and behavioral setting, so that parasites can still find milieus that favor their reproduction and maintenance.

2. In Brazil there is an increasing trend toward obesity (Martins 1999; Mondini and Monteiro 1997; Monteiro et al. 1995a, 1995b, 2000). National surveys show that the prevalence of obesity in adults aged twenty-five to sixty living in metropolitan areas increased from 5.7 percent in 1974–75 to 9.6 percent in 1989 (Monteiro et al. 1995b, 250). Recent studies suggest a changing relationship between obesity and socioeconomic status. In the past, obesity was concentrated in the better-off social sectors of the Brazilian urban population. More recent data have shown a reversal of this situation: increases in obesity tend to be greater in rural than in urban settings and in poor rather than wealthy families (Monteiro et al. 2000). Doak et al. (2000) analyzed data from a nationally representative survey and showed that in 11 percent of the households investigated underweight and overweight coexisted, most commonly underweight children with an overweight young or middle-aged adult, often the mother.

3. We had access to the original anthropometric data collected by Neel et al. (1964), and we carried out a reanalysis, presented in Coimbra et al. 2001. The weight and height measurements in table 9.2 are slightly different from those reported by Neel et al. (1964, 65), possibly because of the use of different age groupings. The measurements presented in table 9.2 include individuals twenty to fifty years of age. In the article by Neel and his associates it is not clear to which age groups the individuals whose measurements are reported belong.

4. A measure frequently used to evaluate the nutritional status of adults is BMI (Gibson 1990, 178–81; WHO 1995, 1998). It is a simple measurement, highly correlated with other estimates of fatness, that also minimizes the effect of stature. Various studies have called attention to the significant relationship between BMI and morbidity and mortality from cardiovascular disease, cancer, and diabetes (Baik et al. 2000; Higgins et al. 1988; Hoffmans et al. 1988; NIH 1985). One calculates BMI by dividing weight in kilograms by stature in meters squared [BMI = (weight in kilograms)/(height in meters)^2]. Thus, an adult who weighs 65 kg and is 1.64 m tall will have a BMI of 24.2 kg/m^2.

5. These three men were twenty, twenty-one, and twenty-seven years of age. It is possible that their high BMI values were less due to fat than to strong, muscular bodies.

6. The Eténitépa data presented in table 9.4 were collected by Gugelmin in 1994 (see Gugelmin 2001); they are not the same as those that she collected in 1990, which are presented in table 9.2.

7. Two investigations were made in the 1960s. One of these is the study conducted by Neel et al. (1964) at São Domingos, and the other is by Niswander et al. (1967), who collected anthropometric data among the Xavánte of Simões Lopes.

8. The geographic location of São José, only 50 km from the town of Primavera do Leste (see fig. 2.2), the availability of relatively easy and cheap transportation, and electricity in the village may all be affecting their dietary habits and lifestyles. During fieldwork at São José, even though Gugelmin (2001) was not able to do a food consumption survey, she observed that the diet was heavy in starchy foods, some of them purchased, like polished rice, macaroni, cooking oil, bread, crackers, and soft drinks (see also Leite 1998).

9. Both mean blood pressure readings and the percentage of hypertensive individuals among the Xavánte of Etéñitépa are consistently higher than those observed in other studies of Brazilian Indians (Coimbra et al. 2001). Note that when these studies were done most groups still lived according to "traditional" subsistence regimens. Surveys carried out in the upper Xingu (Baruzzi and Franco 1981; Lima 1950; Mancilha-Carvalho et al. 1989), and among the Yanomámi (Crews and Mancilha-Carvalho 1993; Mancilha-Carvalho et al. 1989; Oliver et al. 1975), Karajá (Lowenstein 1961), and Kayapó (Ayres and Salzano 1972) showed mean systolic and diastolic pressure levels close to or under 100 to 105 mmHg and 65 to 70 mmHg, respectively.

10. Blood pressure levels were classified in five categories, according to the "Sixth Report of the Joint National Committee on Prevention, Detection, Evaluation, and Treatment of High Blood Pressure" (Joint National Committee 1997): excellent (SBP $<$ 120 mmHg and DBP $<$ 80 mmHg); normal (SBP $<$ 130 mmHg and DBP $<$ 85 mmHg); high-normal (SBP from 130 to 139 mmHg or DBP from 85 to 89 mmHg); hypertension stage 1 (SBP from 140 to 159 mmHg or DBP from 90 to 99 mmHg); hypertension stage 2 (SBP from 160 to 179 mmHg or DBP from 100 to 109 mmHg); hypertension stage 3 (SBP above 180 mmHg or DBP above 110 mmHg).

11. The long-term consequences of diabetes may include cardiovascular disease, renal failure, retinopathy leading to blindness, gangrene of the extremities (often leading to limb amputation), and premature death.

12. Most of the studies on diabetes in indigenous peoples have been carried out in North America (Ghodes 1986; Narayan 1996; Szathmáry 1994; Young 1993), Australia and New Guinea (King 1992; O'Dea 1991), and the Pacific Islands (Baker and Crews 1986; Zimmet et al. 1990).

Studes in human biology and population genetics have led to the postulation of evolutionary models to explain the high frequency of NIDDM in indigenous populations. James Neel's "thrifty genotype hypothesis" (Neel 1962, 1982b) has been particularly influential. According to this model, during most of human evolution genes promoting rapid insulin response to a plasma glucose stimulus and favoring the ability to store fat were positively selected, as they conferred survival advantage in times of food shortages. This model assumes that for our hunter-gatherer ancestors excess glucose was not a regular condition but occurred periodically under feasting conditions. However, altered conditions in more recent times, including the replacement of low-fat, low-calorie diets with high-fat, high-calorie diets, made the quick insulin response maladaptive. Under these new conditions, the thrifty genotype renders its owners more susceptible to obesity and diabetes. Other evolutionary models have been proposed, including the New World syndrome of metabolic diseases by Weiss et al. (1984) and the low carbohydrate adaptation model by Szathmáry (1990). As Szathmáry (1994, 464) pointed out, "these models all propose that the genes predisposing Native Americans to NIDDM today were selectively advantageous in their pre-modern environment."

Some authors have proposed a novel hypothesis called the "thrifty phenotype hypothesis" (Hales and Barker 1992; Pettitt et al. 1988; Phillips et al. 1994; Ravelli et al. 1998). According to this, poor nutrition and growth in utero and/or early infancy carry a significantly increased risk of NIDDM and cardiovascular disease later in life, especially when obesity in adulthood is preceded by undernutrition in childhood. In adult life, endocrine functions of the pancreas might be impaired by inadequate early nutrition, thereby greatly increasing NIDDM susceptibility (see also Szathmáry 1994).

13. Anthropologist Mariana Ferreira conducted ethnographic research among the Teréna of São Paulo, and she refers to cases of diabetes among them, which she ascribes to poor diet and living conditions (1999).

14. The nutritional consequences of dietary change among Xavánte have been extensive. Vieira-Filho et al. (1997) report two serious cases of beriberi (deficiency of vitamin B_1), involving muscular atrophy, among Xavánte from Sangradouro and São Marcos, which they ascribe to an impoverished diet based on polished rice. According to these authors, the Xavánte from Sangradouro and São Marcos are exchanging the rice they grow in their fields for rice industrially polished in nearby towns, which they have grown to prefer.

15. It should be pointed out that these cases of diabetes mellitus are not from a representative epidemiological survey and that the document gives no information regarding the diagnostic standards used (including laboratory technique used to determine blood glucose levels and the cutoff point for hyperglycemia). We thank Luciene Guimarães de Sousa, from FUNASA in Barra do Garça for providing us with a copy of this document.

16. Esparza et al. (2000) and Ravussin et al. (1994) carried out studies comparing Pima who live in the southwestern United States with those who live in northern Mexico; these studies were designed to elucidate the relationship between lifestyle changes, with their impact on nutrition, and the occurrence of chronic, nontransmissible diseases. In general they concluded that obesity and diabetes were less frequent among Pima communities in Mexico, which had a more "traditional" lifestyle characterized by a diet that included less animal fat and more complex carbohydrates and by greater energy expenditure in physical labor. It seems probable that the determining factors observed among the Pima, as well as the differences between the communities, may be useful in explaining what is happening to the Xavánte de Etéñitépa and São José.

References

Adams, K., and D. Price, eds. 1994. The demography of small-scale societies: Case studies from lowland South America. *South American Indian Studies,* no. 4.

Alcântara, V. M., M. A. C. Lourenço, F. M. Salzano, M. L. Petzl-Ehler, C. E. A. Coimbra Jr., R. V. Santos, and E. A. Chautard-Freire-Maia. 1995. Butyrylcholinesterase polymorphisms (BCHE and CHE2 loci) in Brazilian Indian and admixed populations. *Human Biology* 67:717–26.

Alencastre, J. M. P. d'. 1857. Memória chronológia, histórica e geográphica da Provincia do Piauhy. *Revista do Instituto Histórico e Geographico Brasileiro* 20:5–58.

———. 1864. Annaes da Provincia de Goyaz. *Revista do Instituto Histórico e Geographico Brasileiro* 27:5–186.

Almeida, R. H. 1997. *O Diretório dos Índios: Um Projeto de "Civilização" no Brasil no Século XVIII.* Brasília: Editora da Universidade de Brasília.

Almeida, S. P. 1998. *Cerrado: Aproveitamento Alimentar.* Planaltina: Empresa Brasileira de Pesquisa Agropecuária.

Almeida, S. P., C. E. B. Proença, S. M. Sano, and J. F. Ribeiro. 1998. *Cerrado: Espécies Vegetais Úteis.* Planaltina: Empresa Brasileira de Pesquisa Agropecuária.

Almudena, C. 1995. Sacode a tribo: Sepultura (Adöoru em Xavante) grava música e clipe em aldeia indígena. *Folha de São Paulo,* 13 November, section 6, 1.

Alter, M. J. K., P. J. Coleman, W. J. Alexander, E. Kramer, J. K. Miller, E. Mandel, S. C. Hadler, and H. S. Margolis. 1989. Importance of heterosexual activity in the transmission of hepatitis B and non-A, non-B hepatitis. *Journal of the American Medical Association* 262:1201–5.

Alter, H. J., R. H. Purceli, J. L. Gerin, W. T. London, P. M. Kaplan, V. J. McAuliffe, J. Wagner, and P. V. Holland. 1977. Transmission of hepatitis B to chimpanzees by hepatitis B surface antigen-positive saliva and semen. *Infection and Immunity* 16:928–33.

Alvard, M. 1993. A test of the ecologically noble savage hypothesis: Interspecific prey choice by neotropical hunters. *Human Ecology* 21:355–87.

———. 1995. Intraspecific prey choice by Amazonian hunters. *Current Anthropology* 36:789–818.

Alvarez, R. R., I. Campbell, H. Friedman, M. L. Bertoli, G. B. M. Gama, and L. A. Diaz. 1991. Dermatoses entre os Xavánte da Área Indígena Pimentel Barbosa, Mato Grosso (Brasil). *Cadernos de Saúde Pública* 7:581–84.

Alves, R. E. 2000. Xavantes abrem “embaixada” em São Paulo. *Folha de São Paulo,* 2 April, sec. 1, 12.

Amarante, J. M., J. F. Porto, and F. A. Silva. 1996. Controle da tuberculose em área indígena: Experiência de uma nova abordagem em Água Branca, MT, maio de 1996. *Revista de Saúde do Distrito Federal* 7:25–32.

Andrade, F. M., C. E. A. Coimbra Jr., R. V. Santos, A. Goicoechea, F. R. Carnese, F. M. Salzano, and M. H. Hutz. 2000. High heterogeneity of apolipoprotein E gene frequencies in South American Indians. *Annals of Human Biology* 27:29–34.

Anhalt, G. J., C. Martins, E. Rivitti, and L. A. Diaz. 1989. Endemic pemphigus foliaceus (fogo selvagem). *Advances in Dermatology* 4:73–93.

Anonymous, 1980. Xavantes colhem 7 mil sacas de arroz. *Jornal de Brasília,* 14 June, National sec., 5.

———. 1995. “Heavy” índio. *Jornal do Brasil,* 18 November, sec. B, 2.

———. 2000. Protesto de Xavantes. *Jornal do Brasil,* 24 April, sec. 1, 4.

Arantes, R. 1998. Saúde oral de uma comunidade indígena do Brasil Central: Uma abordagem epidemiológica e bioantropológica. M.Sc. thesis, Rio de Janeiro, Escola Nacional de Saúde Pública, Fundação Oswaldo Cruz.

Arantes, R., R. V. Santos, and C. E. A. Coimbra Jr. 2001. Saúde bucal na população indígena Xavánte de Pimentel Barbosa, Mato Grosso, Brasil. *Cadernos de Saúde Pública* 17:375–84.

Araújo, F. D. 1995. A review of *Caryocar brasiliense* (Caryocaraceae): An economically valuable species of the Central Brazilian cerrados. *Economic Botany* 49:40–48.

Araújo, J. D. 1992. Polarização epidemiológica no Brasil. *Informe Epidemiológico do SUS* 1:5–16.

Arnt, R. A., and S. Schwartzman. 1992. *Um Artificio Orgânico: Transição na Amazônia e Ambientalismo (1985–1990).* Rio de Janeiro: Rocco.

Askew, G. P., D. J. Moffatt, R. F. Montgomery, and P. L. Searl. 1970a. Soil landscapes in northeastern Mato Grosso. *Geographic Journal* 136:211–27.

———. 1970b. Interrelationships of soil and vegetation in the savanna-forest boundary zone of northeastern Mato Grosso. *Geographic Journal* 136:370–76.

Aspelin, P. L. 1975. External articulation and domestic production: The artifact trade of the Mamaindê of northwestern Mato Grosso, Brazil. Ph.D. diss., Cornell University.

Aston, D. L., and A. P. Thorley. 1970. Leishmaniasis in Central Brazil: Results from a Montenegro skin test among Amerindians in the Xingu National Park. *Transactions of the Royal Society of Tropical Medicine and Hygiene* 64:671–78.

Athias, R., and M. Machado. 2001. A saúde indígena no processo de implantação dos Distritos Sanitários: Temas críticos e propostas para um diálogo interdisciplinar. *Cadernos de Saúde Pública* 17:425–31.

Auad, A. 1972. Pênfigo foliáceo sul-americano no estado de Goiás. *Revista de Patologia Tropical* 1:293–346.

Ávila-Desser, N. 1993. Projeto de assistência médico-sanitária e de formação de agentes comunitários de saúde para tribos Xavante e Bororo: Estado de Mato Grosso, Brasil, Nova Xavantina and Barra do Garças, Medecins sans Frontiers. Unpubished Report.

Ayres, M., and F. M. Salzano. 1972. Health status of Brazilian Cayapo Indians. *Tropical and Geographical Medicine* 24:178–85.

Ayres de Casal, M. [1817] 1945–47. *Corografia Brasílica.* Rio de Janeiro: Imprensa Nacional. Facsimile of the 1817 edition.

Azevedo, M. M. 1994. Demografia dos povos indígenas do Alto Rio Negro. *Revista Brasileira de Estudos Populacionais* 11:235–44.

———. 1997. Fontes de dados sobre as populações indígenas brasileiras da Amazônia. *Cadernos de Estudos Sociais* 13:163–77.

Baik, I., A. Ascherio, E. B. Rimm, E. Giovannucci, D. Spiegelman, M. J. Stampfer, and W. C. Willett. 2000. Adiposity and mortality in men. *American Journal of Epidemiology* 152:264–71.

Baker, P. T., and D. E. Crews. 1986. Mortality patterns and some biological predictors. In *The Changing Samoans: Behavior and Health in Transition,* P. T. Baker, J. M. Hanna, and T. S. Baker, eds., 93–122. New York: Oxford University Press.

Baker, S. J. 1978. Nutritional anaemia: A major controllable public health problem. *Bulletin of the World Health Organization* 56:659–75.

Baldy, J. L. S., and H. V. Lopes. 1976. Estafilococcias. In *Doenças Infecciosas e Parasitárias,* R. Veronesi, ed., 6th ed., 296–305. Rio de Janeiro: Guanabara Koogan.

Balée, W. 1994. *Footprints of the Forest.* New York: Columbia University Press.

Baptista, F. M. 1999. Crônica de uma fraude anunciada: Licenciamento da Hidrovia Araguaia-Tocantins tem denúncias de irregularidades, mas governo faz vista grossa. *Parabólicas* (São Paulo) 54:15.

Barata, F. J. R. 1848. Memória em que se mostram algumas providencias tendentes ao melhoramento da agricultura e commércio da Capitania de Goyaz. *Revista do Instituto Histórico e Geographico Brasileiro* 11:336–65.

Barbira-Scazzocchio, F., ed. 1980. *Land, People, and Planning in Contemporary Amazonia.* Cambridge: Center of Latin American Studies, Cambridge University.

Barreto, M. L., and E. H. Carmo. 1995. Mudanças em padrões de morbimortalidade: Conceitos e métodos. In *Velhos e Novos Males da Saúde no Brasil: A Evolução do País e de suas Doenças,* C. A. Monteiro, ed., 17–30. São Paulo: Hucitec.

———. 1998. Tendências recentes das doenças crônicas no Brasil. In *O Adulto Brasileiro e as Doenças da Modernidade: Epidemiologia das Doenças Crônicas Não-Transmissíveis,* I. Lessa, ed., 15–42. São Paulo: Hucitec.

Barreto, M. L., E. H. Carmo, C. V. Noronha, R. B. B. Neves, and P. C. Alves. 1993. Mudanças os padrões de morbi-mortalidade: Uma revisão crítica das abordagens epidemiológicas. *Physis* 3:126–46.

Bartlett, A. V., E. Hurtado, D. G. Schroeder, and H. Mendez. 1992. Association of indicators of hygiene behavior with persistent diarrhea of young children. *Acta Paediatrica* 38 (sup.): 66–67.

Baruzzi, R. G. 1970. Contribution to the study of toxoplasmosis epidemiology: Serologic survey among the Indians of the Upper Xingu River, Central Brazil. *Revista do Instituto de Medicina Tropical de São Paulo* 12:93–104.

Baruzzi, R. G., R. M. Castro, C. D'Andretta Jr., S. Carvalhal, O. L. Ramos, and P. L. Pontes. 1973. Occurrence of Lobo's blastomycosis among Caiabi Brazilian Indians. *International Journal of Dermatology* 12:95–98.

Baruzzi, R. G., and L. J. Franco. 1981. Amerindians of Brazil. In *Western Diseases: Their Emergence and Prevention,* H. C. Trowell and D. P. Burkitt, eds., 138–53. London: Edward Arnold.

Baruzzi, R. G., L. J. Franco, J. R. Jardim, A. Masuda, C. Naspitz, E. R. Paiva, and N. Ferreira-Novo. 1976. The association between splenomegaly and malaria in Indians from the Alto Xingu, Central Brazil. *Revista do Instituto de Medicina Tropical de São Paulo* 18:322–48.

Baruzzi, R. G., C. S. Lacaz, and F. A. A. Souza. 1979. História natural da doença de Jorge Lobo: ocorrência entre os índios Caiabí (Brasil Central). *Revista do Instituto de Medicina Tropical de São Paulo* 21:302–38.

Baruzzi, R. G., C. F. Marcopito, and M. Iunes. 1978. Programa médico preventivo da Escola Paulista de Medicina no Parque Nacional do Xingu. *Revista de Antropologia* (São Paulo) 21:155–70.

Baruzzi, R. G., C. F. Marcopito, M. L. C. Serra, F. A. A. Souza and C. Stabile. 1977. The Kren-Akorore: A recently contacted indigenous tribe. In *Health and Disease in Tribal Societies,* K. Elliot and J. Whelan, eds., 179–211. Amsterdam: Elsevier.

Becker, B. K. 1982. *Geopolítica da Amazônia.* Rio de Janeiro: Zahar.

Benavides, M. 1996. Amazon indigenous peoples: New challenges for political participation and sustainable development. *Cultural Survival Quarterly* 20:50–53.

Bensabath, G., S. C. Hadler, M. C. P. Soares, H. Fields, and J. E. Maynard. 1987. Epidemiologic and serologic studies of acute viral hepatitis in Brazil's Amazon. *Bulletin of the Pan American Health Organization* 21:16–27.

Berman, S. 1991. Epidemiology of acute respiratory infections in children of developing countries. *Reviews of Infectious Diseases* 13 (Sup. 6): S454–62.

Bern, C., J. Martines, I. de Zoysa, and R. I. Glass. 1992. The magnitude of the global problem of diarrhoeal disease: A ten-year update. *Bulletin of the World Health Organization* 70:705–14.

Bertrand, W. E., and B. F. Walmus. 1983. Maternal knowledge, attitudes, and practice as predictors of diarrhoeal in young children. *International Journal of Epidemiology* 12:205–10.

Bevilaqua, L. R. M., V. S. Mattevi, G. M. Ewald, F. M. Salzano, C. E. A. Coimbra Jr., R. V. Santos, and M. H. Hutz. 1995. Beta-globin gene cluster haplotype distribution in five Brazilian Indian tribes. *American Journal of Physical Anthropology* 98:395–401.

Black, F. L. 1966. Measles endemicity in insular populations: Critical community size and its evolutionary implication. *Journal of Theoretical Biology* 11:207–11.

———. 1975. Infectious diseases in primitive societies. *Science* 187:515–18.

———. 1990. Infectious disease and evolution of human populations: The example of South American forest tribes. In *Disease in Populations in Transition: Anthropological and Epidemiological Perspectives,* A. C. Swedlund and G. J. Armelagos, eds., 55–74. New York: Bergin and Garvey.

———. 1992. Why did they die? *Science* 258:1739–49.

Black, F. L., W. J. Hierholzer, J. F. Lian-Chen, L. L. Berman, Y. Gabbai, and F. P. Pinheiro. 1982. Genetic correlates of enhanced measles susceptibility in Amazon Indians. *Medical Anthropology* 6:37–46.

Black, F. L., W. J. Hierholzer, F. P. Pinheiro, A. S. Evans, J. P. Woodall, E. M. Opton, J. E. Emmons, B. S. West, G. Edsall, W. G. Downs, and G. D. Wallace. 1974. Evidence for persistence of infectious agents in isolated human populations. *American Journal of Epidemiology* 100:230–50.

Black, F. L., J. P. Pandey, and R. A. Capper. 1986. Hepatitis B epidemiology and its relation to immunogenetic traits in South American Indians. *American Journal of Epidemiology* 123:336–43.

Black, F. L., F. L. Pinheiro, O. Oliva, W. J. Hierholzer, R. V. Lee, J. E. Briller, and V. A. Richards. 1978. Birth and survival patterns in numerically unstable proto agricultural societies in the Brazilian Amazonia. *Medical Anthropology* 2:95–127.

Bloch, K. V., E. S. F. Coutinho, M. E. C. Lôbo, J. E. P. Oliveira, and A. Milech. 1993. Pressão arterial, glicemia capilar, e medidas antropométricas em uma população Yanomámi. *Cadernos de Saúde Pública* 9:428–38.

Blumberg, B. S., and J. E. Hesser. 1975. Anthropology and infectious disease. In *Physiological Anthropology,* A. Damon, ed., 260–98. New York: Oxford University Press.

Bobadilla, J. L., J. Frenk, R. Lozano, T. Frejka, and C. Stern. 1993. The epidemiologic transition and health priorities. In *Disease Control Priorities in Developing Countries,* D. T. Jamison, W. H. Mosley, A. R. Measham, and J. L. Bobadilla, eds., 51–63. New York: Oxford University Press.

Bobadilla, J. L., and C. A. Possas. 1993. Health policy issues in three Latin American countries: Implications of the epidemiological transition. In *The Epidemiological Transition: Policy and Planning Implications for Developing Countries,* J. N. Gribble and S. H. Preston, eds., 145–69. Washington, D.C.: National Academy Press.

Bogdawa, H. M., M. H. Hutz, F. M. Salzano, and T. A. Weimer. 2000. Diversity of two short tandem repeat loci (CD4 and F13A1) in three Brazilian ethnic groups. *Human Biology* 72:1045–53.

Bortolini, M. C., C. Baptista, S. M. Callegari-Jacques, T. A. Weimer, and F. M. Salzano. 1998. Diversity in protein, nuclear DNA, and mtDNA in South Amerinds: Agreement or discrepancy? *Annals of Human Genetics* 62:133–45.

Boxer, C. R. 1962. *The Golden Age of Brazil, 1695–1750: Growing Pains of a Colonial Society.* Berkeley and Los Angeles: University of California Press.

Brabin, L., and B. J. Brabin. 1985. Cultural factors and transmission of hepatitis B virus. *American Journal of Epidemiology* 122:725–30.

Brainard, J. M. 1991. *Health and Development in a Rural Kenyan Community.* New York: Peter Lang.

Brandford, S., and O. Glock. 1985. *The Last Frontier: Fighting over Land in the Amazon.* London: Zed.

Brieger, F. G., J. Gurgel, E. Paterniani, A. Blumenschein, and M. R. Alleoni. 1958. *Races of Maize in Brazil and Other Eastern South American Countries.* Publication 593. Washington, D.C.: National Science Foundation.

Briend, A. 1990. Is diarrhea a major cause of malnutrition among under-fives in developing countries? A Review of available evidence. *European Journal of Clinical Nutrition* 44:611–28.

Brown, M. F. 1993. Facing the state, facing the world: Amazonia's native leaders and the new politics of identity. *L'Homme* 33:307–26.

Buchillet, D. 1992. Nobody is there to hear: Desana therapeutic lamentations. In *Portals of Power: Shamanism in South America,* E. J. M. Langdon and G. Baer, eds., 211–30. Albuquerque: University of New Mexico Press.

Bunker, S. G. 1985. *Underdeveloping the Amazon: Extraction, Unequal Exchange, and the Failure of the Modern State.* Urbana-Champaign: University of Illinois Press.

Burkitt, D. P. 1973. Some diseases characteristic of modern Western civilization. *British Medical Journal* 1:274–78.

Buss, P., and P. Gadelha. 1996. Health care systems in transition: Brazil. Part 1: An outline of Brazil's health care system reforms. *Journal of Public Health and Medicine* 18:289–95.

Caldeiron, S. S. 1993. *Recursos Naturais e Meio Ambiente: Uma Visão do Brasil.* Rio de Janeiro: Instituto Brasileiro de Geografia e Estatística.

Callegari-Jacques, S. M., F. M. Salzano, T. A. Weimer, M. H. L. P. Franco, M. A. Mestriner, M. H. Hutz, and L. Schüler. 1996. The Wai Wai Indians of South America: History and genetics. *Annals of Human Biology* 23: 189–201.

Campbell, H., J. R. M. Armstrong, and P. Bypass. 1989. Indoor air pollution in developing countries and acute respiratory infection in children. *Lancet* 1:1012.

Capelli, J. C. S., and S. Koifman. 2001. Avaliação do estado nutricional da comunidade indígena Parkatêjê, Bom Jesus do Tocantins, Pará, Brasil. *Cadernos de Saúde Pública* 17:433–37.

Cardoso, A. M., I. E. Mattos, and R. J. Koifman. 2001. Prevalência de fatores de risco para doenças cardiovasculares na população Guarani-Mbyá do Estado do Rio de Janeiro. *Cadernos de Saúde Pública* 17:345–54.

Cardoso, J. L. C. 1979. *Relato de Viagem à Reserva Xavante do Rio Culuene (Setembro de 1978).* São Paulo. Mimeo.

Carneri, I., N. Nutels, and J. A. Miranda. 1963. Epidemia de leishmaniose tegumentar entre os índios Waurá do Parque Nacional do Xingu (Estado de Mato Grosso, Brasil). *Revista do Instituto de Medicina Tropical de São Paulo* 5:271–72.

Carone, E. 1976. *O Estado Novo (1937–1945).* Rio de Janeiro: Difel.

Carvalho, F. 2000. Pela reeducação dos brancos: Índios xavante e mehinaku mostram rituais de suas tribos em espetáculos itinerantes. *Jornal do Brasil,* 29 April, B2.

Carvalho, J. A. M. 1997–98. Demographic dynamics in Brazil: Recent trends and perspectives. *Brazilian Journal of Population Studies* 1:5–24.

Castelnau, F. de. 1850–61. *Expedition dans les Parties Centrales de l'Amérique du Sud, de Rio de Janeiro a Lima, et de Lima au Pará, 1843–47.* Part 1: *Histoire du Voyage,* 6 vols. Paris: A. Bertrand.

Castro, E. A., and J. B. Kauffman. 1998. Ecosystem structure in the Brazilian cerrado: A vegetation gradient of aboveground biomass, root mass, and consumption by fire. *Journal of Tropical Ecology* 14:263–83.

Castro, E. V. 1996. Images of nature and images of society in Amazonian ethnology. *Annual Review of Anthropology* 25:179–200.

Castro, E. V., and L. M. M. de Andrade. 1988. Hidroelétricas do Xingu: O estado contra as sociedades indígenas. In *As Hidroelétricas do Xingu e os Povos Indígenas,* L. A. O. Santos and L. M. M. de Andrade, eds., 7–23. São Paulo: Comissão Pró-Índio de São Paulo.

Castro, R. M., J. T. Roscoe, and S. A. P. Sampaio. 1983. Brazilian pemphigus foliaceus. *Clinical Dermatology* 1:22–41.

CEDI (Centro Ecumênico de Documentação e Informação). 1991. *Povos Indígenas no Brasil, 1987–88–89–90.* Aconteceu special series 18. São Paulo: CEDI.

CENEPI (Centro Nacional de Epidemiologia). 1992. *Guia de Controle da Leishmaniose Tegumentar Americana.* Brasília: CENEPI, Ministério da Saúde.

Cerna, M., M. Falco, H. Friedman, E. Raimondi, A. Maccagno, M. Fernandez-Viña, and P. Stastny. 1993b. Differences in HLA Class II alleles of isolated South American Indian populations from Brazil and Argentina. *Human Immunology* 37:213–20.

Cerna, M., M. Fernandez-Viña, H. Friedman, J. R. Moraes, M. E. Moraes, L. Diaz, and P. Stastny. 1993a. Genetic markers for susceptibility to endemic Brazilian pemphigus foliaceus (fogo selvagem) in Xavante Indians. *Tissue Antigens* 42:138–40.

Cerqueira, R. 1982. South American landscapes and their mammals. In *Mammalian Biology in South America,* M. A. Mares and H. H. Genoways, eds. 53–75. Pymatuning Symposium in Ecology, Special Publications, no. 6. Linesville, Pa.: Pymatuning Laboratory of Ecology, University of Pittsburgh.

Chagnon, N. A. 1974. *Studying the Yanomamo.* New York: Holt, Rinehart and Winston.

Chaim, M. M. 1974. *Os Aldeamentos Indígenas na Capitania de Goiás: Sua Importância na Política do Povoamento (1749–1811).* Goiânia: Oriente.

———. 1983. *Aldeamentos Indígenas (Goiás 1749–1811).* São Paulo: Nobel.

Chandra, R. K. 1979. Nutritional deficiency and susceptiblity to infection. *Bulletin of the World Health Organization* 57:167–77.

Chandra, R. K., and N. S. Scrimshaw. 1980. Immunocompetence in nutritional assessment. *American Journal of Clinical Nutrition* 33:2694–97.

Chen, L. C., A. Chowdhury, and S. L. Huffman. 1980. Anthropometric assessment of energy-protein malnutrition and subsequent risk of mortality among pre-school-aged children. *American Journal of Clinical Nutrition* 33:1836–45.

Chen, L. C., and N. S. Scrimshaw, eds. 1983. *Diarrhea and Malnutrition: Interactions, Mechanisms, and Interventions.* New York: Plenum.

Clarke, D. H., and J. Casals. 1958. Techniques for haemagglutination and haemagglutination-inhibition with arthropod-borne virus. *American Journal of Tropical Medicine and Hygiene* 10:227–49.

Cleary, D. 1990. *Anatomy of the Amazon Gold Rush.* Iowa City: University of Iowa Press.

Cleve, H., and J. Constans. 1988. The mutants of the vitamin-D-binding protein: More than 120 variants of the GC/DBP system. *Vox Sanguinis* 54:215–25.

Cohen, M. N. 1989. *Health and the Rise of Civilization.* New Haven: Yale University Press.

Cohen, M. N., and G. J. Armelagos, eds. 1984. *Paleopathology at the Origins of Agriculture.* New York: Academic Press.

Coimbra, C. E. A., Jr. 1987. O sarampo entre sociedades indígenas brasileiras e algumas considerações sobre a prática da saúde pública entre essas populações. *Cadernos de Saúde Pública* 3:22–37.

———. 1988. Human settlements, demographic patterns, and epidemiology in lowland Amazonia: The case of Chagas's disease. *American Anthropologist* 90:82–97.

———. 1989a. From shifting cultivation to coffee farming: The impact of change on the health and ecology of the Suruí Indians in the Brazilian Amazon. Ph.D. diss., Indiana University.

———. 1989b. Avaliação dos serviços de saúde das administrações regionais da FUNAI em Rondonópolis, Barra do Garças e Nova Xavantina (Mato Grosso). Report presented to the World Bank, PRODEAGRO Project, Cuiabá and Washingotn, D.C.

———. 1995. Epidemiological factors and human adaptation in Amazonia. In *Indigenous Peoples and the Future of Amazonia: An Ecological Anthropology of an Endangered World,* L. E. Sponsel, ed., 167–81. Tucson: University of Arizona Press.

———. 1998. Minorias étnico-raciales, desigualdad y salud: Consideraciones teóricas preliminares. In *Salud, Cambio Social y Política: Perspectivas desde América Latina,* M. N. Bronfman and R. Castro, eds., 151–61. Mexico City: Edamex.

Coimbra, C. E. A., Jr., M. M. Borges, N. M. Flowers, R. V. Santos, and R. Piazza. 1992. Sero-epidemiological survey for Chagas' disease among the Xavánte Indians of Central Brazil. *Annals of Tropical Medicine and Parasitology* 86:567–68.

Coimbra, C. E. A., Jr., D. Chor, R. V. Santos, and F. M. Salzano. 2001. Blood pressure levels in Xavánte adults from the Pimentel Barbosa Indian Reserve, Mato Grosso, Brazil. *Ethnicity and Disease* 11:232–40.

Coimbra, C. E. A., Jr., M. F. F. Cruz, C. T. Daniel-Ribeiro, R. V. Santos, and R. M. Passo. 1995. Sero-epidemiology of malaria in the Xavánte and Tupí-

Mondé from Mato Grosso and Rondônia, Brazil. *Revista da Sociedade Brasileira de Medicina Tropical* 28 (Sup. 1): 155.

Coimbra, C. E. A., Jr., and R. V. Santos. 1991. Parasitismo intestinal entre o grupo indígena Zoró, Estado de Mato Grosso (Brasil). *Cadernos de Saúde Pública* 7:100–103.

———. 1994a. Epidemiologic profile of Amazonian Amerindians from Brazil, with special emphasis on the Xavánte from Mato Grosso and on groups from Rondônia. Report presented to the World Bank, Rio de Janeiro and Washington, D.C.

———. 1994b. Ocupação do espaço, demografia e epidemiologia na América do Sul: A doença de Chagas entre as populações indígenas. In *Saúde e Povos Indígenas,* R. V. Santos and C. E. A. Coimbra Jr., eds., 43–62. Rio de Janeiro: Editora Fiocruz.

———. 2000. Saúde, minorias e desigualdade: Algumas teias de inter-relações, com ênfase nos povos indígenas no Brasil. *Ciência e Saúde Coletiva* 5:125–32.

Coimbra, C. E. A., Jr., R. V. Santos, and R. Tanus. 1985a. Estudos epidemiológicos entre grupos indígenas de Rondônia, part 1: *Staphylococcus* sp. na boca e nariz entre os Suruí e Karitiana. *Revista do Instituto de Medicina Tropical de São Paulo* 27:13–19.

Coimbra, C. E. A., Jr., R. V. Santos, R. Tanus, and M. T. Inham. 1985b. Estudos epidemiológicos entre grupos indígenas de Rondônia, part 2: Bactérias intestinais e gastrinterites entre os Suruí e Karitiána. *Revista da Fundação SESP* 30:111–19.

Coimbra, C. E. A., Jr., R. V. Santos, and A. C. F. Valle. 1996a. Cutaneous leishmaniasis in Tupí-Mondé Amerindians from the Brazilian Amazonia. *Acta Tropica* 61:201–11.

Coimbra, C. E. A., Jr., R. V. Santos, C. F. T. Yoshida, M. I. Baptista, N. M. Flowers, and A. C. F. Valle. 1996b. Hepatitis B epidemiology and cultural practices in Amerindian populations of Amazonia: The Tupí-Mondé and the Xavánte from Brazil. *Social Science and Medicine* 42:1738–43.

Cole, M. M. 1986. *The Savannas: Biogeography and Geobotany.* London: Academic Press.

Collins, K. J., and J. S. Weiner. 1977. *Human Adaptability: A History and Compendium of Research in the International Biological Programme.* London: Taylor and Francis.

Comas, J. 1966. *Manual de Antropologia Física.* Mexico City: Universidad Nacional Autónoma de México.

Conklin, B. A. 1997. Body paint, feathers, and VCRs: Aesthetics and authenticity in Amazonian activism. *American Ethnologist* 24:711–37.

Conklin, B. A., and L. R. Graham. 1995. The shifting middle ground: Amazonian Indians and eco-politics. *American Anthropologist* 97:695–710.

Cook, N. D. 1998. *Born to Die: Disease and New World Conquest, 1492–1650.* Cambridge: Cambridge University Press.

Coope, E., and D. F. Roberts. 1971. Dermatoglyphic studies of populations in

Latin America. In *The Ongoing Evolution of Latin American Populations,* F. M. Salzano, ed., 405–53. Springfield: Charles C. Thomas.

Costa, C. C. C. da, J. P. Lima, L. D. Cardoso, and V. Q. Henriques. 1981. *Fauna do Cerrado: Lista Preliminar de Aves, Mamíferos e Répteis.* Rio de Janeiro: Fundação Instituto Brasileiro de Geografia e Estatística.

Costa, D. C. 1987. Política indigenista e assistência à saúde: Noel Nutels e o Serviço de Unidades Sanitárias Aéreas. *Cadernos de Saúde Pública* 4:388–401.

Coudreau, H. A. 1897. *Voyage au Tocantins-Araguaya.* Paris: A. Lahure.

Coutinho, L. M. 1978. O conceito de cerrado. *Revista Brasileira de Botânica* 1:17–23.

Crews, D. E., and J. J. Mancilha-Carvalho. 1993. Correlates of blood pressure in Yanomami Indians of northwestern Brazil. *Ethnicity and Disease* 3:362–71.

Crocker, J. C. 1985. *Vital Souls: Bororo Cosmology, Natural Symbolism, and Shamanism.* Tucson: University of Arizona Press.

Crocker, W., and J. Crocker. 1994. *The Canela: Bonding through Kinship, Ritual, and Sex.* Fort Worth: Harcourt Brace.

Crompton, D. W. T., and L. S. Stephenson. 1990. Hookworm infection, nutritional status and productivity. In *Hookworm Disease: Current Status and New Directions,* G. A. Schad and K. S. Warren, eds., 231–64. London: Taylor and Francis.

Crosby, A. W. 1972. *The Columbian Exchange: Biological and Cultural Consequences of 1492.* Westport: Greenwood.

Cunha, M. C. 1992a. Introdução a uma história indígena. In *História dos Índios no Brasil,* M. C. Cunha, ed., 9–24. São Paulo: Companhia das Letras.

———, ed. 1992b. *História dos Índios no Brasil.* São Paulo: Companhia das Letras.

———. 1993. Les études Gé. *L'Homme* 33:77–93.

Da Rocha, F. J., R. S. Spielman, and J. V. Neel. 1974. A comparison of gene frequency and anthropometric distance matrices in seven villages of four Indian tribes. *Human Biology* 46:295–310.

Davis, S. 1977. *Victims of the Miracle: Development and the Indians of Brazil.* New York: Cambridge University Press.

Dégallier, N., A. P. A. T. Rosa, P. F. C. Vasconcelos, J. P. Hervé, G. C. Sá-Filho, J. F. S. T. Rosa, E. S. T. Rosa, and S. G. Rodrigues. 1992. Modifications of arbovirus transmission in relation to construction of dams in Brazilian Amazonia. *Ciência e Cultura* 44:124–35.

DeMaeyer, E., and M. Adiels-Tegman. 1985. The prevalence of anemia in the World. *World Health Statistics Quarterly* 38:302–16.

Denevan, W. M., ed. 1992. *The Native Populations of the Americas in 1492.* 2d ed., Madison: University of Wisconsin Press.

Denny, F. W., and W. A. Clyde. 1983. Acute respiratory tract infections: An overview. *Pediatric Research* 17:1026–29.

Descola, F., and A. C. Taylor, eds. 1993. La remontée de l'"Amazone: Anthropologie et histoire des sociétés Amazoniennes. *L'Homme,* special issue, 33:126–28.

Diamond, S. 1993. *In Search of the Primitive: A Critique of Civilization.* New Brunswick, N.J.: Transaction Publishers.

Dias, B. F. S. 1994. Conservação da natureza no cerrado brasileiro. In *Cerrado: Caracterização Ocupação, e Perspectivas,* M. N. Pinto, ed., 586–605. Brasília: Editora da Universidade de Brasília.

Diaz, L. A., S. A. P. Sampaio, E. A. Rivitti, C. R. Martins, P. R. Cunha, C. Lombardi, F. A. Almeida, R. M. Castro, M. L. Macca, and C. Lavrado. 1989. Endemic pemphigus foliaceus (fogo selvagem), part 2: Current and historical epidemiologic studies. *Journal of Investigative Dermatology* 92:4–12.

Dickie, E. R., R. M. Knight Jr., and C. Merten. 1982. Ethnographic observations on child care and the distribution of hepatitis B virus in the nuclear family. *Medical Anthropology* 6:21–36.

Doak, C. M., L. S. Adair, C. A. Monteiro, and B. M. Popkin. 2000. Overweight and underweight coexist within households in Brazil, China, and Russia. *Journal of Nutrition* 130:2965–71.

Dobyns, H. F. 1983. *Their Number Become Thinned: Native American Population Dynamics in Eastern North America.* Knoxville: University of Tennessee Press.

Donnelly, C. J., L. A. Thomson, H. M. Stiles, C. Brewer, J. V. Neel, and J. A. Brunelle. 1977. Plaque, caries, periodontal diseases, and acculturation among Yanomami Indians, Venezuela. *Community Dentistry and Oral Epidemiology* 5:30–39.

Duchiade, M. P. 1995. População brasileira: Um retrato em movimento. In: *Os Muitos Brasis: Saúde e População na Década de 80,* M. C. S. Minayo, ed., 14–56. São Paulo: Hucitec.

Dufour, D. 1992. Nutritional ecology in the tropical rain forests of Amazonia. *American Journal of Human Biology* 4:197–207.

Early, J. D., and T. N. Headland. 1998. *Population Dynamics of a Philippine Rain Forest People.* Gainesville: University Press of Florida.

Early, J. D., and J. F. Peters. 1990. *The Population Dynamics of the Mucajai Yanomama.* San Diego: Academic Press.

Eaton, D. P., L. A. Diaz, G. Hans-Filho, V. Santos, V. Aoki, H. Friedman, E. A. Rivitti, S. A. P. Sampaio, M. S. Gottlieb, G. J. Giudice, A. Lopez, E. W. Cupp, and the Cooperative Group on Fogo Selvagem Research. 1998. Comparison of black fly species (Diptera: Simuliidade) on an Amerindian reservation with a high prevalence of fogo selvagem to neighboring disease-free sites in the state of Mato Grosso do Sul, Brazil. *Journal of Medical Entomology* 35:120–31.

Eiten, G. 1972. The cerrado vegetation of Brazil. *Botanical Review* 38:201–341.

———. 1975. The vegetation of the Serra do Roncador. *Biotropica* 7:112–35.

———. 1994. Vegetação. In *Cerrado: Caracterização, Ocupação, e Perspectivas,* M. N. Pinto, ed., 2d ed., 17–73. Brasília: Editora da Unviersidade de Brasília.

Emmons, L. H. 1997. *Neotropical Rainforest Mammals.* Chicago: University of Chicago Press.

Eschwege, L. W. 1830. *Brasilien: Die Neue Welt.* Braunschweig: Friedrich Bieweg.

Esparza, J., C. Fox, I. T. Harper, P. H. Bennett, L. O. Schulz, M. E. Valencia, and E. Ravussin. 2000. Daily energy expenditure in Mexican and USA Pima Indians: Low physical activity as a possible cause of obesity. *International Journal of Obesity and Related Metabolic Disorders* 24:55–59.

Eveland, W. C., W. J. Oliver, and J. V. Neel. 1971. Characteristics of *E. coli* serotypes in the Yanomama, a primitive Indian tribe of South America. *Infection and Immunity* 4:753–56.

Eveleth, P. B., and J. V. Tanner. 1990. *Worldwide Variation in Human Growth.* 2d ed. Cambridge: Cambridge University Press.

Everhart, J. E., D. J. Pettitt, P. H. Bennett, and W. C. Knowler. 1992. Duration of obesity increases the incidence of NIDDM. *Diabetes* 41:235–40.

Fausto, B. 1994. *História do Brasil.* São Paulo: Editora da Universidade de São Paulo.

Feachem, R. G. A. 1984. Interventions for the control of diarrhoeal diseases among young children: Promotion of personal and domestic hygiene. *Bulletin of the World Health Organization* 62:467–76.

Feachem, R. G. A., R. C. Hogan, and M. H. Merson. 1983. Diarrhoeal disease control: Reviews of potential interventions. *Bulletin of the World Health Organization* 61:637–40.

Fearnside, P. M. 1986. *Human Carrying Capacity of the Brazilian Rainforest.* New York: Columbia University Press.

Ferguson, R. B. 1990. Blood of the Leviathan: Western contact and warfare in Amazonia. *American Ethnologist* 17:237–57.

Ferreira, M. K. L. 1999. Pobreza causa diabetes em índios. *Parabólicas* (São Paulo) 53:11.

Ferreira, M. R. 1946. *Terrras e Índios do Alto Xingu.* São Paulo: Melhoramentos.

Ferri, M. G., ed. 1963. *Simpósio sobre o Cerrado.* São Paulo: Edgard Blücher and Editora da Universidade de São Paulo.

———, ed. 1971. *III Simpósio sobre o Cerrado.* São Paulo: Edgard Blücher and Editora da Universidade de São Paulo.

———, ed. 1977a. *IV Simpósio sobre o Cerrado: Bases para Utilização Agropecuária.* Belo Horizonte and São Paulo: Itatiaia and Editora da Universidade de São Paulo.

———. 1977b. Ecologia dos cerrados. In *IV Simpósio sobre o Cerrado: Bases para Utilização Agropecuária,* M. G. Ferri, ed., 15–33. Belo Horizonte and São Paulo: Itatiaia and Editora da Universidade de São Paulo.

Fisher, W. H. 1994. Megadevelopment, environmentalism, and resistance: The institutional context of Kayapó politics in Central Brazil. *Human Organization* 53:220–32.

Fleming, P. 1934. *Brazilian Adventure.* New York: Scribner's.

Fleming-Moran, M., and C. E. A. Coimbra Jr. 1990. Blood pressure studies among Amazonian native populations: A review from an epidemiological perspective. *Social Science and Medicine* 31:593–601.

Fleming-Moran, M., R. V. Santos, and C. E. A. Coimbra Jr. 1991. Blood pressure levels of the Suruí and Zoró Indians of the Brazilian Amazon: Group- and sex-specific effects resulting from body composition, health status, and age. *Human Biology* 63:835–61.

Flowers, N. M. 1983a. Forager-farmers: The Xavante Indians of Central Brazil. Ph.D. diss., City University of New York.

———. 1983b. Seasonal factors in subsistence, nutrition, and child growth in a Central Brazilian Indian community. In *Adaptive Responses of Native Amazonians,* R. B. Hames and W. T. Vickers, eds., 357–90. New York: Academic Press.

———. 1994a. Demographic crisis and recovery: A case study of the Xavante of Pimentel Barbosa. *South American Indian Studies* 4:18–36.

———. 1994b. Subsistence strategy, social organization, and warfare in Central Brazil. In *Amazonian Indians from Prehistory to the Present,* A. C. Roosevelt, ed., 249–69. Tucson: University of Arizona Press.

———. 1994c. Crise e recuperação demográfica: Os Xavánte de Pimentel Barbosa, Mato Grosso. In *Saúde e Povos Indígenas,* R. V. Santos and C. E. A. Coimbra Jr., eds., 213–42. Rio de Janeiro: Editora Fiocruz.

Flowers, N. M., D. R. Gross, M. Ritter, and D. Werner. 1982. Variation in swidden practices in four Central Brazilian Societies. *Human Ecology* 10:203–17.

Flowers, N. M., S. A. Gugelmin, and R. V. Santos. 1998. Settlement pattern as economic and political strategy: The Xavánte of Central Brazil. *South American Indian Studies* 5:18–28.

Fonseca, S. 1948. *Frente a Frente com os Xavante.* 2d. ed. Rio de Janeiro: Pongetii.

Franco, L. J. 1981. Aspectos metabólicos da população indígena do Alto Xingu. D.Sc. diss., São Paulo, Escola Paulista de Medicina.

———. 1992. Diabetes in Brazil: Review of recent survey data. *Ethnicity and Disease* 2:158–65.

Freire, J. R. [1790] 1951. *Relação da Conquista do Gentio Xavante.* São Paulo: Faculdade de Filosofia, Ciências e Letras, Universidade de São Paulo. originally published in Lisbon by Typographya Nunesianna.

Freitas-Filho, A. S. 1955. Inquérito médico sanitário entre os índios Xavánte. In *Relatório de Atividades do Serviço de Proteção aos Índios Durante o Ano de 1954,* M. F. Simões, ed., 145–72. Rio de Janeiro: Serviço do Proteção aos Índios.

Freitas-Filho, A. S., and N. B. Oliveira. 1955. Estudo sôbre o estado nutritivo dos Xavantes. *Revista Brasileira de Medicina* 12:565–67.

Freitas, F. G., and C. O. Silveira. 1977. Principais solos sob vegetação de cerrado e sua aptidão agrícola. In *IV Simpósio sobre o Cerrado,* M. G. Ferri, 155–94. Belo Horizonte and São Paulo: Itatiaia and Editora da Universidade de São Paulo.

Frenk, J., J. L. Bobadilla, J. Sepúlveda, and M. L. Cervantes. 1989. Health transition in middle-income countries: New challenges for health care. *Health Policy and Planning* 4:29–39.

Frenk, J., J. L. Bobadilla, C. Stern, T. Frejka, and R. Lozano. 1994. Elements for a theory of health transition. In *Health and Social Change in International Perspectives,* L. C. Chen, A. Kleinman, and N. C. Ware, eds., 26–49. Boston: Harvard School of Public Health and Harvard University Press.

Frenk, J., T. Frejka, J. L. Bobadilla, C. Stern, R. Lozano, J. Sepúlveda, and M. José. 1991. La transición epidemiológica en América Latina. *Boletin de la Oficina Sanitaria Panamericana* 111:485–96.

Friedman, H., I. Campbell, R. Alvarez, I. Ferrari, C. E. A. Coimbra Jr., J. R. Moraes, N. M. Flowers, P. Stastny, M. Fernandez-Viña, M. O. Alcalá, and L. A. Diaz. 1995. Endemic pemphigus foliaceous (fogo selvagem) in Native Americans from Brazil. *Journal of the American Academy of Dermatology* 32:949–56.

Friedman, H., C. E. A. Coimbra Jr., R. Alvarez, I. Campbell, L. A. Diaz, N. M. Flowers, R. V. Santos, M. L. Bertoli, G. B. N. Gama, and M. O. Alcalá. 1992. Pênfigo foliáceo endêmico (fogo-selvagem) no grupo indígena Xavánte, Mato Grosso, Brasil. *Cadernos de Saúde Pública* 8:331–34.

Frikel, P. 1978. Áreas de arboricultura pré-agrícola na Amazônia: Notas preliminars. *Revista de Antropologia* (São Paulo) 21:45–52.

FUNASA (Fundação Nacional de Saúde). 2000a. *Informe sobre Saúde Indígena: Edição Brasil.* Brasília: FUNASA.

———. 2000b. Casos de diabetes mellitus conhecidos no âmbito do Distrito Sanitário Especial Indígena Xavánte. Memoradum no. 347, distrito Sanitário Especial Indígena Xavánte, DSEIXAV/00, 11 October. Barra do Garças: FUNASA.

Gaiger, J. M. G. 1989. *Direitos Indígenas na Constituição Brasileira de 1988.* Brasília: Conselho Indigenista Missionário.

Gardner, G. [1846] 1970. *Viagem ao Interior do Brasil, 1836–1841.* Belo Horizonte and São Paulo: Itatiaia and Editora da Universidade de São Paulo.

Garfield, S. W. 1996. "Civilized" but discontent: The Xavante Indians and government policy in Brazil, 1937–1988. Ph.D. diss., Yale University.

———. 1997. The roots of a plant that today is Brazil: Indians and the nation-state under the Brazilian Estado Novo. *Journal of Latin American Studies* 29:747–68.

Gaspar, P. A., M. H. Hutz, F. M. Salzano, and T. A. Weimer. 2000. TP53 polymorphisms and haplotypes in South Amerindians and neo-Brazilians. *Annals of Human Biology* 28:184–94.

Gershowitz, H., P. C. Junqueira, F. M. Salzano, and J. V. Neel. 1967. Further studies on the Xavante Indians, part 3: Blood groups and ABH-Le^{a} secretor types in the Simões Lopes and São Marcos Xavantes. *American Journal of Human Genetics* 19:502–13.

Gershowitz, H., and J. V. Neel. 1978. The immunoglobulin allotypes (Gm and Km) of twelve Indian tribes of Central and South America. *American Journal of Physical Anthropology* 49:289–301.

Géry, J. 1984. The fishes of Amazonia. In *The Amazon: Limnology and Landscape Ecology of a Mighty Tropical River and Its Basin,* H. Sioli, ed., 353–70. Dordrecht: Dr. W. Junk.

Ghodes, D. M. 1986. Diabetes in American Indians: A growing problem. *Diabetes Care* 9:609–13.

Giaccaria, B., and A. Heide. 1975. *Jerônimo Conta: Mitos e Lendas.* Campo Grande: Casa da Cultura.

———. 1984. *Xavante (Auwẽ Uptabi: Povo Autêntico).* 2d ed. São Paulo: Editorial Dom Bosco.

Gibson, R. S. 1990. *Principles of Nutritional Assessment.* New York: Oxford University Press.

Giraldin, O. 1997. *Cayapó e Panará: Luta e Sobrevivência de um Povo Jê no Brasil Central.* Campinas: Editora da UNICAMP.

Giugliano, L. G., M. G. P. Bernardi, J. C. Vasconcelos, C. A. Costa, and R. Giugliano. 1986. Longitudinal study of diarrhoeal incidence in a periurban community in Manaus (Amazon, Brazil). *Annals of Tropical Medicine and Parasitology* 80:443–50.

Gomes, M. P. 1988. *Os Índios e o Brasil: Ensaio sobre um Holocausto e sobre uma Nova Possibilidade de Convivência.* Petrópolis: Vozes.

Gómez, F., R. R. Galvan, S. Frenk, J. C. Muñoz, R. Chávez, and J. Vasquez. 1956. Mortality in second and third degree malnutrition. *Journal of Tropical Pediatrics* 2:77–83.

Goodland, R., and M. G. Ferri. 1979. *Ecologia do Cerrado.* Belo Horizonte and São Paulo: Itatiaia and Editora da Universidade de São Paulo.

Goodman, A. H., and T. L. Leatherman. 1998. Traversing the chasm between biology and culture: An introduction. In *Building a New Biocultural Synthesis: Political-Economic Perspectives on Human Biology,* A. H. Goodman and T. L. Leatherman, eds., 3–41. Ann Arbor: University of Michigan Press.

Gordon, J. E., I. D. Chitkara, and J. B. Wyon. 1963. Weanling diarrhea. *American Journal of Medical Sciences* 245:129–61.

Graham, L. R. 1987. Uma aldeia por um "projeto." In *Povos Indígenas no Brasil, 85–86,* 348–50. Aconteceu special issue, no. 17. São Paulo: Centro Ecumênico de Documentação e Informação.

———. 1990. The always living: Discourse and the male lifecyle of the Xavánte Indians of Central Brazil. Ph.D. diss., University of Texas.

———. 1993. A public sphere in Amazonia? The depersonalized collaborative construction of discourse in Xavante. *American Ethnologist* 70:717–41.

———. 1995. *Performing Dreams: Discourse of Immortality among the Xavanate of Central Brazil.* Austin: University of Texas Press.

———. 1999. Xavante education fund. *Cultural Survival Quarterly* 22:16–19.

———. 2000. Lessons in Collaboration: The Xavante/WWF Wildlife Management Project in Central Brazil. In *Indigenous Peoples and Conservation Organization: Experiences in Collaboration,* R. Weber, J. Butler, and P. Larson, eds., 47–71. Washington, D.C.: World Wildlife Fund.

Greenberg, J. H. 1956. The general classification of Central and South American Indian languages. In *Men and Cultures,* A. F. C. Wallace, ed., 791–94. Philadelphia: University of Pennsylvania Press.

Grim, C. W., E. B. Broderick, B. Jasper, and K. R. Phipps. 1994. A comparison of dental caries experience in Native American and caucasian children in Oklahoma. *Journal of Public Health Dentristy* 54:220–27.

Gross, D. R. 1979. A new approach to Central Brazilian social organization. In *Brazil, Anthropological Perspectives: Essays in Honor of Charles Wagley,* M. Margolis and W. E. Carter, eds., 321–435. New York: Columbia University Press.

———. 1984. Time allocation: A tool for the study of cultural behavior. *Annual Review of Anthropology* 13:519–58.

Gross, D. R., G. Eiten, N. M. Flowers, F. M. Leoi, M. L. Ritter, and D. W. Werner. 1979. Ecology and acculturation among native peoples of Central Brazil. *Science* 206:1043–50.

Gugelmin, S. A. 1995. Nutrição e alocação de tempo dos Xavánte de Pimentel Barbosa, Mato Grosso. M.Sc. thesis, Rio de Janeiro, Escola Nacional de Saúde Pública, Fundação Oswaldo Cruz.

———. 2001. Antropometria nutricional e ecologia humana dos Xavánte de Sangradouro—Volta Grande, Mato Grosso. D.Sc. diss., Rio de Janeiro, Escola Nacional de Saúde Pública, Fundação Oswaldo Cruz.

Gugelmin, S. A., and R. V. Santos. 2001. Ecologia humana e antropometria nutricional de adultos Xavánte, Mato Grosso, Brasil. *Cadernos de Saúde Pública* 17:313–22.

Gust, I., 1980. Acute viral hepatitis. In *Changing Disease Patterns and Human Behaviour,* N. F. Stanley and R. A. Joske, eds., 85–114. London: Academic Press.

Gustafsson, B. E., C. E. Quensel, L. S. Lanke, C. Lundqvist, H. Grahnén, B. E. Bonow, and B. Krasse. 1954. The Vipeholm dental caries study: The effect of different levels of carbohydrate intake on caries activity in 436 individuals observed for five years. *Acta Odontologica Scandinavica* 11:232–64.

Hadler, S. C., O. H. Fay, F. P. Pinheiro, and J. E. Maynard. 1987. La hepatitis en las Americas: Informe del grupo colaborador de la OPS. *Boletin de la Oficina Sanitaria Panamericana* 103:185–208.

Hales, C. N., and D. J. Baker. 1992. Type 2 (non-insulin-dependent) diabetes mellitus: The thrity phenotype hypothesis. *Diabetologia* 35:595–601.

Hames, R. B., and W. T. Vickers, 1983. Introduction to *Adaptive Responses of Native Amazonians,* R. B. Hames and W. T. Vickers, eds., 1–26. New York: Academic Press.

Handro, W., and M. M. Barradas. 1971. Sôbre os óleos do fruto e da semente do piquí—*Caryocar brasiliense* Camb. In *III Simpósio sobre o Cerrado,* M. G. Ferri, ed., 110–13. São Paulo: Edgard Blücher and Editora da Universidade de São Paulo.

Haridasan, M. 1994. Solos. In *Cerrado: Caracterização, Ocupação, e Perspectivas,* M. N. Pinto, ed., 2d. ed., 321–44. Brasília: Editora da Universidade de Brasília.

Harris, D. R. 1980a. Tropical savanna environments: Definition, distribution, diversity, and development. In *Human Ecology in Savanna Environments,* D. R. Harris, ed., 3–27. New York: Academic Press.

———. 1980b. Commentary: Human occupation and exploitation of savanna environments. In *Human Ecology in Savanna Environments,* D. R. Harris, ed., 31–39. New York: Academic Press.

Haswell-Elkins, M., D. Elkins, and R. M. Anderson. 1989. The influence of individual, social group, and household factors on the distribution of *Ascaris lumbricoides* within a community and implications for control strategies. *Parasitology* 98(1): 125–34.

Hecht, S., and A. Cockburn. 1989. *The Fate of the Forest: Developers, Destroyers, and Defenders of the Amazon.* New York: Verso.

Hees, D. R., M. E. P. C. Sá, and T. C. Aguiar. 1987. A evolução da agricultura na região Centro-Oeste na década de 70. *Revista Brasileira de Geografia* 49:197–257.

Heidrich, E. M., M. H. Hutz, F. M. Salzano, C. E. A. Coimbra Jr., and R. V. Santos. 1995. D1S80 locus variability in three Brazilian ethnic groups. *Human Biology* 67:311–19.

Hemming, J. 1978. *Red Gold: The Conquest of the Brazilian Indians, 1500–1760.* Cambridge: Harvard University Press.

———, ed. 1985. *Change in the Amazon Basin.* Vol. 2. Manchester: Manchester University Press.

———. 1987. *Amazon Frontier: The Defeat of the Brazilian Indians.* Cambridge: Cambridge University Press.

Hershkovitz, F. 1972. The recent mammals of the neotropical region. In *Evolution, Mammals and Southern Continents,* A. Keast, F. C. Erk, and B. Glass, eds., 311–431. Albany: SUNY Press.

Higgins, M., W. Kannel, R. Garrison, J. Pinsky, and J. Stokes III. 1988. Hazards of obesity: The Framingham experience. *Acta Medica Scandinavica* 723 (Sup.): 23–36.

Hill, K., and A. M. Hurtado. 1996. *Aché Life History: The Ecology and Demography of a Foraging People.* Hawthorne: Aldine de Gruyter.

Hillson, S. 1996. *Dental Anthropology.* Cambridge: Cambridge University Press.

Hoehne, F. C. 1940. *Gênero Arachis: Precedido da Chave Geral para as Sub-Famílas, Tribus, e Generos das Leguminosas do Brasil.* Flora Brasílica, no. 25/2. São Paulo: Secretaria da Agricultura, Indústria e Comércio.

Hoffmans, M. D., D. Kromhout, and C. L. Coulander. 1988. The impact of body mass index of 78,612 18-year old Dutch men on 32-year-mortality from all causes. *Journal of Clinical Epidemiology* 41:749–56.

Howell, N. 1979. *Demography of the Dobe !Kung.* New York: Academic Press.

Hoyle, B., M. Yunus, and L. C. Chen. 1980. Breastfeeding and food intake among children with acute diarrhoeal disease. *American Journal of Clinical Nutrition* 33:2365–71.

Hsu, F. L. K. 1964. Rethinking the concept "primitive." *Current Anthropology* 5:169–78.

Hudson-Thomas, M., K. C. Binghan, and W. K. Simmons. 1994. An evaluation of the HemoCue for measuring haemoglobin in field studies in Jamaica. *Bulletin of the World Health Organization* 72:423–26.

Hutz, M. H., S. Almeida, C. E. A. Coimbra Jr., R. V. Santos, and F. M.

Salzano. 2000. Haplotype and allele frequencies for three genes of the dopaminergic system in South American Indians. *American Journal of Human Biology* 12:638–45.

Hutz, M. H., S. M. Callegari-Jacques, M. C. Bortolini, and F. M. Salzano. 1999. Variability in nDNA, mtDNA, and proteins: A test case. In *Genomic Diversity: Applications in Human Population Genetics,* S. S. Papiha, R. Deka, and R. Chakraborty, eds., 23–32. New York: Kluwer and Plenum.

Hutz, M. H., V. S. Mattevi, S. M. Callegari-Jacques, F. M. Salzano, C. E. A. Coimbra Jr., R. V. Santos, R. F. Carnese, A. S. Goicoechea, and C. B. Dejean. 1997. D1S80 locus variability in South American Indians. *Annals of Human Biology* 24:249–55.

Ianelli, R. V. 1997. Epidemiologia da malária em uma população indígena do Brasil Central: Os Xavánte de Pimentel Barbosa. M.Sc. thesis, Rio de Janeiro, Escola Nacional de Saúde Pública, Fundação Oswaldo Cruz.

Ianelli, R. V., N. A. Honório, D. C. Lima, R. Lourenço-de-Oliveira, R. V. Santos, and C. E. A. Coimbra Jr. 1998. Faunal composition and behavior of anopheline mosquitoes in the Xavánte Indian Reservation of Pimentel Barbosa, Central Brazil. *Parasite* 5:197–202.

Ianelli, R. V., J. P. Silva, and S. M. Agostini. 1995. Parasitoses intestinais nos índios Xavánte de Parabubure, Mato Grosso, Brasil. *Cadernos de Saúde Pública* 11:630.

International Rivers Network. 1999a. Hidrovia e posição indígena. http//:www.irn.org (accessed 4 August 1999).

———. 1999b. Judge suspends hidrovia. http//:www.irn.org (accessed 29 October, 1999).

ISA (Instituto Socioambiental). 1996. *Povos Indígenas do Brasil, 1991–1995.* São Paulo: ISA.

Jackson, J. 1975. Recent ethnography of indigenous Northern lowland South America. *Annual Review of Anthropology* 4:307–40.

———. 1995. Culture, genuine and spurious: The politics of Indianness in the Vaupés, Colombia. *American Ethnologist* 22:3–27.

James, G. D., and P. T. Baker. 1990. Human population biology and hypertension: Evolutionary and ecological aspects of blood pressure. In *Hypertension: Pathophysiology, Diagnosis, and Management,* J. H. Laragh and B. M. Brenner, eds., 137–45. New York: Raven.

Jeliffe, D. B., and E. F. P. Jeliffe. 1977. "Breast is best": Modern meanings. *New England Journal of Medicine* 297:912–15.

Johns, W. L., and S. M. Lewis. 1989. Primary health screening by hemoglobinometry in a tropical community. *Bulletin of the World Health Organization* 67:627–33.

Johnson, A. 1975. Time allocation in a Machiguenga community. *Ethnology,* 14:301–10.

Joly, A. B. 1970. *Conheça a Vegetação Brasileira.* São Paulo: Polígono and Editora da Universidade de São Paulo.

Juruna, M., A. Hohfeldt, and A. Hoffman. 1982. *O Gravador do Juruna.* Porto Alegre: Mercado Aberto.

Kameyama, I. 1985. Parasitoses intestinais entre os índios do Parque Nacional do Xingu: Alguns aspectos epidemiológicos e ecológicos. M.Sc. thesis, São Paulo, Faculdade de Saúde Pública, Universidade de São Paulo.

Kaplan, G. A., T. W. Wilson, and R. D. Cohen. 1994. Social functioning and overall mortality: Prospective evidence from the Kuopio Ischemic Heart Disease Risk Factor Study. *Epidemiology* 5:494–500.

Karasch, M. 1992. Catequese e cativeiro: Política indigenista em Goiás, 1780–1889. In *História dos Índios no Brasil,* M. C. Cunha, ed., 397–412. São Paulo: Companhia das Letras.

Kaufman, L., A. F. Vargas, C. E. A. Coimbra Jr., R. V. Santos, F. M. Salzano, and M. H. Hutz. 1999. Apoliprotein B genetic variability in Brazilian Indians. *Human Biology* 71:87–98.

Kasongo Project Team. 1983. Anthropometric assessment of young children's nutritional status as an indicator of subsequent risk of dying. *Journal of Tropical Pediatrics* 29:69–75.

Kerr, W. E., and D. A. Posey. 1984. Informações adicionais sobre a agricultura dos Kayapó. *Interciencia* 9:392–400.

Kerr, W. E., D. A. Posey, and W. Walter-Filho. 1978. Cupá ou cipó babão, alimento de alguns índios amazônicos. *Acta Amazonica* 8:702–5.

Khan, M. V. 1982. Interruption of shigellosis by hand washing. *Transactions of the Royal Society of Tropical Medicine and Hygiene* 76:164–68.

King, H. 1992. The epidemiology of diabetes mellitus in Papua New Guinea and the Pacific: Adverse consequences of natural selection in the face of sociocultural change. In *Human Biology in Papua New Guinea: The Small Cosmos,* R. D. Attenborough and M. P. Alpers, eds., 363–72. Oxford: Clarendon.

Knowler, W. C., P. H. Bennett, R. F. Hamman, and M. Miller. 1978. Diabetes incidence and prevalence in Pima Indians: A 19-fold greater incidence than in Rochester, Minnesota. *American Journal of Epidemiology* 108:497–505.

Knowler, W. C., D. J. Petitt, M. F. Saad, M. A. Charles, R. G. Nelson, B. V. Howard, C. Bogardus, and P. H. Bennett. 1991. Obesity in the Pima Indians: Its magnitude and relationship with diabetes. *American Journal of Clinical Nutrition* 53 (Sup.): S1543–51.

Koining, H. W., K. R. Smith, and J. R. Last. 1985. Biomass fuel consumption and health. *Bulletin of the World Health Organization* 6:11–26.

Kunitz, S. J. 1994a. Disease and the destruction of indigenous populations. In *Companion Encylcopedia of Anthropology,* T. Ingold, ed., 297–325. London: Routledge.

———. 1994b. *Disease and Social Diversity: The European Impact on the Health of Non-Europeans.* New York: Oxford University Press.

Kvitko, K., J. C. B. Nunes, T. A. Weimer, F. M. Salzano, and M. H. Hutz. 2000. Cytochrome P4501A1 polymorphisms in South American Indians. *Human Biology* 72:1039–43.

Langdon, E. J. M. 1992a. Shamanism and anthropology. In *Portals of Power: Shamanism in South America,* E. J. M. Langdon and G. Baer, eds., 1–21. Albuquerque: University of New Mexico Press.

———. 1992b. Shamanic power in Siona religion and medicine. In *Portals of*

Power: Shamanism in South America, E. J. M. Langdon and G. Baer, eds., 41–61. Albuquerque: University of New Mexico Press.

———. 2000. Salud y pueblos indígenas: Los desafios en el cambio de siglo. In *Salud y Equidad: Una Mirada desde las Ciencias Sociales,* R. Briceño-León, M. C. S. Minayo, and C. E. A. Coimbra Jr., eds., 107–117. Rio de Janeiro: Editora Fiocruz.

Langendorfer, A., W. Davenport, W. T. London, B. S. Blumberg, and S. Mazzur. 1984. Sex-related differences in transmission of hepatitis B infection in a Melanesian population. *American Journal of Physical Anthropology* 64:243–54.

Larsen, C. S., and G. R. Milner, eds. 1994. *In the Wake of Contact: Biological Responses to Conquest.* New York: Wiley-Liss.

Lawrence, D. N., R. R. Facklam, F. O. Sottnek, G. A. Hancock, J. V. Neel, and F. M. Salzano. 1979. Epidemiologic studies among Amerindian populations of Amazônia, part 1: Pyoderma—prevalence and associated pathogens. *American Journal of Tropical Medicine and Hygiene* 28:548–58.

Leacock, S. 1964. Economic life of the Maué Indians. *Boletim do Museu Paraense Emílio Goeldi (Antropologia)* 19: 1–30.

Leeuwenberg, F. 1995. Diagnóstico de caça e manejo da fauna cinegética com os Índios Xavante, Aldeia Etenhiritipá. report to the World Wildlife Fund, Brasilia.

———. 1997. Manejo da fauna cinegética na reserva indígena Xavante de Pimentel Barbosa, Estado de Mato Grosso. In *Manejo e Conservação de Vida Silvestre no Brasil,* C. Valladares-Padua and R. E. Bodmer, eds., 233–38. Brasília: Conselho Nacional de Desenvolvimento Científico e Tecnológico and Sociedade Civil Mamirauá.

Leeuwenberg, F., and J. G. Robinson. 2000. Traditional management of hunting by a Xavante community in Central Brazil: The search for sustainability. In *Hunting for Sustainability in Tropical Forests,* J. G. Robinson and E. L. Bennett, eds., 375–94. New York: Columbia University Press.

Leeuwenberg, F., and M. Salimon. 1999. *Para Sempre A'uwê: Os Xavante na Balança das Civilizações.* Brasília: UNICEF.

Leite, M. 1998. Avaliação do estado nutricional da população Xavánte de São José, Terra Indígena Sangradouro-Volta Grande, Mato Grosso. M.Sc. thesis, Rio de Janeiro, Escola Nacional de Saúde Pública, Fundação Oswaldo Cruz.

Lerrer, Débora F. 2001. Alto risco ambiental para baixo retorno financeiro. *Planeta na Web.* http://www.terra.com.br/planetanaweb/flash/reconectando/agrandeteia/araguaia.htm (accessed December 20, 2001).

Leslie, J. 1989. Women's time: A factor in the use of child survival technologies. *Health Policy and Planning* 4:1–16.

Levine, R. M. 1970. *The Vargas Regime: The Critical Years, 1934–1938.* New York: Columbia University Press.

Lima, A. C. S. 1995. *Um Grande Cerco de Paz: Poder Tutelar, Indianidade, e Formação do Estado no Brasil.* Petrópolis: Vozes.

Lima, P. 1950. Níveis tensionais dos índios Kalapalo e Kamaiurá. *Revista Brasileira de Medicina* 7:787–88.

Linhares, A. C. 1992. Epidemiologia das infecções diarréicas entre populações indígenas da Amazônia. *Cadernos de Saúde Pública* 8:121–28.

Linhares, A. C., J. D. M. Pereira, C. M. Nakauth, and Y. B. Gabbay. 1986a. Rotavirus infection in wild marsupials (*Didelphis marsupialis*) of the Amazon region. *Transactions of the Royal Society of Tropical Medicine and Hygiene* 80:20–24.

Linhares, A. C., F. P. Pinheiro, R. B. Freitas, Y. B. Gabbay, J. A. Shirley, and G. M. Beards. 1981. An outbreak of rotavirus diarrhea among a nonimmune, isolated South American Indian community. *American Journal of Epidemiology* 113:703–10.

Linhares, A. C., E. V. Salbe, Y. B. Gabbay, and N. Rees. 1986b. Prevalence of rotavirus antibodies among isolated South American Indian communities. *American Journal of Epidemiology* 123:699–709.

Lins, Z. C. 1970. Studies on enteric bacterias in the Lower Amazon Region, part 1: Serotypes of *Salmonella* isolated from wild forest animals in Pará State, Brazil. *Transactions of the Royal Society of Tropical Medicine and Hygiene* 64:439–43.

Lombardi, C., P. C. Borges, A. Chaul, S. A. Sampaio, E. A. Rivitti, H. Friedman, C. R. Martins, J. A. Sanches, P. R. Cunha, R. G. Hoffman, and the Cooperative Group on Fogo Selvagem Research. 1992. Environmental risk factors in endemic pemphigus foliaceus (fogo selvagem). *Journal of Investigative Dermatology* 98:847–50.

Loureiro, E. C. B. 1985. Ocorrência do gênero *Salmonella* em animais domésticos da ordem Edentata na região amazônica, norte do Estado do Pará, Brasil. *Revista Latinoamericana de Microbiologia* 27:31–43.

Loureiro, E. C. B., and Z. C. Lins. 1976. Observações sobre a incidência de enterobactérias patogênicas em índios da tribo Parakanã. *Revista da Fundação SESP* 22:43–44.

Lourenço-de-Oliveira, R. 1989. Some observations on the mosquitoes on Indian settlements in Xingu National Park, Mato Grosso State, Brazil, with emphasis on malaria vectors. *Revista Brasileira de Biologia* 49:393–97.

Lowe-McConnell, R. H. 1975. *Fish Communities in Tropical Freshwaters: Their Distribution, Ecology, and Evolution.* London: Longmans.

Lowenstein, F. W. 1961. Blood pressure in relation to age and sex in the tropics and subtropics. *Lancet* 1:389–92.

MacCluer, J. W., J. V. Neel, and N. A. Chagnon. 1971. Demographic structure of a primitive population: A simulation. *American Journal of Physical Anthropology* 35:193–207.

Magalhães, J. V. C. de. 1934. *Viagem ao Araguaya.* São Paulo: Companhia Editora Nacional.

Mahar, D. J. 1979. *Frontier Development Policy in Brazil: A Study of Amazonia.* New York: Praeger.

Mancilha-Carvalho J. J., R. G. Baruzzi, P. F. Howard, N. Poulter, M. P. Alpers,

L. J. Franco, L. F. Marcopito, V. J. Spooner, A. R. Dyer, P. Elliott, and J. Stamler. 1989. Blood pressure in four remote populations in the INTERSALT study. *Hypertension* 14:238–46.

Mancilha-Carvalho, J. J., J. A. C. Lima, J. V. Carvalho, and C. A. M. Souza. 1985. Blood pressure is directly related to the degree of acculturation among primitive Yanomamo Indians. *Circulation* 72 (Sup.): 296.

Marcellino, A. J., F. J. da Rocha, and F. M. Salzano. 1978. Size and shape differences among six South American Indian tribes. *Annals of Human Biology* 5:69–74.

Marchetti, D., and A. D. Machado, eds. 1979. *V Simpósio sobre o Cerrado.* Brasília: Editerra.

Mares, M. A., J. K. Braun, and D. Gettinger. 1989. Observations on the distribution and ecology of the mammals of the cerrado grasslands of Central Brazil. *Annals of the Carnegie Museum* 58:1–60.

Marmot, M. G. 1985. Psychosocial factors and blood pressure. *Preventive Medicine* 14:451–65.

Martines, J., M. Phillips, and R. G. A. Feachem. 1993. Diarrheal disease. In *Disease Control Priorities in Developing Countries,* D. T. Jamison, W. H. Mosley, A. R. Measham, and J. L. Bobadilla, eds., 91–116. New York: Oxford University Press.

Martins, C. 1999. Obesity in Brazil: An overview. *Nutrition* 15:960–61.

Martius, C. F. P., von. 1867. *Beitrage zur Ethnographie und Sprachenkunde Amerika's zumal Brasiliens.* 2 vols. Leipzig: Friedrich Fleischer.

Martorell, R. 1980. Interrelationships between diet, infectious disease, and nutritional status. In *Social and Biological Predictors of Nutritional Status, Physical Growth, and Behavioral Development,* L. S. Greene and F. Johnston, eds., 81–106. New York: Academic Press.

Martorell, R., and J. P. Habicht. 1986. Growth in early childhood in developing countries. In *Human Growth: A Comprehensive Treatise,* F. Falkner and J. M. Tanner, eds., 241–62. 2d ed. New York: Plenum.

Martorell, R., C. Yarbrough, S. Yarbrough, and R. E. Klein. 1980. The impact of ordinary illness on the dietary intake of malnourished children. *American Journal of Clinical Nutrition* 33:345–50.

Marzochi, M. C. A. 1992. Leishmanioses no Brasil: As leishmanioses tegumentares. *Jornal Brasileiro de Medicina* 63:82–104.

Mason, J. A. 1950. The languages of South American Indians. In *Handbook of South American Indians,* J. H. Steward, ed., 5:157–317. Washington, D.C.: Government Printing Office.

Matta, R., and R. B. Laraia. 1979. *Índios e Castanheiros: A Empresa Extrativa e os Índios do Meio Tocantins.* Rio do Janeiro: Paz e Terra.

Mattevi, V. S., C. E. A. Coimbra Jr., R. V. Santos, F. M. Salzano, and M. H. Hutz. 2000. Association of the low-density lipoprotein receptor gene with obesity in Native American populations. *Human Genetics* 106:546–52.

Mattos, R. J. da C. 1874. Chorographia historica da Provincia de Goyaz. *Revista do Instituto Histórico e Geographico Brasileiro* 37 (1): 213–398.

———. 1875. Chorographia historica da Provincia de Goyaz. *Revista do Instituto Histórico e Geographico Brasileiro* 38 (1): 5–150.

Maybury-Lewis, D. 1967. *Akwẽ-Shavante Society.* Oxford: Clarendon.

———., ed. 1979. *Dialectical Societies: The Gê and Bororo of Central Brazil.* Cambridge: Harvard University Press.

———. 1983. The Shavante struggle for their lands. *Cultural Survival Quarterly* 7:54–55.

———. 1985. Brazilian Indianist policy: Some lessons from the Shavante project. In *Native Peoples and Economic Development: Six Case Studies from Latin America,* T. MacDonald, ed., 75–86. Cambridge, Mass.: Cultural Survival.

———. 1988. *The Savage and the Innocent.* 2d ed. Boston: Beacon.

Mazzur, S. 1973. Menstrual blood as a vehicle of Australia-antigen transmission. *Lancet* 1:749–51.

McGarvey, S. T., and D. E. Schendel. 1986. Blood pressure of Samoans. In *The Changing Samoans: Behavior and Health in Transition,* P. T. Baker, J. M. Hanna, and T. S. Baker, eds., 350–93. New York: Oxford University Press.

McGrath, J. W. 1991. Biological impact of social disruption resulting from epidemic disease. *American Journal of Physical Anthropology* 84:407–19.

McKusick, V. A. 1988. *Mendelian Inheritance in Man: Catalogs of Autosomal Dominant, Autosomal Recessive, and X-linked Phenotypes.* Baltimore: Johns Hopkins University Press.

Melatti, J. C. 1967. *Índios e Criadores: A Situação dos Krahó na Área Pastoril do Tocantins.* Rio de Janeiro: Instituto de Ciências Sociais, Universidade Federal do Rio de Janeiro.

Mello, D. A. 1981. Aspectos do ciclo silvestre do *Trypanosoma cruzi* em regiões de cerrado. *Memórias do Instituto Oswaldo Cruz* 76:227–46.

Mello Jorge, M. H. P., and S. L. D. Gotlieb. 2000. *As Condições de Saúde no Brasil: Retrospecto de 1979 a 1995.* Rio de Janeiro: Editora Fiocruz.

Menezes, C. 1982. Os Xavante e o movimento de fronteira no leste matogrossense. *Revista de Antropologia* (São Paulo) 25:63–87.

———. 1984. Missionários e índios em Mato Grosso (Os Xavante da reserva de São Marcos). D.Sc. diss., São Paulo, Faculdade de Filosofia, Letras e Ciências Humanas, Universidade de São Paulo.

Menezes, M. L. P. 1999. *Parque Indígena do Xingu: A Construção de um Território Estatal.* Campinas: Editora da Unicamp.

Merchan-Hamann, E. 1997. Diagnóstico macrorregional da situação das endemias das regiões Norte e Nordeste. *Informe Epidemiológico do SUS* 6:43–114.

Milne, A., G. K. Allwood, C. D. A. Moyes, N. E. Pearce, and K. Newell. 1987. A seroepidemiological study of the prevalence of hepatitis B infections in a hyperendemic New Zealand community. *International Journal of Epidemiology* 16:84–90.

Minayo, M. C. S., ed. 1995. *Os Muitos Brasis: Saúde e População na Década de 80.* São Paulo: Hucitec.

Mondini, L., and C. A. Monteiro. 1997. The stage of nutrition transition in different Brazilian regions. *Archivos Latinoamericanos de Nutrición* 47 (2, sup. 1): 17–21.

Monteiro, C. A. 1995, ed. *Velhos e Novos Males da Saúde no Brasil: A Evolução do País e de suas Doenças.* São Paulo: Hucitec.

Monteiro, C. A., M. H. D. Benício, W. L. Conde, and B. M. Poplin. 2000. Shifting obesity trends in Brazil. *European Journal of Clinical Nutrition* 54:342–46.

Monteiro, C. A., M. H. D. Benício, and N. C. Gouveia. 1992. Saúde e nutrição das crianças brasileiras no final da década de 80. In *Perfil Estatístico de Crianças e Mães no Brasil,* M. F. G. Monteiro, and R. Cervini, eds., 19–42. Rio de Janeiro: IBGE.

Monteiro, C. A., M. H. D. Benício, R. F. Iunes, N. C. Gouveia, and M. A. A. Cardoso. 1995b. Da desnutrição para a obesidade: A transição nutricional no Brasil. In *Velhos e Novos Males da Saúde no Brasil: A Evolução do País e de suas Doenças,* C. A. Monteiro, ed., 247–55. São Paulo: Hucitec.

Monteiro, C. A., L. Mondini, A. L. Souza, and B. M. Popkin. 1995a. The nutrition transition in Brazil. *European Journal of Clinical Nutrition* 49: 105–13.

Morais, F. 1994. *Chatô. O Rei do Brasil.* São Paulo: Companhia das Letras.

Moran, E. F. 1981. *Developing the Amazon.* Bloomington: Indiana University Press.

———. 1983. Growth without development: Past and present development efforts in Amazonia. In *The Dilemma of Amazonian Development,* E. F. Moran, ed., 3–23. Boulder; Westview.

———. 1996. An agenda for anthropology. In *Transforming Societies, Transforming Anthropology,* E. F. Moran, ed., 1–24. Ann Arbor: University of Michigan Press.

Morse, R. M. 1965. Introduction to *The Bandeirantes: The Historical Role of the Brazilian Pathfinders,* R. M. Morse, ed., 3–36. New York: Knopf.

Mota, J. L. 1955. A epidemia de sarampo no Xingu. In *Relatório das Atividades do Serviço de Proteção aos Índios Durante o Ano de 1954,* M. F. Simões, ed., 131–44. Rio de Janeiro: Serviço de Proteção aos Índios.

Mourant, A. E., A. C. Kopec, and K. Domaniewska-Sobczak. 1976. *The Distribution of the Human Blood Groups and Other Polymorphisms.* London: Oxford University Press.

Munhoz, S., Jr., C. J. F. Fontes, and S. M. P. Meirelles. 1997. Avaliação do programa de controle da hanseníase em municípios mato-grossenses, Brasil. *Revista de Saúde Pública* 31:282–87.

Murphy, R. F. 1960. *Headhunter's Heritage: Social and Economic Change among the Mundurucu Indians.* Berkeley: University of California Press.

Nabarro, D., P. Howard, C. Cassels, M. Pant, A. Wijga, and N. Padfield. 1988. The importance of infections and environmental factors as possible determinants of growth retardation in children. In *Linear Growth Retardation in Less Developed Countries,* J. C. Waterlow, ed., 165–83. New York: Raven.

Narayan, K. M. V. 1996. Diabetes mellitus in Native Americans: The problem and its implications. In *Changing Numbers, Changing Needs: American Indian Demography and Public Health,* G. D. Sandefur, R. R. Rindfuss, and B. Cohen, eds., 262–88. Washington, D.C.: National Academy Press.

Nasser, D. 1944. Enfrentando os Chavantes! *O Cruzeiro* (Rio de Janeiro), 24 June, 47–60.

Neel, J. V. 1962. Diabetes mellitus: A "thrifty" genotype rendered detrimental by progress? *American Journal of Human Genetics* 14:353–62.

———. 1968. The American Indian in the International Biological Program. In *Biomedical Challenges Presented by the American Indian,* 47–54. Scientific Publications, no. 165. Washington, D.C.: Pan American Health Organization.

———. 1970. Lessons from a "primitive" people. *Science* 170:815–22.

———. 1974. Control of disease among Amerindians in cultural transition. *Bulletin of the Pan American Health Organization* 8:205–11.

———. 1977. Health and disease in unacculturated Amerindian populations. In *Health and Disease in Tribal Societies,* K. Elliott and J. Whelan, eds., 155–77. Amsterdam: Elsevier.

———. 1978. Rare variants, private polymorphisms, and locus heterozygosity in Amerindian populations. *American Journal of Human Genetics* 30:465–90.

———. 1982a. Infectious disease among Amerindians. *Medical Anthropology* 6:47–55.

———. 1982b. The thrifty genotype revisited. In *The Genetics of Diabetes Mellitus,* J. Kobberling and R. Tattersall, eds., 283–93. New York: Academic Press.

———. 1991. Symposium on South American Indians: Introduction. *Human Biology* 63:737–41.

———. 1994. *Physician to the Gene Pool: Genetic Lessons and Other Stories.* New York: Wiley.

Neel, J. V., A. H. P. Andrade, G. E. Brown, E. E. Warren, J. Goobar, W. A. Sodfam, G. H. Stollerman, E. B. Weinstein, and H. H. Wheeler. 1968a. Further studies of the Xavante Indians, part 9: Immunologic status with respect to various diseases and organisms. *American Journal of Tropical Medicine and Hygiene* 17:486–98.

Neel, J. V., W. R. Centerwall, N. A. Chagnon, and H. L. Casey. 1970. Notes on the effects of measles and measles vaccine in a virgin-soil population of South American Indians. *American Journal of Epidemiology* 91:418–29.

Neel, J. V., M. Layrisse, and F. M. Salzano. 1977. Man in the tropics: The Yanomama Indians. In *Population Structure and Human Variation,* G. A. Harrison, ed., 109–42. Cambridge: Cambridge University Press.

Neel, J. V., W. M. Mikkelsen, D. L. Rucknagel, E. D. Weinstein, R. A. Goyer, and S. H. Abadie. 1968b. Further studies of the Xavante Indians, part 8: Some observations on blood, urine, and stool specimens. *American Journal of Tropical Medicine and Hygiene* 17:474–85.

Neel, J. V., and F. M. Salzano. 1967a. Further studies on the Xavante Indians, part 10: Some hypothesis-generalizations resulting from these studies. *American Journal of Human Genetics* 19:554–74.

———. 1967b. A prospectus for genetic studies on the American Indians. In *The Biology of Human Adaptability,* P. T. Baker and J. S. Weiner, eds., 245–74. Oxford: Clarendon.

Neel, J. V., F. M. Salzano, P. C. Junqueira, F. Keiter, and D. Maybury-Lewis. 1964. Studies on the Xavante Indians of the Brazilian Mato Grosso. *American Journal of Human Genetics* 16:52–140.

Neel, J. V., and K. M. Weiss. 1975. The genetic structure of a tribal population, the Yanomama Indians, part 12: Biodemographic studies. *American Journal of Physical Anthropology* 42:25–51.

Neves, W. A., F. M. Salzano, and F. J. da Rocha. 1985. Principal-components analysis of Brazilian Indian anthropometric data. *American Journal of Physical Anthropology* 67:13–17.

Newbrum, E. 1982. Sugar and dental caries: A review of human studies. *Science* 217:418–23.

NIH (National Institutes of Health, Consensus Development Panel on the Health Implications of Obesity). 1985. Health implications of obesity. *Annals of Internal Medicine* 103:1073–77.

Nimer, E. 1989. *Climatologia do Brasil.* 2d ed. Rio de Janeiro: Instituto Brasileiro de Geografia e Estatística.

Nimuendaju, C. 1942. *The Šerente.* Los Angeles: Frederick Webb Hodge Anniversary Publication Fund.

———. 1946. *The Eastern Timbira.* Berkeley: University of California Press.

Niswander, J. D. 1967. Further studies on the Xavante Indians, part 7: The oral status of the Xavantes of Simões Lopes. *American Journal of Human Genetics* 19:543–53.

Niswander, J. D., F. Keiter, and J. V. Neel. 1967. Further studies on the Xavante Indians, part 2: Some anthropometric, dermatoglyphic, and nonquantitative morphological traits of the Xavantes of Simões Lopes. *American Journal of Human Genetics* 19:490–501.

Novais, Washington. 2001. A lambaçada hidrovia. *Estado de São Paulo,* October 26. <http://www.estado. estadao.com/editorias/01/10/26/aberto001.htm> (accessed December 20, 2001).

Nutels, N. 1952. Plano para uma campanha de defesa do índio brasileiro contra a tuberculose. *Revista Brasileira de Tuberculose* 20:3–28.

———. 1955. Cadastro torácico entre índios do Brasil Central. In *Relatório das Atividades do Serviço de Proteção aos Índios Durante o Ano de 1954,* M. F. Simões, ed., 125–30. Rio de Janeiro: Serviço de Proteção aos Índios.

———. 1960. Resumo das atividades durante o ano passado, Serviço Nacional de Tuberculose. *Revista do Serviço Nacional de Tuberculose* 4:27–35.

———. 1968. Medical problems of newly contacted Indian groups. In *Biomedical Challenges Presented by the American Indian,* 68–76. PAHO Scientific Publications, no. 165. Washington, D.C.: Pan American Health Organization.

Nutels, N., and L. V. Duarte. 1961. Cadastro tuberculínico na área indígena. *Revista do Serviço Nacional de Tuberculose* 3:259–70.

O'Dea, K. 1991. Westernization, insulin resistance, and diabetes in Australian aborigines. *Medical Journal of Australia* 155:258–64.

Oliveira, J. P. 1994. Pardos, mestiços ou caboclos: Os índios nos censos nacionais no Brasil (1972–1980). *Horizontes Antropológicos* 6:60–83.

———. 1999. *Ensaios em Antropologia Histórica.* Rio de Janeiro: Editora da UFRJ.

Oliveira, L. R. C. 1981. Colonização e diferenciação: Os colonos de Canarana. M.A. thesis, Museu Nacional, Universidade Federal do Rio de Janeiro.

Oliveira, R. C. 1960. *O Processo de Assimilação dos Terena.* Rio de Janeiro: Museu Nacional.

———. 1972. *O Índio e o Mundo dos Brancos: Uma Interpretação Sociológica da Situação dos Tukúna.* São Paulo: Pioneira.

Oliveira, S. F., D. E. B. Lopes, S. E. B. Santos, and J. F. Guerreiro. 1998. The Awá-Guajá Indians of the Brazilian Amazon: Demographic data, serum protein markers, and blood groups. *Human Heredity* 48:163–68.

Oliver, W. J., E. L. Cohen, and J. V. Neel. 1975. Blood pressure, sodium intake, and sodium related hormones in the Yanomamo Indians, a "no-salt" culture. *Circulation* 52:146–51.

Omran, A. R. 1971. The epidemiologic transition: A theory of the epidemiology of population change. *Milbank Memorial Fund Quarterly* 49:509–38.

Oricchio, L. Z. 1997. Índios produzem programa para a televisão. *O Estado de São Paulo,* 24 September, sec. 2, 1.

Page, L. B. 1974. Hypertension and artherosclerosis in primitive and acculturating societies. In *International Symposium on Hypertension,* J. C. Hunt, ed., 1–12. New York: Health Learning Systems.

PAHO (Pan American Health Organization). 1980. Diarrhoeal diseases in the Americas. *Epidemiological Bulletin* 1:1–4.

———. 1997. Health of indigenous peoples. *Pan American Journal of Public Health* 2:357–62.

Palmore, J. A., and R. W. Gardner. 1996. *Measuring Mortality, Fertility, and Natural Increase: A Self-Teaching Guide to Elementary Measures.* Honolulu: East-West Center.

Pappiani, A. 1992. Programa de índio. *Caminhos da Terra* (São Paulo) 1:28–33.

Patarra, N. L. 1995. Mudanças na dinâmica demográfica. In *Velhos e Novos Males da Saúde no Brasil: A Evolução do País e de Suas Doenças,* C. A. Monteiro, ed., 61–78. São Paulo: Hucitec.

PCDCh/MT (Programa de Controle da Doença de Chagas/Mato Grosso). 1999. Interrupção da Transmissão Vetorial da Doença de Chagas por *Triatoma infestans:* Relatório de Avaliação, Estado de Mato Grosso. Cuiabá, Fundação Nacional de Saúde. Photocopy.

Pelletier, D. L., E. A. Frongillo Jr., D. G. Schroeder, and J. P. Habicht. 1995. The effects of malnutrition on child mortality in developing countries. *Bulletin of the World Health Organization* 73:443–48.

Penna, G. O. 1994. Hanseníase: Epidemiologia e controle. *Revista da Sociedade Brasileira de Medicina Tropical* 27 (sup. 3): 37–44.

Pereira, P. 1997a. Meninos são enviados para viver com brancos. *O Estado de São Paulo,* 9 November, A18.

———. 1997b. Xavantes contam sua história da pacificação. *O Estado de São Paulo,* 9 November, A17.

Petersen, E. E., R. Clemens, H. L. Bock, K. Friesse, and G. Hess. 1992. Hepatitis B and C in heterosexual patients with various sexually transmitted diseases. *Infection* 20:128–31.

Petersen, N. J., D. B. Barrett, W. W. Bond, K. R. Berquiot, M. S. Favero, T. R. Bender, and J. E. Maynard. 1976. Hepatitis B surface antigen in saliva, impetiginous lesions, and the environment in two remote Alaskan villages. *Applied Environmental Microbiology* 32:572–74.

Pettitt, D. J., K. A. Aleck, H. R. Baird, M. J. Carraher, P. H. Bennett, and W. C. Knowler. 1988. Congenital susceptibility to NIDDM: Role of intrauterine environment. *Diabetes* 37:622–28.

Phillips, D. I., S. Hirst, P. M. Clark, C. N. Hales, and C. Osmond. 1994. Fetal growth and insulin secretion in adult life. *Diabetologia* 37:592–96.

Picchi, D. 1991. The impact of an industrial agricultural project on the Bakairi Indians of Central Brazil. *Human Organization* 50:26–38.

———. 2000. *The Bakairí Indians of Brazil: Politics, Ecology, and Change.* Prospect Heights, Ill.: Waveland Press.

Pine, R. H., I. R. Bishop, and R. L. Jackson. 1970. Preliminary list of mammals of the Xavantina/Cachimbo expedition (Central Brazil). *Transactions of the Royal Society of Tropical Medicine and Hygiene* 64:668–70.

Pinstrup-Andersen, P., S. Burger, J. P. Habicht, and K. Peterson. 1993. Protein-energy malnutrition. In *Disease Control Priorities in Developing Countries,* D. T. Jamison, W. H. Mosley, A. R. Measham, and J. L. Bobadilla, eds., 391–420. New York: Oxford University Press.

Pinto, M. N., ed. 1994. *Cerrado: Caracterização, Ocupação e Perspectivas.* 2d ed. Brasília: Editora da Universidade de Brasília.

Pohl, J. E. [1837] 1951. *Viagem no Interior do Brasil.* 2 vols. Rio de Janeiro: Instituto Nacional do Livro, Ministério da Educação. Originally published as *Reise im Innern von Brasilien.* 2 vols. Vienna: Zweyter Theil, 1832–37.

Popkin, B. M. 1993. Nutritional patterns and transition. *Population and Development Review* 19:138–57.

———. 1994. The nutrition transition in low income countries: An emerging crisis. *Nutrition Reviews* 52:285–98.

Popkin, B. M., and C. M. Doak. 1998. The obesity epidemic is a worldwide phenomenon. *Nutrition Reviews* 56, no. 4 (pt. 1): 106–14.

Pose, S. B. 1993. Avaliação das condições de saúde bucal dos índios Xavánte do Brasil Central. M.Sc. thesis, Rio de Janeiro, Escola Nacional de Sáude Pública, Fundação Oswaldo Cruz.

Posey, D. A. 1988. Kayapó Indians natural resource management. In: *People of the Tropical Rain Forest* (J. S. Denslow and C. Padoch, eds.), pp. 89–90. Berkeley: University of California Press.

Prado, C., Jr. 1963. *Formação do Brasil Contemporâneo: Colônia.* São Paulo: Brasiliense.

Prata, P. R. 1992. A transição epidemiológica no Brasil. *Cadernos de Saúde Pública* 8:168–75.

Price, D. 1989. *Before the Bulldozer: The Nambiquara Indians and the World Bank.* Cabin John, Md.: Seven Locks Press.

———. 1994. Notes on Nambiquara demography. *South American Indian Studies* 4:63–76.

Queiroz, J. F. 1980. Xavánte: Como manter o orgulho de ser índio na trilha do trator. *Interior* (Brasília) 7:20–31.

Rachou, R. G. 1958. Anofelinos do Brasil: Comportamento das espécies vetoras de malária. *Revista Brasileira de Malariologia e Doenças Tropicais* 10:145–81.

Ramos, A. R. 1984. Frontier expansion and Indian peoples in the Brazilian Amazon. In *Frontier Expansion in Amazonia,* M. Schmink and C. H. Wood, eds., 83–104. Gainesville: University of Florida Press.

———. 1998. *Indigenism: Ethnic Politics in Brazil.* Madison: University of Wisconsin Press.

Ratter, J. A., P. W. Richards, G. Argent, and D. R. Gifford. 1973. Observations on the vegetation of northeastern Mato Grosso, part 1: The woody vegetation types of the Xavantina-Cachimbo Expedition area. *Philosophical Transactions of the Royal Society of London* 266B:449–92.

Ravagnani, O. M. 1978. A experiência Xavante com o mundo dos brancos. D.Sc. diss., São Paulo, Escola de Sociologia e Política.

Ravelli, A. C., J. H. van der Meulen, R. P. Michels, C. Osmond, D. J. Baker, C. N. Hales, and O. P. Bleker. 1998. Glucose tolerance in adults after prenatal exposure to famine. *Lancet* 351:173–77.

Ravussin, E., M. E. Valencia, J. Esparza, P. H. Bennett, and L. O. Schulz. 1994. Effects of a traditional lifestyle on obesity in Pima Indians. *Diabetes Care* 17:1067–74.

Rees, R. G. P., and A. J. Shelley. 1977. Estimativa quantitativa de resposta à clortetraciclina em um caso grave de desinteria por *Balantidium coli. Acta Amazonica* 7:47–49.

Ribeiro, D. 1956. Convívio e contaminação: Efeitos dissociativos da depopulação provocada por epidemias em grupos indígenas. *Sociologia* (São Paulo) 18:3–50.

———. 1977. *Os Índios e a Civilização.* 2d ed. Petrópolis: Vozes.

Ribeiro, F. P. 1841. Memória sobre as nações gentias que habitam o Continente do Maranhão (escripta em 1819). *Revista do Instituto Histórico e Geographico Brasileiro* 3:184–196, 297–321, 442–56.

———. 1849. Descripção do território de Pastos Bons, nos sertões do Maranhão, pelo Major Francisco de Paula Ribeiro. *Revista do Instituto Histórico e Geographico Brasileiro* 12:41–86.

———. 1870. Roteiro da viagem que fez o Capitão Francisco de Paula Ribeiro às fronteiras da Capitania do Maranhão e de Goyaz em 1815 em serviço da S. M. Fidelissima. *Revista do Instituto Histórico e Geographico Brasileiro* 10:5–80.

Ricardo, C. A. 1996. A sociodiversidade nativa contemporânea no Brasil. In

Povos Indígenas no Brasil, 1991/1995, C. A. Ricardo, ed., i–xii. São Paulo: Instituto Socioambiental.

RIHGB (Revista do Instituto Histórico e Geographico Brasileiro). 1918. Subsídios para a história da Capitania de Goyaz (1756–1806). *Revista do Instituto Histórico e Geographico Brasileiro* 84:41–294.

Rizzini, C. T. 1979. *Tratado de Fitogeografia do Brasil.* Vol. 2. São Paulo: Hucitec and Editora da Universidade de São Paulo.

Rizzini, C. T., and H. P. Heringer. 1961. Underground organs of plants from southern Brazilian savannas, with special reference to xylopodium. *Phyton* 17:105–24.

Roche, M., and M. Layrisse. 1966. The nature and causes of "hookworm anemia." *American Journal of Tropical Medicine and Hygiene* 15:1029–1102.

Rodrigues, A. D. 1986. *Línguas Brasileiras: Para o Conhecimento das Línguas Indígenas.* São Paulo: Edições Loyola.

Rodriguez-Delfin, L. A., F. M. Salzano, S. M. Callegari-Jacques, and J. V. Neel. 1994. Blood polymorphisms and morphological variability in Brazilian Indians. *American Journal of Human Biology* 6:619–25.

Rondon, C. M. S. 1946. *Índios do Brasil.* Vol. 1: *Centro, Noroeste, e Sul de Mato Grosso.* Rio de Janeiro: Conselho Nacional de Proteção aos Índios.

Roosevelt, A. C. 1992. Arqueologia amazônica. In *História dos Índios no Brasil,* M. C. Cunha, ed., 53–86. São Paulo: Companhia das Letras.

———. 1994. Amazonian anthropology: Strategy for a new synthesis. In *Amazonian Indians from Prehistory to the Present: Anthropological Perspectives,* A. C. Roosevelt, ed., 1–29. Tucson: University of Arizona Press.

Ross, J. B. 1984. Effects of contact on revenge hostilities among the Achuara Jivaro. In *Warfare, Culture, and Environment,* R. B. Ferguson, ed., 83–109. Orlando: Academic Press.

Rowland, M. G. M., R. A. E. Barrel, and R. G. Whitehead. 1978. The weanling's dilemma: Bacterial contamination in traditional Gambian weaning foods. *Lancet* 1:316–18.

Roychoudhury, A. K., and M. Nei. 1988. *Human Polymorphic Genes: World Distribution.* New York: Oxford University Press.

Sá, C. 1983. Observações sobre a habitação em três grupos indígenas brasileiros. In *Habitações Indígenas,* S. C. Novaes, ed., 103–45. São Paulo: Nobel and Editora da Universidade de São Paulo.

Sabroza, P., H. Kawa, and W. S. Q. Campos. 1995. Doenças transmissíveis: Ainda um desafio. In *Os Muitos Brasis: Saúde e População na Década de 80,* M. C. S. Minayo, ed., 177–244. São Paulo: Hucitec.

Sahlins, M. 1999. What is anthropological enlightenment? Some lessons of the twentieth century. *Annual Review of Anthropology* 28:i–xxiii.

Saint-Hilaire, A. de 1847–48. *Voyage aux Sources du Rio de S. Francisco et dans la Province de Goyaz.* 2 vols. Paris: A. Bertrand.

Salzano, F. M. 1985. Changing patterns of disease among South American Indians. In *Diseases of Complex Etiology in Small Populations,* R. Chakraborty and E. J. E. Szathmáry, eds., 301–23. New York: Liss.

———. 1991. Interdisciplinary approaches to the human biology of South Amerindians. *Human Biology* 63:875–82.

Salzano, F. M., and S. M. Callegari-Jacques. 1988. *South American Indians: A Case Study in Human Evolution.* Oxford: Clarendon.

Salzano, F. M., S. M. Callegari-Jacques, M. A. Mestriner, T. A. Weimer, M. H. L. P. Franco, L. Schüler, M. L. Harada, H. Schneider, M. H. Hutz, and M. J. M. Freitas. 1990. Reconstructing history: The Amazonian Mura Indians. *Human Biology* 62:619–35.

Salzano, F. M., S. M. Callegari-Jacques, T. A. Weimer, M. H. L. P. Franco, M. H. Hutz, and M. L. Petzl-Erler. 1997a. Electrophoretic protein polymorphisms in Kaingang and Guarani Indians of southern Brazil. *American Journal of Human Biology* 9:505–12.

Salzano, F. M., M. H. L. P. Franco, T. A. Weimer, S. M. Callegari-Jacques, M. A. Mestriner, M. H. Hutz, N. M. Flowers, R. V. Santos, and C. E. A. Coimbra Jr. 1997b. The Brazilian Xavante Indians revisited: New protein genetic studies. *American Journal of Physical Anthropology* 104:23–34.

Salzano, F. M., J. V. Neel, and D. Maybury-Lewis. 1967. Further studies on the Xavante Indians, part 1: Demographic data on two additional villages—genetic structure of the tribe. *American Journal of Human Genetics* 19:463–89.

Salzano, F. M., and C. V. Tondo. 1968. Hemoglobin types of Brazilian Indians. *American Journal of Physical Anthropology* 28:355–59.

Santos, A. K., M. Ishak, S. Santos, J. F. Guerreiro, and R. Ishak. 1995. A possible correlation between the host genetic background in the epidemiology of hepatitis B virus in the Amazon region of Brazil. *Memórias do Instituto Oswaldo Cruz* 90:435–41.

Santos, F. B., M. Hutz, R. V. Santos, C. E. A. Coimbra Jr., F. M. Salzano, and S. D. J. Pena. 1995. Further evidence for the existence of a major founder Y chromosome haplotype in Amerindians. *Brazilian Journal of Genetics* 18:669–72.

Santos, L. G. 1991. O sonho de Sibupá. In *Povos Indígenas no Brasil, 1987/88/89/90,* 73–74. Aconteceu Especial, no. 18. São Paulo: Centro Ecumênico de Documentação e Informação.

Santos, R. 1980. *História Econômica da Amazônia.* São Paulo: T. A. Queiroz.

Santos, R. V. 1993. Crescimento fisico e estado nutricional de populações indígenas brasileiras. *Cadernos de Saúde Pública* 9 (sup. 1): 46–57.

Santos, R. V., and C. E. A. Coimbra Jr. 1996. Socioeconomic differentiation and body morphology in the Suruí of Southwestern Amazonia. *Current Anthropology* 37:851–56.

———. 1998. On the (un)natural history of the Tupí-Mondé Indians: Bioanthropology and change in the Brazilian Amazonia. In *Toward a Biocultural Synthesis: Political-Economic Perspectives in Biological Anthropology,* A. Goodman and T. Leatherman, eds., 269–94. Ann Arbor: University of Michigan Press.

Santos, R. V., C. E. A. Coimbra Jr., N. M. Flowers, and J. P. Silva. 1995.

Intestinal parasitism in the Xavánte Indians, Central Brazil. *Revista do Instituto de Medicina Tropical de São Paulo* 37:145–48.

Santos, R. V., C. E. A. Coimbra Jr., and A. C. Linhares. 1991. Estudos epidemiológicos entre grupos indígenas de Rondônia, part 4: Inquérito sorológico para rotavírus entre os Suruí e Karitiána. *Revista de Saúde Pública* 25:230–32.

Santos, R. V., N. M. Flowers, C. E. A. Coimbra Jr., and S. A. Gugelmin. 1997. Tapirs, tractors, and tapes: The changing economy and ecology of the Xavánte Indians of Central Brazil. *Human Ecology* 25:545–66.

Schmink, M., and C. H. Wood, eds. 1984. *Frontier Expansion in Amazonia.* Gainesville: University of Florida Press.

Schmink, M., and C. H. Wood. 1987. The "political ecology" of Amazonia. In *Land at Risk in the Third World: Local-Level Perspectives,* P. D. Little and M. M. Horowitz, eds., 38–57. Boulder: Westview.

———. 1992. *Contested Frontiers in Amazonia.* New York: Columbia University Press.

Scott, G. R., and C. G. Turner. 1988. Dental anthropology. *Annual Review of Anthropology* 17:99–126.

Scrimshaw, N. S. 1977. Effect of infection on nutrient requirements. *American Journal of Clinical Nutrition* 30:739–43.

Scrimshaw, N. S., C. E. Taylor, and J. E. Gordon. 1968. *Interactions of Nutrition and Infection.* WHO Monographs, no. 57. Geneva: World Health Organization.

Seeger, A. 1981. *Nature and Society in Central Brazil: The Suya Indians of Mato Grosso.* Cambridge: Harvard University Press.

Sereburã, Hipru, Rupawê, Serezabdi, and Sereñimirãmi. 1998. *Wamrêmé Za'ra–Nossa Palavra: Mito e História do Povo Xavante.* Paulo Supretaprã Xavante and Jurandir Siridiwê Xavante, trans. São Paulo: Senac.

Shreffler, D. C., and A. G. Steinberg. 1967. Further studies on the Xavánte Indians, part 4: Serum protein groups and the SC1 trait of saliva in the Simões Lopes and São Marcos Xavantes. *American Journal of Human Genetics* 19:514–23.

Sick, H. 1955. O aspecto fitofisionômico da paisagem do médio Rio das Mortes, Mato Grosso e a avifauna da região. *Arquivos do Museu Nacional* (Rio de Janeiro) 42:542–76.

———. 1965. A fauna do cerrado. *Arquivos de Zoologia* (São Paulo) 12:71–93.

Silva, A. L. 1983. Xavante: Casa-aldeia-chão-terra-gente. In *Habitações Indígenas,* S. C. Novaes, ed., 33–55. São Paulo: Nobel and Editora da Universidade de São Paulo.

———. 1986. *Nomes e Amigos: Da Prática Xavante a uma Reflexão sobre os Jê.* São Paulo: Faculdade de Filosofia, Letras e Ciências Humanas, Universidade de São Paulo.

———. 1992. Dois séculos e meio de história Xavante. In *História dos Índios no Brasil,* M. C. Cunha, ed., 357–78. São Paulo: Companhia das Letras.

Silva, H. B. da. 1935. *Nos Sertões do Araguaia.* São Paulo: Edições Cultura Brasileira.

Silva, M. F. 1994. A demografia e os povos indígenas no Brasil. *Revista Brasileira de Estudos Populacionais* 11:261–64.

Simões, C. C. S., and C. A. Monteiro. 1995. Tendência secular e diferenciais regionais da mortalidade infantil no Brasil. In *Velhos e Novos Males da Saúde no Brasil: A Evolução do País e de Suas Doenças,* C. A. Monteiro, ed., 153–56. São Paulo: Hucitec.

Sinnett, P. F., I. H. Kevau, and D. Tyson. 1992. Social change and the emergence of degenerative cardiovascular disease in Papua New Guinea. In *Human Biology in Papua New Guinea: The Small Cosmos,* R. D. Attenborough and M. P. Alpers, eds., 373–86. Oxford: Clarendon.

Sixth Report of the Joint National Committee on Prevention, Detection, Evaluation, and Treatment of High Blood Pressure. 1997. The Sixth Report of the Joint National Committee on Prevention, Detection, Evaluation, and Treatment of High Blood Pressure. *Archives of Internal Medicine* 157: 2413–45.

Skidmore, T. 1967. *Politics in Brazil, 1930–1964: An Experiment in Democracy.* New York: Oxford University Press.

Smith, A. 1971. *Mato Grosso: Last Virgin Land.* New York: Dutton.

Smith, N. J. H. 1981. *Rainforest Corridors: The Transamazon Colonization Scheme.* Berkeley: University of California Press.

Smouse, P. E., V. Vitzthum, and J. V. Neel. 1981. The impact of random and lineal fission on the genetic divergence of small human groups: A case study among the Yanomama. *Genetics* 98:179–97.

Soares, M. C. P., and G. Bensabath. 1988. Possível participação do vegetação na transmissão do virus das hepatitis B e Delta na microrregião do Purus, Amazonas, Brasil. In *Resumos do 4° Encontro Nacional de Virologia,* 143. São Lourenço: Sociedade Brasileira de Virologia.

Souza, C. F. M., Jr. 1994. On Brazil and its Indians. In *Indigenous Peoples and Democracy in Latin America,* D. L. van Cott, ed., 213–33. New York: St. Martin's.

Souza, L. 1953a. *Os Xavantes e a Civilização (Ensaio Histórico).* Rio de Janeiro: Instituto Brasileiro de Geografia e Estatística.

———. 1953b. *Entre os Xavante do Roncador.* Rio de Janeiro: Serviço de Documentação, Minstério da Educação e Saúde.

Souza, L. G. 1999. Perfil demográfico dos Xavánte de Sangradouro-Volta Grande, Mato Grosso, 1993–1997. M.Sc. thesis, Rio de Janeiro, Escola Nacional de Saúde Pública, Fundação Oswaldo Cruz.

Souza, L. G., and R. V. Santos. 1999. *Mortalidade, Fecundidade e Padrão de Assentamento dos Xavánte de Sangradouro-Volta Grande, Mato Grosso (1993–1997).* Working Papers, no. 2. Porto Velho: Centro de Estudos em Saúde do Índio de Rondônia and Rio de Janeiro: Escola Nacional de Saúde Pública.

———. 2001. Perfil demográfico da população indígena Xavánte de Sangradouro-Volta Grande, Mato Grosso (1993–1997). *Cadernos de Saúde Pública* 17 17:355–65.

Soyano, A., J. Malavé, R. Walder, Z. Layrisse, and M. Layrisse. 1976. Hepatitis

B antigen in an isolated Indian population (Yanomama Indians), southern Venezuela. *Revista Brasileira de Pesquisas Médicas e Biológicas* 9:247–53.

SPI (Serviço de Proteção aos Índios). 1944. Instruções ao Inspetor Especializado Francisco F. Soares de Meireles, para os trabalhos de pacificação dos índios Chavantes. Rio de Janeiro: Museu do Índio. Setor de Documentação, microfilm 381/ photos 277–99.

Spielman, R. S., S. S. Fajans, J. V. Neel, S. Pek, J. C. Floyd, and W. J. Oliver. 1982. Glucose tolerance in two unacculturated Indian tribes of Brazil. *Diabetologia* 23:90–93.

Spix, J. B. von, and C. F. P. von Martius. 1824. *Travels in Brazil in the Years 1817–1820.* 2 vols. London: Longman, Hurst, Rees, Orme, Brown, and Green.

Sreebny, L. M. 1983. Cereal availability and dental caries. *Community Dentistry and Oral Epidemiology* 11:148–55.

Stansfield, S. K., and D. S. Shepard. 1993. Acute respiratory infections. In *Disease Control Priorities in Developing Countries,* D. T. Jamison, W. H. Mosley, A. R. Measham, and J. L. Bobadilla, eds., 67–90. New York: Oxford University Press.

Stark, A. E., F. M. Salzano, and F. J. da Rocha. 1990. Marital correlation for anthropometric characteristics in Brazilian Indians. *Annals of Human Biology* 17:417–22.

Stauffer, D. H. 1955. The origin and establishment of Brazil's Indian Service. Ph.D. diss., University of Texas.

Stearman, A. M. 2000. A pound of flesh: Social changes and modernization as factors in hunting sustainability among neotropical indigenous societies. In *Hunting for Sustainability in Tropical Forests,* J. G. Robinson and E. L. Bennett, eds., 233–50. New York: Columbia University Press.

Steward, J. 1949. South American cultures; An interpretative summary. In *Handbook of South American Indians,* J. H. Steward, ed., 5:669–772. Washington, D.C.: Government Printing Office.

Steward, J., and L. Faron. 1959. *Native Peoples of South America.* New York: McGraw-Hill.

Switkes, G. 1998. Resumo de informação histórica sobre a Hidrovia Tocantins-Araguaia. Cuiabá: International Rivers Network. Manuscript.

Szathmáry, E. J. E. 1990. Diabetes in Amerindian populations: The Dogrib studies. In *Disease in Populations in Transition: Anthropological and Epidemiological Perspectives,* A. C. Swedlund and G. J. Armelagos, eds., 75–103. New York: Bergin and Garvey.

———. 1994. Non-insulin dependent diabetes mellitus among aboriginal North Americans. *Annual Review of Anthropology* 23:457–82.

Szmuness, W., M. I. Much, A. M. Prince, J. H. Hoofnagle, C. E. Cherubin, E. J. Harley, and G. H. Block. 1975. On the role of sexual behavior in the spread of hepatitis B infection. *Annals of Internal Medicine* 83:489–95.

Tanis, R., R. E. Ferrell, J. V. Neel, and M. Morrow. 1974. Albumin Yanomama-2, a "private" polymorphism of serum albumin. *Annals of Human Genetics* 38:179–90.

Taplin, D., and T. L. Meinking. 1988. Infestations. In *Pediatric Dermatology,* L. A. Schachner and R. C. Hansen, eds., 2:1465–93. New York: Churchill Livingstone.

Tashian, R. E., G. J. Brewer, H. Lehmann, D. A. Davies, and D. L. Rucknagel. 1967. Further studies on the Xavánte Indians, part 5: Genetic variability in some serum and erythrocyte enzymes, hemoglobin, and the urinary excretion of beta-aminoisobutyric acid. *American Journal of Human Genetics* 19:524–31.

Tavares, E. F., J. P. B. Vieira-Filho, A. Andriolo, and L. J. Franco. 1999. Anormalidades de tolerância à glicose e fatores de risco cardiovascular em uma tribo indígena aculturada da região amazônica brasileira. *Arquivos Brasileiros de Endocrinologia e Metabologia* 43 (sup. 1): S235.

Tavener, C. J. 1973. The Karajá and the Brazilian frontier. In *People and Cultures of Native South America,* D. R. Gross, ed., 433–59. Garden City, N.Y.: Natural History Press.

Terborgh, J. 1975. Faunal equilibria and the design of wildlife preserves. In *Tropical Ecological Systems,* F. B. Goley and E. Medina, eds., 369–80. New York: Springer Verlag.

Thompson, E. A. 197. Fission models of population variability. *Genetics* 93:479–95.

Thornton, R. 1987. *American Indian Holocaust and Survival.* Norman: University of Oklahoma Press.

Tills, D., A. C. Kopec, and R. E. Tills. 1983. *The Distribution of the Human Blood Groups and Other Polymorphisms.* Supplement 1. Oxford: Oxford University Press.

Tsega, E., B. Mengesha, B. G. Hansson, J. Lindberg, and E. Nordenfelt. 1986. Hepatitis A, B, and delta infection in Ethiopia: A serologic survey with demographic data. *American Journal of Epidemiology* 123:344–51.

Tsubouchi, J., M. Tsubouchi, R. J. Maynard, P. K. Domoto, and P. Weinstein. 1995. A study of dental caries and risk factors among Native American infants. *Journal of Dentistry for Children* 67:283–87.

Tupasi, T. E., M. A. Velmonte, M. E. G. Sanvictores, L. Abraham, L. E. Leon, S. A. Tan, C. A. Miguel, and M. C. Saniel. 1988. Determinants of morbidity and mortality due to acute respiratory infections: Implications for intervention. *Journal of Infectious Diseases* 157:615–23.

Turner, T. S. 1993. The role of indigenous peoples in the environmental crisis: The example of the Kayapó of the Brazilian Amazon. *Perspectives in Biology and Medicine* 36:526–45.

———. 1995. An indigenous people's struggle for socially equitable and ecologically sustainable production: The Kayapo revolt against extractivism. *Journal of Latin American Anthropology* 1:98–121.

———. 2000. Indigenous rights, environmental protection, and the struggle over forest resources in the Amazon: The case of the Brazilian Kayapo. In *Earth, Air, Fire, and Water: Humanistic Studies of the Environment,* J. K. Conway, K. Keniston, and L. Marx, eds., 226–61. Amherst: University of Massachusetts Press.

Urban, G. 1985. Developments in the situation of Brazilian tribal populations from 1876 to 1982. *Latin American Research Review* 20:7–25.

———. 1992. A história da cultura brasileira segundo as línguas nativas. In *História dos Índios no Brasil,* M. C. Cunha, ed., 87–102. São Paulo: Companhia das Letras.

Vale, A. 1980a. Índios insistem em mudanças: Xavantes deixam o DF prometendo voltar para tirar o presidente da Funai. *Correio Braziliense,* 13 May, National sec., 10.

———. 1980b. Xavantes vão à Funai "Cansados de esperar." *Correio Braziliense,* 24 May, National sec., 3.

Valente, R. 1999. Estudo ambiental é posto sob suspeita. *Jornal do Brasil,* 23 August, sec. 1, 8–9.

Van Lerberghe, W. 1988. Linear growth retardation and mortality. In *Linear Growth Retardation in Less Developed Countries,* J. C. Waterlow, ed., 245–60. New York: Raven.

Van Loon, H., V. Saverys, J. P. Vuyesteke, R. F. Vlietinck, and R. Eeckels. 1986. Local versus universal growth standards: The effect of using NCHS as universal reference. *Annals of Human Biology* 13:347–57.

Variyam, E. P., and J. G. Banwell. 1982. Hookworm disease: Nutritional implications. *Reviews of Infectious Diseases,* 4:830–35.

Vasconcelos, P. F. C., J. F. S. T. Rosa, S. G. Guerreiro, N. Dégallier, E. S. T. Rosa, and A. P. A. T. Rosa. 1989. Primeiro registro de epidemias causadas pelo vírus Oropouche nos estados do Maranhão e Goiás, Brasil. *Revista do Instituto de Medicina Tropical de São Paulo* 31:271–78.

Vasconcelos, P. F. C., J. F. S. T. Rosa, A. P. A. T. Rosa, N. Dégallier, F. P. Pinheiro, and G. S. Sá-Filho. 1991. Epidemiologia das encefalites por arbovirus na Amazônia brasileira. *Revista do Instituto de Medicina Tropical de São Paulo* 33:465–76.

Vaughan, J. P. 1978. A review of cardiovascular diseases in developing countries. *Annals of Tropical Medicine and Parasitology* 72:101–9.

Velho, O. G. 1976. *Capitalismo Autoritário e Campesinato: Um Estudo Comparativo a Partir da Fronteira em Movimento.* São Paulo: Difel.

———. 1981. *Frentes de Expansão e Estrutura Agrária: Estudo do Processo de Penetração numa Área da Transamzônica.* 2d ed. Rio de Janeiro: Zahar.

Vellar, O. D. 1964. Acute viral hepatitis in Norwegian track finders: An epidemiological study in Norway. *Acta Medica Scandinavica* 176:651–55.

Velloso, W. 1957. Report from Walter Velloso to Lincoln Pope. 18 October. Microfilm 381, Rio de Janeiro, Museu do Índio.

Verano, J. W., and D. H. Ubelaker, eds., 1992. *Disease and Demography in the Americas.* Washington, D.C.: Smithsonian Institution Press.

Veras, R. P. 1988. Considerações acerca de um país jovem que envelhece. *Cadernos de Saúde Pública* 4:382–97.

Verdesio, J. J. 1994. As perspectivas ambientais do cerrado brasileiro. In *Cerrado: Caraterização, Ocupaçao, e Perspectivas,* M. N. Pinto, ed., 586–605. Brasília: Editora da Universidade de Brasília.

Vianna, H. 1996. O futuro do samba. *Folha de São Paulo,* 14 April, Entertainment sec. 5.

Vianna, U. 1927. Akuen ou Xerente. *Revista do Instituto Histórico e Geographico Brasileiro* 155:33–48.

Victora, C. G., S. R. Huttly, S. C. Fuchs, L. C. Nobre, and F. C. Barros. 1992. Deaths due to dysentery, acute and persistent diarrhoea among Brazilian infants. *Acta Paediatrica* 381 (sup.): 7–11.

Victora, C. G., P. G. Smith, J. P. Vaughan, L. C. Nobre, C. Lombardi, A. M. Teixeira, S. M. Fuchs, L. B. Moreira, L. P. Gigante, and F. C. Barros. 1987. Evidence for a strong protective effect of breast-feeding against infant deaths due to infectious diseases in Brazil. *Lancet* 2:319–22.

Vieira, C. O. C. 1951. Notas sobre os mamíferos obtidos pela expedição do Instituto Butantan ao Rio das Mortes e Serra do Roncador. *Papéis Avulsos do Departmento de Zoologia* (São Paulo) 10:105–25.

Vieira-Filho, J. P. B. 1975. Análise das glicemias dos índios das aldeias Suruí, Gaviões, e Xikrín. *Revista da Associação Médica Brasileira* 21:253–55.

———. 1977. O diabetes mellitus e as glicemias de jejum dos índios Caripuna e Palikur. *Revista da Associação Média Brasileira* 23:175–78.

———. 1981. Problemas de aculturação alimentar dos Xavantes e Bororo. *Revista de Antropologia* (São Paulo) 24:37–40.

———. 1996. Emergência do diabetes melito tipo II entre os Xavantes. *Revista da Associação Médica Brasileira* 42:61–62.

———. 2000. Genótipo econômico e produção de insulina. In *Diabetes, Saúde, and Cia,* Aventis Pharma, ed., 2. São Paulo: Aventis Pharma.

Vieira-Filho, J. P. B., A. S. B. Oliveira, M. R. D. Silva, A. L. Amaral, and R. T. Schultz. 1997. Polineuropatia nutricional entre índios Xavantes. *Revista da Associação Média Brasileira* 43:82–88.

Vieira-Filho, J. P. B., E. M. K. Russo, and N. Ferreira-Novo. 1983. A hemoglobina glicosilada (HbA_1) dos índios Xavantes. *Arquivos Brasileiros de Endocrinologia e Metabologia* 27:153–55.

Vieira-Filho, J. P. B., E. M. K. Russo, and Y. Juliano. 1984. A hemoglobina glicosilada (HbA1) dos índios Bororo. *Arquivos Brasileiros de Endocrinologia e Metabologia* 28:87–90.

Villas-Bôas, O., and C. Villas-Bôas. 1994. *A Marcha para o Oeste: A Epopéia da Expedição Roncador-Xingu.* São Paulo: Globo.

Wagley, C. 1977. *Welcome of Tears: The Tapirapé Indians of Central Brazil.* New York: Oxford University Press.

Waldman, E. A., L. J. Silva, and C. A. Monteiro. 1995. Trajetória das doenças infecciosas: Da eliminação da poliomielite à reintrodução da cólera. In *Velhos e Novos Males da Saúde no Brasil: A Evolução do País e de suas Doenças,* C. A. Monteiro, ed., 195–244. São Paulo: Hucitec.

Ward, R. H., F. M. Salzano, S. L. Bonatto, M. H. Hutz, C. E. A. Coimbra Jr., and R. V. Santos. 1996. Mitochondrial DNA polymorphism in three Brazilian Indian tribes. *American Journal of Human Biology* 8:317–23.

Waterlow, J. C. 1992. *Protein-Energy Malnutrition.* London: Edward Arnold.

Weinstein, E. D., J. V. Neel, and F. M. Salzano. 1967. Further studies on the Xavante Indians, part 6: The physical status of the Xavantes of Simões Lopes. *American Journal of Human Genetics* 19:532–42.

Weiss, K. M., R. E. Ferrell, and C. L. Hanis. 1984. A New World syndrome of metabolic diseases with a genetic and evolutionary basis. *Yearbook of Physical Anthropology* 27:153–78.

Weitkamp, L. R., E. M. McDermid, J. V. Neel, J. M. Fine, C. Petrim, L. Bonazzi, V. Ortali, F. Porta, R. Tanis, D. J. Harris, T. Peters, G. Ruffini, and E. Johnston. 1973. Additional data on the population distribution of human serum albumin genes: Three new variants. *Annals of Human Genetics* 37:19–26.

Werner, D. W. 1983. Fertility and pacification among the Mekranoti of Central Brazil. *Human Ecology* 11:227–45.

West, K. M. 1974. Diabetes in American Indians and other native populations of the New World. *Diabetes* 23:841–55.

Whitehead, N .L. 1988. *Lords of the Tiger Spirit: A History of the Caribs in Colonial Venezuela and Guyana, 1498–1820.* Royal Institute of Linguistics and Anthropology, Caribbean Studies Series, no. 10. Dordrecht: Foris and Leiden.

———. 1993. Native American cultures along the Atlantic littoral of South America, 1499–1650. *Proceedings of the British Academy* 81:197–231.

WHO (World Health Organization). 1972. *Nutritional Anaemias.* Technical Reports, no. 503. Geneva: WHO.

———. 1975. *Control of Nutritional Anaemia with Special Reference to Iron Deficiency.* Technical Reports, no. 580. Geneva: WHO.

———. 1987. *Oral Health Surveys: Basic Methods.* Geneva: WHO.

———. 1988. *Persistent Diarrhoea in Children in Developing Countries.* Report of a WHO meeting, document CDD/88.27. Geneva: WHO.

———. 1995. *Physical Status: The Use and Interpretation of Anthropometry.* Technical Reports, no. 854. Geneva: WHO.

———. 1998. *Obesity: Preventing and Managing the Global Epidemic.* Report of a WHO consultation on obesity. Document WHO/NUT/NCD/98.1 Geneva: WHO.

Williams, A. O., and A. I. O. Williams. 1972. Carrier stage prevalence of hepatitis associated antigen (Au/Sh) in Nigeria. *American Journal of Epidemiology* 96:227–30.

Williamson, L. 1995. Xavantes: Novos tempos, antigos rituais. *Horizonte Geográfico* (São Paulo) 8:52–61.

Wood, J. W. 1990. Fertility in anthropological populations. *Annual Review of Anthropology* 19:211–42.

Woodward, M., and A. R. P. Walker. 1994. Sugar consumption and dental caries: Evidence from 90 countries. *British Dental Journal* 176:297–302.

Wright, R. M. 1988. Anthropological presuppositions of indigenous advocacy. *Annual Review of Anthropology* 17:364–90.

Wüst, I. 1994. The eastern Bororo from an archaeological perspective. In *Ama-*

zonian Indians from Prehistory to the Present: Anthropological Perspectives, A. C. Roosevelt, ed., 315–42. Tucscon: University of Arizona Press.

Young, T. K. 1993. Diabetes mellitus among Native Americans in Canada and the United States: An epidemiological review. *American Journal of Human Biology* 5:399–413.

Zimmet, P., G. Dowse, C. Finch, S. Serjeantson, and H. King. 1990. The epidemiology and natural history of NIDDM: Lessons from the South Pacific. *Diabetes/Metabolism Reviews* 6:91–124.

Zuckerman, A. J. 1983. Aspects of the ecology of viral hepatitis. *Ecology and Disease* 2:135–44.

Index

Page numbers for photographs and figures are in italics. Page numbers for tables are in bold type.